D1605031

Enders A. Robinson
Member of the National Academy of Engineering

EINSTEIN'S RELATIVITY IN METAPHOR AND MATHEMATICS

PRENTICE HALL
Englewood Cliffs, New Jersey 07632

Library of Congress Cataloging-in-Publication Data

Robinson, Enders A.
    Einstein's relativity in metaphor and mathematics / Enders A. Robinson.
      p. cm.
    Bibliography: p.
    Includes index.
    ISBN 0-13-246497-7
    1. Special relativity (Physics)   2. Space and time.   I. Title.
QC173.65.R63   1990
530.1'1—dc20   89-8754
                                                                                  CIP

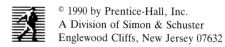

Editorial/production supervision and
   interior design: KERRY REARDON
Cover design: BRUCE KENSELAR
Manufacturing buyer: RAY SINTEL

© 1990 by Prentice-Hall, Inc.
A Division of Simon & Schuster
Englewood Cliffs, New Jersey 07632

All rights reserved. No part of this book may be
reproduced, in any form or by any means,
without permission in writing from the publisher.

Printed in the United States of America

10 9 8 7 6 5 4 3 2 1

ISBN 0-13-246497-7

PRENTICE-HALL INTERNATIONAL (UK) LIMITED, *London*
PRENTICE-HALL OF AUSTRALIA PTY. LIMITED, *Sydney*
PRENTICE-HALL CANADA INC., *Toronto*
PRENTICE-HALL HISPANOAMERICANA, S.A., *Mexico*
PRENTICE-HALL OF INDIA PRIVATE LIMITED, *New Delhi*
PRENTICE-HALL OF JAPAN, INC., *Tokyo*
SIMON & SCHUSTER ASIA PTE. LTD., *Singapore*
EDITORA PRENTICE-HALL DO BRASIL, LTDA., *Rio de Janeiro*

CONTENTS

FOREWORD   *ix*

PREFACE   *xi*

Part I   Metaphor   1

    INTRODUCTION   1

        Jet Lag   2

        Poetic Metaphors   6

    1   *FROM COPERNICUS TO EINSTEIN*   11

        Parallel Paths to Gravitation   12

        The Velocity of Light   20

        Macroworld and Microworld   24

        Natural Units   27

        Newton's Laws of Motion   29

        Inertial Frames of Reference   30

        The Relativity Principle and the Light Principle   34

Relativity as a Shear in Space-Time   36

Time-Space Interval Between Events   39

The Lorentz Transform and the Galilean Transform   40

2 SPACE AND TIME   42

The Same and the Ultimate   42

Space and Time   49

The Fable of the Two Eskimos   56

The Fable of the Muon and the Mountain   63

3 MAXWELL'S EQUATIONS AND RELATIVITY   69

Kepler's Equation   70

Olympic Metaphor of Relativity   72

Mathematical Model of Relativity   78

Wave Motion   81

The Wave Equation   82

Maxwell's Equation   85

Physical Theory of Relativity   88

Appendix: Derivation of Kepler's Equation   91

4 SYMMETRY OF THE POINCARÉ GROUP   96

Poincaré Symmetry of Space-Time   96

The Fable of the King's Messengers   99

Relativistic Doppler Factor   101

Distance and Time   104

Physical Velocity and Apparent Velocity   106

Two-Way Velocity of Light   112

The Electromagnetic Spectrum   116

The Doppler Effect   119

Newton's Failure and Maxwell's Success   126

Light Speed and Light Color   132

Space-Time Symmetry   136

Appendix 1: Derivation of Physical Velocity Addition Formula from Lorentz Transform   144

Appendix 2: Derivation of the Relativistic Doppler Factor from Lorentz Transform   145

5 MICHELSON-MORLEY EXPERIMENT   147

The Aether   147

The Michelson-Morley Experiment   151

The Null Result and the Aether   155

The Recognition Principle   156

Appendix: Light Clocks and the FitzGerald Length Contraction   159

6 WRIGHT'S EQUATION AND THE LORENTZ TRANSFORM   165

Trigonometry   165

Round Globe   168

Flat Map   170

Toscanelli's Map and the Discoveries of Columbus   171

Mercator, Dee, and Wright   176

Mercator Projection   181

Transverse Mercator Projection   188

The King Is Right   191

Round World   194

Flat World   197

Lorentz Transform   201

Appendix 1: Hipparchus' Stereographic Projection   203

Appendix 2: Wright's Equation   207

7 EINSTEIN, MASS, AND ENERGY   *209*

    Mass   209

    Conservation of Momentum   211

    Newtonian Dynamics   212

    Conservation of Energy   214

    The Muon and the Mountain Revisited   217

    Relativistic Mass   220

    Relativistic Energy   223

    $E = mc^2$ 227

    Invariants and Symmetry   231

    The Conservation Laws   233

Part II Mathematics   *239*

8 HYPERBOLIC TRIGONOMETRY   *239*

    Euclidean and Non-Euclidean Geometry   240

    Analytic Geometry   243

    Pythagorean Theorem   248

    Circle and Hyperbola   251

    Trigonometry   253

    Hyperbolic Trigonometry   256

    Linear Transformations   260

    Rotations and Shears   265

    Rotation of Space Coordinates   271

    Shear of Space-Time Coordinates   274

    Rotation Versus Shear   278

    Group Theory   281

9 CATENARY AND RELATIVITY   *286*

    Science as Fable   286

    Metaphor of Ants   287

CONTENTS vii

    Time-Space Interval   289

    Lorentz Transform   290

    Velocity   292

    Physical Velocity Addition Formula   294

    Larmor Time Dilation   294

    Appendix: Derivation of the Catenary   294

10 THE MINKOWSKI WORLD   *298*

    Time-Space Interval as the Metric   298

    World Lines   301

    Relativity of Simultaneity   305

    Einstein's Train   306

    Relativity of the Same Place   309

    Timelike and Spacelike Events   310

    Proper Time and Proper Length   313

    Minkowski Geometry   315

    FitzGerald Length Contraction and Larmor Time Dilation   321

11 LEONHARD EULER AND AFFINE GEOMETRY   *324*

    Affine Geometry   324

    Minkowski Plane   330

    Time and Distance by Radar Measurements   334

    Minkowski Coordinate Systems   337

    Gudermannian Space-Time Diagram   340

    Proper and Improper Times   343

    FitzGerald Length Contraction   346

12 THE GUDERMANN WORLD   *350*

    Hyperbolic Sine and Cosine   350

    The Gudermannian   353

Reason for the Gudermannian 356

Geometric Interpretation for the Timelike Hyperbola 358

Geometric Interpretation for the Spacelike Hyperbola 361

A Matter of Perspective 364

Example of the Gudermann World 369

13 CONCLUSION   373

The Long Road with No Turning 373

So Close, Yet So Far 377

Three Flashes of Insight 380

CHRONOLOGICAL TABLE   *385*

PLATE CREDITS   *387*

POETRY CREDITS   *388*

BIBLIOGRAPHY   *389*

INDEX   *395*

FOREWORD

Modern Physics still basks in the satisfaction of some profound intellectual prizes—all captured a little more than a lifetime ago. Among these, relativity, in both special and general cases, still demands top billing for its fresh insights into natural realms hidden from our direct experience, as well as its remarkable triumph of deduction over empiricism. With relativity—and with Einstein—physics became the science of our times.

That was not always so. At the turn of this century, physics was undergoing some troubling trials indeed. The Newtonian paradigm, with its elegant and common-sense approach to nature, was being challenged by some rather irritating experimental findings. Some of these were to be later explained by quantum theory, but the Michelson-Morley experiment, with its ether-destroying results, didn't make sense at all. Fitzgerald tried to fix Newtonian physics—so did Poincaré. Einstein brought the entire house down; ether was a figment of an incorrect paradigm.

Oddly, the correct paradigm, relativistic mechanics, didn't replace the Newtonian approach as much as absorb it in one great gulp. Newtonian physics became a special case of special relativity. In the history of science, that is a most curious state of affairs. Imagine a heliocentric view of the solar system absorbing, rather than replacing, a geocentric one, or Darwinism becoming LaMarckianism under just the right conditions. One best paradigm topples the others—at least that is the normal course of scientific revolutions.

But then, Einstein was no ordinary scientist. Indeed, relativity emerged almost full blown from that creative mind not once, but twice! First the mathematically simple but conceptually boggling special relativity and then, much later, the mathematically abstract and bizarre general relativity (for which the physicist Einstein had to become the mathematician Einstein). For more than a decade, Einstein was perhaps the only one who truly understood relativity.

Today, every young scientist worth his or her salt (as well as more than a few curious nonscientists) discovers the intrigue of relativity. The general case needs non-Euclidean geometries and concurrent, abstract mathematics in order to gain a real mastery—not so with special relativity. Here the ideas can be described in simple mathematical terms requiring, at the most, knowledge of hyperbolic trigonometric functions. Still, there are only two principle paths for this fascinating information: a totally watered-down layman's exposition or text book offerings. Either way, the ideas of relativity lose out; the first to a barrage of incompleteness and the second to the tendency to make basic physics a mathematical recipe: "*The steps are left for completion by the interested student.*" Let's hope that the student or reader is still interested after that!

So along comes this marvelous book. Robinson has enticed you to look at an old subject in some very new ways. In addition to the usual physical and mathematical rigor (I promise that you'll like it) you'll be feted to the development of a grand idea, whose precursors were as tantalizing as they were incomplete; Robinson will show you just how close relativity came into being with Kepler, Galileo and even Newton. You'll learn about Wright; you'll revisit Flatland—sans politics—and you'll see the resurrection of Gudermann. Who but Robinson would show you just how close the science of maps (cartography) is to special relativity, or how a simple tale of jet lag shows that relativity may be more commonsensical than you think? And who but Robinson would show you that the *shear transform*, well known to the mechanical engineer, is the same construction as relativistic time dilation and dimension distortion? Where will you get as satisfying a background on the concepts of relativity as this elegant book?

Robinson's RELATIVITY is not a picture book, a problem book, or a graduate text. Instead, it is an exploration in which you'll find out things that Einstein didn't even know. Enjoy it.

Nathan Cohen
Director
Science and Engineering Program
Boston University

PREFACE

This book is about space and time. The concept of space is an abstraction; in itself, space is literally nothing at all. The idea of space forms a fertile ground for many kinds of speculation. It gives rise to totally different interpretations—religious, philosophical, scientific. It presents to the mind an image of such vastness that the imagination is extended to the limit to conceive it. The history of ideas about space is fascinating, as it demonstrates the changes in attitudes and perception throughout the whole period of recorded human history.

The concept of time, as treated in art and poetry, is always mysterious, and philosophical treatments are not much clearer. When we turn to science, we find that questions about time raise deep issues. A formal definition of time already presents major obstacles, but it is usually defined in terms of the clocks used to measure it. This method originates with the use of the Earth as a clock to define one year as the period of revolution about the sun. Time passes in such an unforgiving way that it is considered unidirectional. We cannot turn back a clock, for some other clock will register the inexorable flow that we vainly try to circumvent.

The special theory of relativity unites the concepts of space and time. Relativity has given us a new concept called *space-time*. This idea has stimulated the imaginations of layperson and scientist alike, but it is a difficult concept. It must be explained by means of language. What language should be used? The language of the layperson is *metaphor*, the language of the scientist is *mathematics*. To describe relativity theory we must use both languages, for one is not enough to teach this baffling subject. The challenge is basically to overcome the imposing difficulties of relativity by appealing to the power of the imagination.

Before the advent of the special theory of relativity in 1905, it was possible for physics to be explained to the layperson in verbal terms that, although fuzzy, made sense. Henceforth, this was impossible, for the characteristic quality of the special theory was that it seemed to violate all the principles of common sense. Previously, science could usually illustrate its most abstruse discoveries by means of a mechanical model; afterward, models were mathematical. The full sense of dismay felt by those in the classical tradition may be illustrated by Lord Kelvin's remark that he could understand something only if he could make a mechanical model of it. The mathematization of nature, however, had its compensations. In particular, it opened up the whole area of theoretical physics to abstract mathematics, where symmetries could be seen and exploited.

The so-called affront to common sense is the *second postulate: the velocity of light is constant in all inertial frames*. In deference to Kelvin, there is no mechanical model that can be made to illustrate this behavior. It was Einstein who elevated this statement about light to the status of a postulate. (Webster defines a postulate as a proposition that is taken for granted or put forth as axiomatic; that is, a postulate is something assumed without proof as being self-evident or generally accepted.) It was Einstein's boldness in taking this step that transformed the theory of relativity into one of the cornerstones upon which the edifice of modern physics is built.

A metaphor is defined as a figure of speech in which a word denoting one object or idea is used in place of another to suggest a likeness between them. It has been said that modern physics can only be understood by use of metaphors. The work of the scientist, like that of a poet, is to create and extend our concept of the universe by visualizing and expressing nature in terms of analogies. The words of Pythagoras in the fifth century B.C. summed up this idea: "There is geometry in the humming of the strings. There is music in the spacings of the spheres." Here Pythagoras was referring to the spherical shells on which the ancient Greeks thought the planets moved in the heavens. However, the metaphor finds its worth today, for the same mathematical eigenvalues that govern the sweet music of his instrument, the lyre, determine the spacings of spherical shells of the electrons surrounding the atomic nucleus. The musical metaphor of Pythagoras mathematically describes the quantum nature of physical reality.

The use of metaphor in relativity has a long and distinguished history, for Einstein freely used this language as well as the language of mathematics to convey his ideas. The metaphors of Einstein appear in every relativity book. One of Einstein's favorite metaphors makes use of a moving train. He first formulated this metaphor while riding to work in Bern, Switzerland, as a young man. Of course, in his imagination he replaced the actual slow-moving train by one traveling at close to the speed of light. Today, of course, the tendency is to replace his train with a space vehicle, but still his metaphor remains intact. Another metaphor is Einstein's vision of himself riding upon a light beam. This metaphor conveys the essential idea that there is no rest frame for light, a difficult concept at best, and one that can best be grasped in this way. Everyone is familiar with the metaphor of the twins. One twin takes a trip and returns home younger than her stay-at-home sibling. These metaphors and more form the backbone of the theory of relativity.

This book is divided into two parts: Part I, Mataphor, and Part II, Mathematics. The reader can start reading the book either at Part I or Part II, depending upon his or her interests, as each part is self-contained and can be used independently of the other.

In Part I, we give two historical metaphors. The first shows that special relativity first appeared in the sixteenth century in cartography with the development of Mercator's map of the world. This book establishes that Wright's equation given in 1599 for the Mercator projection is equivalent to the Lorentz transformation of Einstein's theory. The second historical metaphor shows that special relativity again manifested itself in the seventeenth century with Kepler's discovery that the planets travel in elliptic orbits. Although nearly every writer compares the revolution caused by the Copernicus heliocentric theory in the seventeenth century to the space-time revolution brought on by relativity theory in the twentieth century, this book shows for the first time that both theories have the same mathematical structure. Kepler's formula for elliptical orbits is identical to Einstein's formula relating velocities in two relativistic frames of reference. Thus through metaphor, we come to the surprising conclusion that Mercator, Kepler, and Einstein mathematically were all dealing with the same theory of relative perceptions.

In addition to the historical metaphors, the book develops several new mythical metaphors in the form of fables and stories. They are the fable of the two Eskimos, the fable of the muon and the mountain, the Olympic metaphor of relativity, the fable of the king's messengers, the story of the flat world, and the metaphor of ants. Each of these brings an understanding of relativity in terms of things that can be visualized. All these metaphors occur in Part I except for the metaphor of the ants, which is relegated to Part II because it makes use of hyperbolic functions.

Most of the developments occurring in physics require the use of mathematics. Mathematical symbols are used to define physical concepts. Mathematical operations are then carried out, and finally physical conclusions are drawn by implicit comparison or analogy. Unfortunately, mathematics involves much manipulation of symbols, and this presents a problem to many people who want to understand physics but do not want to suffer the pains that strict mathematical discipline requires. However, there is no way to avoid the use of mathematics in physics. Also, the difficulties that a person experiences with mathematics are usually at a basic level. High school algebra, geometry, and trigonometry are so important that they cannot be studied too much. This book makes use of this high school mathematics and expands it in directions that are not covered in high school because of limited time. Often a student asks what is the point of learning high school mathematics. This book tries to show that these mathematical skills let him or her probe entirely new fields of thought such as the theory of relativity. Part II of this book carries out this program.

Let us now indicate why the second postulate is so difficult to grasp. A body subject to no external forces moves by inertia at a constant speed in a straight line. This body, as well as all other objects moving along with it at the same speed in the same direction, constitute an *inertial frame* in which the body and these other objects are at rest. An example is Einstein's train and the passengers seated on it. They regard the train and themselves as being at rest. They cannot tell whether the train is stopped, or is moving uniformly with respect to the embankment, unless they look out the windows and see for themselves.

The second postulate requires that each inertial frame has its own unique time and space coordinates. If two frames are in uniform motion with respect to each other, then each frame has a different perception of time and space. While the train is stopped, workers on the embankment perceive both time and space in the same way as the passengers on the train. While the train is in motion, the workers' perception of time and space is different from that of the passengers.

Let us explain why. Because the workers and the passengers each regard themselves as being at rest, we would expect their spatial coordinates to differ. This is not in conflict with classical theory. However, the second postulate states that the velocity of light must be the same in both frames. This is in conflict with classical theory. Because the velocity of light is a ratio of space over time, the conflict can only be resolved in this way. The time coordinates in the two frames must also differ. This has to be so to keep the critical ratio (i.e., the velocity of light) constant. Therefore time is no longer an absolute quantity that is the same everywhere but, instead, time is perceived differently in different intertial frames. This is the most surprising thing about relativity theory.

The implications of the second postulate are astounding. The second postulate says that a particle of light, a photon, has the same constant velocity, 300,000 kilometers per second, in each and all inertial frames. It follows, therefore, that there is no inertial frame in which light has a velocity of zero. In other words, there is no frame in which a photon is at rest.

As we have just seen, a photon has no rest frame. But an inertial frame is required to assign coordinates to the perceptions of time and space. Thus a photon has no concept, no idea, no inkling of time and space. In our inertial frame, we can see a photon aging in our time, and traveling through our space, but the photon itself has no such perception. A photon operates outside the domain of time and space. To a photon the time clocked in any inertial frame is null, and the space meted out is void. The null and void of the photon corresponds to all time and all space in our perspective. Thus *the photon is at the same place at all times (semper), and at the same time at all places (ubique)*.

The words of Sir Isaac Newton apply to light. In terms of his *general scholium* in the *Principia*, a particle of light, a photon, is

> eternal, infinite, absolutely perfect. Its duration reaches from eternity to eternity, its presence from infinity to infinity. It is not eternity and infinity, but eternal and infinite. It is not time and space, but it endures and is present. It endures forever,

and is everywhere present, and by existing always (semper) and everywhere (ubique) it constitutes time and space.

Newton's is the most beautiful description yet written of a fundamental principle of relativity theory, the *principle of light*, which exhibits the complete symmetry of time and space.

As we have just seen, Newton's general scholium is a clear and precise statement of what we call the principle of light. This principle says that light is perfect in the sense of the symmetry of space-time. Being perfect, light can only travel at the ultimate speed in the universe, no more and no less. This means that the speed of light is the same constant value in every inertial frame. But this is the second postulate. Thus the principle of light is the second postulate.

The conclusion reached in this book, therefore, is that the *general scholium of Sir Isaac Newton is the same as the second postulate of relativity theory of Albert Einstein. Both are the principle of light.* The trains of ideas of these two great thinkers are now seen to be closer than was previously believed. They converge on the concept of perfection, such as can be found in the nature of light. Of course, Newton and Einstein were also the ones who first realized, each in his own age, that light, although it travels as a wave, is made up of small individual particles, the photons.

Einstein once said:

> I want to know how God created this world. I am not interested in this or that phenomenon. I want to know His thoughts, the rest are details.

This book will not help the reader in this respect, but it will bring the reader closer to the thoughts of Newton and Einstein. Still one must grapple with the metaphors and struggle with the mathematics. There is no other way. It is not easy, but it is there.

Enders A. Robinson
Lincoln, Massachusetts

PART I

METAPHOR

INTRODUCTION

If you can look into the seeds of time
And say which grain will grow and which will not,
Speak then to me.

William Shakespeare (1564–1616)
Macbeth, Act 1, Scene 3

Plate 0.1. Nicolaus Copernicus (1473–1543)

JET LAG

In its relative motion around the Earth, the sun travels 360° of longitude each day of 24 hours. In other words, the speed of the sun around the Earth is 360/24 or 15° of longitude per hour. As a result, the surface of the Earth is divided into 24 time zones of 15° each. The time of the initial or zero zone is based upon the central meridian of the town of Greenwich (a suburb of London, England). Each of the zones eastward and westward is designated by a number representing the hours by which its local time differs from Greenwich time, as illustrated in the table.

Location	Distance from Greenwich (Degrees West)	Zone (Hours Earlier)
London	0°	0
New York	75°	5
Los Angeles	120°	8
Hawaii	150°	10

Let us illustrate the use of this table. The table says that New York is 75° west of London and that New York time is 5 hours earlier than Greenwich time. Thus when it is 10 hours Greenwich time, it is only 5 hours New York time. Similarly, when it is 10 hours Greenwich time, it is only 2 hours Los Angeles time. Also, when it is 10 hours Greenwich time, it is 0 hours Hawaii time.

Jet Lag

Assume that we are at the London airport. Four airplanes are on the ground. Airplane A is not going to fly. The other three take off at 0 hours (noon) Greenwich time, each with instruction to fly for 10 hours westward and then land. The speeds of the airplanes are given in the table.

Airplane	Speed	Speed of Sun	Relative Speed β
A	0.0° per hour	15° per hour	0.0
B	7.5° per hour	15° per hour	0.5
C	12.0° per hour	15° per hour	0.8
D	15.0° per hour	15° per hour	1.0

Note that the relative speed β is defined as the speed of the airplane divided by the speed of the sun.

According to Robert Elston, the ground controller in London, the following takes place. After 10 hours airplane A is still on the ground. It has flown 0° in 10 hours. In 10 hours, airplane B travels 75° and lands in New York. In the same 10 hours, airplane C travels 120° and lands in Los Angeles, and airplane D travels 150° and lands in Hawaii.

London Ground Controller Data			
Airplane	Destination	Flying Time	Distance
A	London	10 hours	0
B	New York	10 hours	75°
C	Los Angeles	10 hours	120°
D	Hawaii	10 hours	150°

However, for the passengers the situation is quite different. Except for the distance from the airplane door to the passenger seat, the passenger covers no distance. The passenger looks up at the clock (which the flight attendant keeps set at the local time) and notes the time of takeoff and landing. Because the passenger travels no distance (within the airplane), this time difference is called her proper time.

Airplane A travels no distance, and in 10 hours the local Greenwich time goes from 0 hours to 10 hours. Thus, the proper time for airplane A is 10 hours. In 10 hours airplane B travels to New York and crosses five time zones. It takes off at 0 hours Greenwich time and lands at 5 hours New York time. The proper time is the difference between these two times, that is, 5 hours minus 0 hours, or 5 hours. Airplane C leaves London at 0 hours Greenwich time and lands in Los Angeles at 2 hours Los Angeles time, so its proper time is 2 hours. Airplane D leaves London at 0 hours Greenwich time and lands in Hawaii at 0 hours Hawaii time, so its proper time is 0 hours.

Airline Passenger Data			
Airplane	Destination	Proper Time	Distance
A	London	10 hours	0°
B	New York	5 hours	0°
C	Los Angeles	2 hours	0°
D	Hawaii	0 hours	0°

Let us comment on this table. The airplane passenger gets on the plane and gets seated comfortably in her seat. As far as she knows, the airplane is always at rest because her chair is just as stable as the one in her living room at her home in London. At home she uses her sundial to tell the passage of time. So also she does on the airplane. In airplane A she might as well be at home. In airplane B the sundial registers a 5-hour change in the passage to New York. In airplane C the sundial shows a 2-hour change in the passage to Los Angeles. Because airplane D is flying at the speed of the sun, the sun is always overhead and the sundial registers no change on the whole trip to Hawaii.

Because on plane D the passenger sees the zenith (or noon) sun all the way, the whole passage from London to Hawaii to her is simultaneous. The whole trip takes place with the sun directly overhead, which is 0 hours (local time) all the way. However, the London ground controller sees this passage quite differently. For him, the passage is spread out over a time duration of 10 hours, from 0 hours (Greenwich time) to 10 hours (Greenwich time). Thus the same passage is simultaneous in the passenger's frame, but not simultaneous in the controller's frame.

The following table summarizes our results.

Airplane	Speed	Flight Time	Flight Distance	Proper Time
A	0°/hr	10 hr	0°	10 hr
B	7.5°/hr	10 hr	75°	5 hr
C	12°/hr	10 hr	120°	2 hr
D	15°/hr	10 hr	150°	0 hr

The flying time is the time measured by the London ground controller. The proper time is the time measured by the airplane passenger. The London ground controller sees airplane C moving at 12° per hour and records the flight time as 10 hours. However, for the passenger only 2 hours have elapsed. Thus the ground controller realizes that the passenger's clock is running slow (in this case, the passenger loses 8 of the 10 hours, because she passed 8 time zones). In fact, the ground controller sees the clock of any westward moving aircraft as running slow. Thus, for plane C, the ground controller must dilate (or increase) the proper time of 2 hours by a magnification factor of $\gamma = 5$ to bring it up to the 10 hours he records for the flight.

The ground controller has a twin, Mary Elston. Both are age 30. If the twin gets on plane B, she will be 30 years plus 5 hours old when she gets off in New York. However, the ground controller, Robert, will be 30 years plus 10 hours old. So now, the traveling twin Mary is younger than the sedentary twin Robert. We can take this to an extreme. The twin Mary instead gets on the Hawaii plane, but now the plane is instructed to fly westward for 10 years. Because the Hawaii plane is flying at the speed of the sun, it is always high noon on the plane, so the traveling twin does not age. Thus, at the completion of the passage, the traveling twin Mary is still 30 years old, but her sedentary brother Robert is 40 years old.

We can take this to a still further extreme. We know that the supersonic plane Concorde travels faster than the sun. A person getting on the Concorde in London will get off the plane at the westward destination at an earlier time than when she got on. This means that she is younger than when she got on the plane. This finally explains why the British built the Concorde airplane and why their movie actors are better than their American counterparts. Each year the British secretly put all their top actors on the Concorde and fly it westward around the world several times. When their actors finally get off in Hollywood, they are all

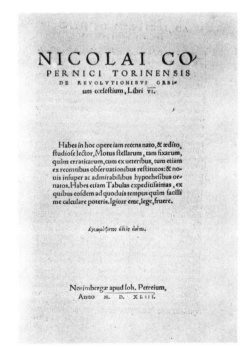

Plate 0.2. Title page of *De Revolutionibus Orbium Coelestium* by Copernicus, 1543.

Plate 0.3. Leaf of Copernicus' *De Revolutionibus*, 1543, showing heliocentric system of planets.

younger, and with this incentive they do a better job at acting. If you believe this, there will not be any trouble in understanding the rest of this book. What we have recounted in this introduction is not the theory of relativity. The theory of relativity is much stranger.

POETIC METAPHORS

The story about jet lag in the preceding section does not depict the theory of relativity. The reason is that the perceived speed of the sun in its apparent daily passage around the Earth is not the ultimate speed in the universe, because, for example, the Concorde airplane can travel at a greater speed. The reference speed in the theory of relativity must be a physical speed that cannot be exceeded by anything. The speed of light satisfies this requirement.

In this book all the metaphors about relativity theory which we have devised are prosaic. However, the most powerful metaphors are poetic, and here one must look to the works of the great poets for elucidation. In *Paradise Lost*, John Milton (1608–1674) writes[1]

---

[1] Milton, John. "Paradise Lost," III, 3–5. In *The Poetical Works of John Milton*, Second Edition, J. Johnson, London, 1809.

> God is light,
> And never but in unapproached light
> Dwelt from eternity.

Milton's "unapproached light" is an expression of the ultimate speed of light, and this poetical metaphor is as relevant as Einstein's metaphor that one can never catch up with a beam of light in order to ride upon it.

In his epic poem Milton meditates on the immensity of the universe and places the Empyrean beyond the bounds of the fixed stars. The name itself, the Empyrean, signifies fire. Because fire is regarded as the best of the elements, and because heavenly fire is better than elemental fire, the Empyrean represents the highest perfection. Milton writes[2]

> Now had the Almighty Father from above,
> From the pure Empyrean where he sits
> High Thron'd above all highth, bent down his eye,
> His own works and their works at once to view.

In this poetic metaphor, Milton uses the Empyrean to signify the perfect quality of light, an essential idea in the theory of relativity.

Let us now turn our imagination to space travel and choose an extreme example, yet well beyond present human ability. A space colonist living on a sister planet of a far-away star wants to make a sentimental journey to the home of her ancestors, her true love, the Earth. Suppose that the star is one million light-years distant from the Earth. Because light requires one million years to travel from the star to Earth, it seems impossible that she can reach her destination in her lifetime. However, if she travels at a great enough speed, at a speed very close to the speed of light, she can reach the Earth in only one day of her lifetime. The reason is that her clock suffers an enormous retardation. In other words, when she travels at this great speed, time for her slows down almost to a stop. Moreover the distance of the journey in her perception will be contracted almost to nothing. Thus for her the trip is not a long one, either in distance or time. By traveling at this great speed, she has come close to the idea expressed by Alexander Pope (1688–1744) who writes[3]

> Ye Gods! Annihilate just space and time
> And make two lovers happy.

Here we have a fundamental idea in relativity theory, namely, that space and time cannot be separated, but are parts of a larger domain known as space-time.

---

[2] Milton, John. "Paradise Lost," III, 56–59. *Ibid.*

[3] Pope, Alexander. *Martinus Scriblerus of the Art of Sinking in Poetry*, Chapter 11, B. Mutte, London, 1727.

As we have just seen, when a space traveler approaches the speed of light very closely, her clock is arrested nearly completely, and she is almost untouched by the ravages of time. Also the distance of the journey to her is contracted almost to nothing. To such a traveler, everything would happen almost at the same time and almost at the same place.

A space traveler, a material thing, can approach but never reach the ultimate speed, the speed of light. However, there are massless (nonmaterial) particles which do travel at this ultimate speed. They are the photons, small particles of light which are just barely detectable, a few at a time, by the human eye. It is said that Newton was the only person who could ever see just one lone photon at a time in the blackness of his darkened chamber at Cambridge University.

A material particle can be accelerated, greatly, but never to the ultimate speed. As a result its space and time can be greatly decimated, but never annihilated. However, a photon travels at the ultimate speed, so for it the destruction of space and time is complete. The passage of time has no reality for a photon. In its make-up, there is nothing that corresponds to past and future. There is no before or after. Its existence is atemporal. Likewise, the extension of space has no reality for a photon. All distances are shrunk to zero. In its framework, there is nothing that indicates here or there. Its existence is aspatial. For a photon, space and time are indeed annihilated.

Suppose a photon is emitted from a source of light, and suppose that it travels for one year to a receiver where it is absorbed. In this description, we are speaking of our own space and our own time. Specifically, the source and receiver are separated by a distance of one light-year and the travel time is one year. But as far as the photon is concerned, both source and receiver are located at the same place, and both emission and absorption happen at the same time. This means that an individual photon is at all places at all times. In other words, a particle of light sees all events in the universe as coincident in space and simultaneous in time. We can say that a photon is in a state of spacelessness and timelessness. In the words of Shakespeare (1564–1616), a photon demonstrates that[4]

> Time, that takes survey of all the world,
> Must have a stop.

Seven hundred years ago Dante Alighieri (1265–1321) wrote *The Divine Comedy*. It is literally about the travels of a pilgrim in three canticles: *Inferno*, *Purgatorio*, and *Paradiso*. Allegorically, it is a search for an understanding of the nature of the universe. In *Paradiso*, the voyage of the pilgrim is a journey of ascending in the universe to the unmoving Empyrean. The pilgrim travels the immaterial path of intellectual light. Throughout the journey the pilgrim is saturated with images of light until the incandescence precipitates an intellectual

---

[4] Shakespeare, William. "Henry IV, Part I," Act 5, Scene 4. From *The Works of William Shakespeare*, Volume 5, Chatterton-Peck Co., New York, 1885.

atmosphere filled with the exultancy of understanding. The Empyrean is[5]

                    pura luce:
luce intellettual, piena d'amore;
      amor di vero ben, pien di letizia;
            letizia che trascende ogne dolzore.

                    [pure light:
light intellectual, full of love;
      love of true good, full of joy;
            joy that transcends every sweetness.]

The pilgrim looks into the absolute light and states his experience[6]

Oh abbondante grazia ond' io presunsi
      ficcar lo viso per la luce etterna,
            tanto che la veduta vi consunsi!

[O abounding grace wherein I presumed
      to fix my look on the eternal light,
            until my vision was absorbed in it!]

Nel suo profondo vide che s'interna
      legato con amore in un volume
            cio che per l'universo si squaderna,

[Within its depths I saw ingathered
      bound together into one volume by the force of love
            that which throughout the universe is scattered in leaves.]

sustanze e accidenti e lor costume
      quasi conflati insieme, per tal modo
            che cio ch'i' dico e un semplice lume.

[matter and events and their relationships
      seeming to be fused together, in such a way
            that what I speak of is one simple light.]

La forma universal di questo nodo
      credo ch'i' vidi, perché piu di largo,
            dicendo questo, mi sento ch'i' godo.

[The universal form of this structure
      I believe that I saw, because more abundantly,
            as I say this, I feel my joy increase.]

---

[5] Alighieri, Dante. "Paradiso," XXX, 39–42. From *La Devina Commedia di Dante Alighieri*, Tom. III, Nella Stamperia de Romanis, Roma, 1822.

[6] Alighieri, Dante. "Paradiso," XXXIII, 82–93. *Ibid.*

In this instant the pilgrim gets a momentary glimpse of the ordering and nature of the universe, and sees the wholly intellectual source of all matter and all events. It is a final and culminating vision of pure light as the unity and totality of the universe. The metaphor compares our spatial-temporal universe to the scattered pages of an unbound book. In the eternal light all the contents of the space-time universe (the dispersed and scattered leaves of the book) are gathered together and bound into one unity (in one volume). Specifically, all matter and events and their relationships are fused together and united in one simple flame of light.

Since light is outside of the dimension of time, there is no before or after in light. Light is the entity for which all times are the same. The happenings of all things in the universe throughout time are simultaneous in this timeless dimension. Since light is outside of the dimensions of space, there is no here or there in light. Light is the entity for which all places are the same. The locations of everything in the universe are coincident in this spaceless dimension. The light is the alpha and omega, the beginning and the ending, which is, and which was, and which is to come.

In this ultimate vision, the pilgrim sees the scattered parts of the universe fused into a flawless simple whole. All the strewn and disjoint pieces are assembled into a perfect entity. Free of the delusive dimensions of space and time, the true nature of the universe is revealed in unity of form without cracks and other defects. All this is revealed as pure light. But a glimpse is all that the pilgrim gets. The vision ends and disappears from the pilgrim's memory. The only evidence that remains behind is the consciousness of joy which he feels. This is Dante's metaphor of Einstein's theory of relativity.

CHAPTER 1

FROM COPERNICUS TO EINSTEIN

I saw Eternity the other night
Like a great Ring of pure and endless light
    All calm, as it was bright
And round beneath it, Time in hours, days, years
    Driven by the spheres
Like a vast shadow moved, in which the world
    And all her train were hurled.

Henry Vaughan (1622–1695)
The World

Plate 1.1. Tycho Brahe (1546–1601)

PARALLEL PATHS TO GRAVITATION

There is a striking parallel in the scientific development of Newton's theory of gravitation and of Einstein's theory of gravitation (also called the general theory of relativity). The corresponding players are

Age of Discovery	*Nuclear Age of Discovery*
Nicolaus Copernicus (1473–1543)	James Clerk Maxwell (1831–1879)
Tycho Brahe (1546–1601)	Albert A. Michelson (1852–1931)
Johannes Kepler (1571–1630)	Hendrik Lorentz (1853–1928)
Galileo Galilei (1564–1642)	Henri Poincaré (1854–1912)
Isaac Newton (1642–1727)	Albert Einstein (1879–1955)

Copernicus wrote in his book *De Revolutionibus Orbium Coelestium*,

At rest, however in the middle of everything is the Sun. For in this most beautiful temple, who would place this lamp in another or better position than that from which it can light up the whole thing at the same time? For the Sun is not inappropriately called by some people the lantern of the universe, its mind by others, and its ruler by others. Hermes the Thrice Great labels it a visible god, and Sophocles's Electra, the all-seeing.

Maxwell in his book *Treatise on Electricity and Magnetism* showed that light must be an electromagnetic wave. With full mathematical support in the form of four equations, now universally known as Maxwell's equations, Maxwell placed light in the central position in modern physics, even as Copernicus had put the sun in this exalted place. Copernicus made the sun perfect in his philosophy, and Maxwell made light perfect in his. Both Copernicus and Maxwell found perfection, one in the sun and the other in light.

Although haughty with the royal family and neglectful of his tenants, Tycho Brahe painstakingly gathered astronomical observations of Copernicus' solar system. Tycho's book *Epistolarum Astronomicarum* was a great descriptive treatise on his instruments and their methods of use.

The passion of Michelson's life was the building of instruments that required great accuracies. In 1881, Michelson constructed an interferometer, a device to split a beam of light in two, send the parts along different paths, and then bring them back together. This instrument was designed to carry out an experiment that Maxwell had suggested two years before. The definitive experiment with his interferometer was performed in 1887 by Michelson and his friend Edward Morley. Both Tycho and Michelson were great instrumentalists and observers, one testing the perfection of the sun and the other of light.

Kepler took Tycho's observations of the planet Mars and tried to reconcile them with some simple geometrical theory of motion. He began by assuming that

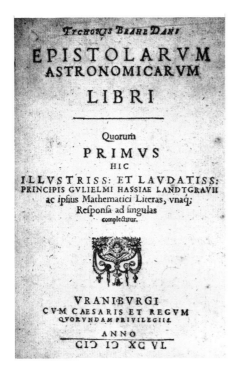

Plate 1.2. Title page of *Epistolarum Astronomicarum* by Tycho Brahe, 1596.

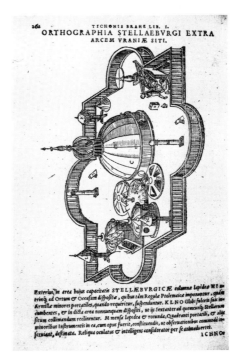

Plate 1.3. The astronomical observatory of Tycho Brahe, 1596. (Turn picture on side.)

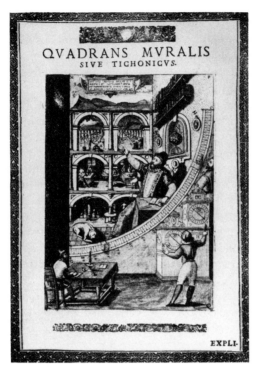

Plate 1.4. Tycho Brahe in his observatory.

the orbit was a circle with the sun slightly off-center, and that Mars did not necessarily move with uniform velocity along this circle. Kepler's first task was to determine the radius of the circle and the direction of the axis connecting perihelion and aphelion. After excruciating trial-and-error calculations, Kepler absentmindedly put down three erroneous figures for three vital longitudes of Mars, never noticing his error. His results, however, were nearly correct because of several mistakes of simple arithmetic committed later, which happened very nearly to cancel out his earlier errors. At the end, he seemed to have achieved his goal, but because two other observations from Tycho's data did not fit, he decided that the sacred concept of circular motion had to go.

At the end of his quest, Kepler obtained the elliptic orbic. He wrote "I thought and searched, until I nearly went mad, why the planet preferred an elliptical orbit." In Copernicus' solar system, the sun was at the center of the circular orbit of a planet (Figure 1.1).

The angle ϕ locates the planet on that circle. In Kepler's solar system, the sun is at one focus of the elliptical orbit of a planet. The angle ϕ' locates the planet on that ellipse. The mean radius of the Kepler ellipse is the same as the radius of the Copernicus circle, and the ellipse has eccentricity β. The formula that relates the circular perception ϕ of Copernicus to the elliptical perception ϕ'

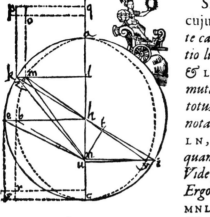

Figure 1.1. Kepler's illustration of his perception *m* and Copernicus' perception *k* of the same planet. This is the theory of relativity. (Courtesy of History of Science Collections of the University of Oklahoma.)

of Kepler was derived by Kepler. It is

$$\sin \phi' = \frac{\sin \phi - \beta}{1 - \beta \sin \phi} \qquad (1.1)$$

Note that both the Copernicus planet and the Kepler planet are the same planet, but the planet is perceived differently in the two different systems.

Lorentz took Michelson's observations and tried to reconcile them with the simple geometrical theory of optical wave propagation. Lorentz's concern over being unable to explain the Michelson-Morley experiment of 1887 is clear from a letter that he wrote to Lord Rayleigh in 1892. Lorentz said "This experiment has been puzzling me for a long time." After many false starts, Lorentz finally decided that the sacred concept of absolute length had to go.

Lorentz struggled long and hard with the observational data of Michelson,

Plate 1.5. Johannes Kepler (1571–1630)

and considered every mathematical scheme. Finally, he adopted an explanation conceived by George Francis FitzGerald (1851–1901). FitzGerald's solution was that the distance covered by the light beam altered with the velocity of motion of the light source in such a way as to allow the beam of light to travel at the same velocity in all directions. The final equations, formulated and named by Poincaré as the Lorentz transformation, relate the coordinates of two inertial frames of reference in relative motion with respect to each other. Let the relative motion between the two frames be given by velocity β. Then the Lorentz transformation yields the physical velocity addition formula:

$$v' = \frac{v - \beta}{1 - \beta v} \qquad (1.2)$$

This equation relates the perception v of the velocity of an object in one frame to the perception v' of the velocity of the same object in the other frame. Although it is the same object, its physical velocity is perceived differently in the two different systems (inertial frames).

Identify sin φ with v and sin φ' with v'. Then Kepler's formula (1.1) is identical with the physical velocity addition formula (1.2) derived from the Lorentz transform. The mathematics of Kepler is identical with the mathematics of Lorentz. Both Kepler and Lorentz were able to glean the mathematics by trial and error, but neither seemed to know the true reason why. We still do not know the reason why in relativity theory, except that it works.

Galileo was a universal genius. Among his contributions to dynamics was his analysis that showed that the path of a projectile is a parabola. It is a striking fact that the conic sections had been studied for 2000 years before two of them almost simultaneously found applications, the ellipse in astronomy and the parabola in physics. As early as 1632 in his *Dialogue Concerning Two Chief World Systems*, Galileo described relative motion. He pointed out that physical experiments on moving bodies made on board ship (in a cabin below decks so the observer could not see out) would not tell the observer whether the ship were stationary or sailing along at a constant speed. For example, if one dropped a coin on the ship, the coin would fall to the same place on the deck whether the ship was at rest or moving at uniform speed. This concept of Galileo marks one of the epochal developments of physics, and it is called the *Galilean principle of relativity*. It was devised by Galileo to answer critics about the laws of physics on the moving Earth. It is one of the basic principles of physics.

Implicit in the application of Galileo's principle of relativity is an assumed infinite signal speed. The Galilean relativity principle is an expression of a symmetry existing in nature. Until the development of special relativity, it was believed that all mechanical systems possessed this symmetry. The branch of mathematics that deals with symmetries is called *group theory*. The idea of a group is one of the great unifying ideas in mathematics. Group theory makes possible the examination of symmetries in nature that are not available by direct observation. As a result, the systematical application of group theory leads to deep physical insights. The importance of abstraction is nowhere more evident than in the concept of a group. Group theory is essential to the development of modern physics, as seen in the fields of spectroscopy, crystallography, atomic and particle physics, as well as relativity theory. In modern mathematical language, the Galilean principle of relativity can be expressed in terms of the mathematical object known as the special Galilean group. This is a symmetry group that encompasses the space-time symmetries imposed by the Galilean principle of relativity under the implicit assumption that there is a signal that travels at an infinite velocity. Because light travels at such a great speed, this still is a reasonable assumption for most Earth-bound purposes.

When Gauss died in 1855, it was generally believed that there would be never again a universalist in mathematics, one who contributes to all branches, pure and applied. Poincaré proved this view wrong. Some calculus problems had been considered as far back as Archimedes, but Newton and Leibniz are the ones who created calculus as a major mathematical discipline. Likewise, some topological problems are found in the works of Euler, Möbius, and Cantor, but it was Poincaré who created topology as a major branch of mathematics. This invention dates from 1895 with the publication of Poincaré's book *Analysis Situs*. Poincaré wrote hastily and extensively, publishing more memoirs per year than any other mathematician. Both Gauss and Poincaré teemed with so many ideas that it was difficult for them to jot the thoughts down on paper, both had a strong preference for general theorems instead of specific cases, and both contributed to a wide variety of branches of science. In mathematics, the concept of a group is fundamental to the fields of differential geometry and topology, and some of the fundamental theorems in this area are due to Poincaré.

Newton had introduced the conceptions of absolute space and absolute time. In a very careful study made in 1899, Poincaré came to the conclusion that there is no place which we can regard as absolutely at rest. Poincaré's analysis had profound implications, which he summed up in his principle of relativity. Since 1900, Poincairé had been speaking in various contexts of this "principle of relativity." The definitive statement came in St. Louis in 1904 where Poincaré delivered an important address. Here he elaborated on the principle of relativity "according to which the laws of physical phenomena must be the same for a stationary observer as for an observer carried along in uniform motion of translation; so that we have not and can not have any means of discussing whether or not we are carried along in such motion." Unlike the Galilean principle of relativity, Poincaré's principle was to apply to all physical phenomena which would include electromagnetic phenomena as well as mechanical phenomena. For this to be possible, Poincaré said that "there would arise an entirely new mechanics, which would be, above all, characterized by the fact that no velocity could surpass that of light, anymore than any temperature could fall below zero absolute, because bodies would impose an increasing inertia to the causes, which would tend to decelerate their motion; and this inertia would become infinite when one approached the velocity of light." Thus, in St. Louis in 1904, Poincaré set forth two principles: the relativity principle and the concomitant principle that the finite speed of light is the ultimate signal speed.

On the basis of these two principles, Poincaré in the following year devised the symmetry group today known as the Poincaré group. (In mathematical terms, it is a ten-parameter Lie group.) The Poincaré group expresses the fundamental space-time symmetry of nature and includes the Lorentz transformation as a subgroup. The scientific developments leading up to Poincaré's discovery were basically inductive methods based on experimental results. However, his mathematics made it now possible to follow a deductive procedure in theoretical physics based on the fundamental assumption that the underlying symmetry of all physical systems is the Poincaré group. For example, in particle physics, new theories usually require complicated structures obeying involved dynamical laws. However, these laws must admit the Poincaré group as a symmetry group, and as a result certain predictions concerning their behavior can be made without a detailed knowledge of these laws. The introduction of group theory represents one of the most fundamental advances in theoretical physics and is essential to new discoveries in modern physics.

Both Galileo and Poincaré each found that perception can only be understood in terms of the principle of relativity. Galileo implicitly assumed an infinite signal speed. As a result, his relativity principle leads to the Galilean group which describes symmetries valid only for classical physics (that is, the physics of systems in which the velocities involved are much less than the velocity of light). Poincaré assumed that the finite speed of light is the ultimate signal speed. As a result, his relativity principle leads to the Poincaré group which describes the space-time symmetries valid for all of physics.

More has been written on Newton and Einstein than any other scientists, so we can only hope to touch upon their greatness here. Both started a new era in physics: Newton initiated Newtonian or classical physics and Einstein modern physics. Among his many contributions, Newton is best remembered for the invention of calculus which forms the basis of modern mathematics, the three laws of motion which make up the essence of mechanics, the inverse-square law of gravitation which unified our concept of the solar system, and his work in optics which first explained the spectral nature of light. In a parallel manner, Einstein is best remembered for his use of Riemannian geometry which forms the basis of modern cosmology, the definition of simultaneity, the concept of relativity of length and time, and the equivalence of mass and energy all of which make up the essence of special relativity, the geometric law of gravitation (or general relativity) which unified our concept of the universe, and his work in optics which first explained the quantum nature of light. Newton's best known equation is his second law of motion, namely,

$$F = \frac{dp}{dt}$$

(which says that force F is equal to the rate of change of momentum p with respect to time t). This equation opened up the mechanical world of the old Industrial Revolution. Einstein's best known equation is his equivalence of energy and mass, namely,

$$E = mc^2$$

(which says that energy E is equal to mass m times the square of the velocity c of light). This equation opened up the nuclear world of the new industrial revolution. Newton and Einstein summed everything up, Newton arriving at the inverse square law of gravitation and Einstein at a geometrical law of gravitation. In this book, we stop short of gravitation, and consider only the developments leading to the special theory of relativity.

In summary,

> Copernicus and Maxwell were the visionaries who provided the impetus by finding perfection in the sun and light, respectively.
>
> Tycho and Michelson were the explorers who collected the essential data by building instruments and making observations.
>
> Kepler and Lorentz were the discoverers who used the data to lead them to the unknown territories thereby making possible the developments of the fundamental formulas of nature. The amazing part of this, here put forward for the first time, is that each discovery resulted in the same mathematical

formula. The greatest physical discovery in the seventeenth century has the same mathematics as the greatest one in the twentieth century. The more things change, the more things stay the same.

Galileo and Poincaré were the inspirers who provided the insight by perceiving the symmetry of nature by means of the principle of relativity, Galileo with an assumed infinite ultimate signal speed, and Poincaré with the finite speed of light as the ultimate signal speed.

Newton and Einstein were the explainers, the grand synthesizers, who produced the dénouements, each with a law of universal gravitation.

THE VELOCITY OF LIGHT

As we know today, light is a stream of particlelike concentrations of electromagnetic energy. These particles are called photons. Photons can exist only in motion at the speed designated by the letter c. The symbol c denotes the *velocity of light*. In fact, the usage of the letter c comes from the Latin word *celeritas*, which means velocity. Furthermore, photons propagate not in the fashion of material objects such as baseballs but, instead, photons propagate in a wavelike fashion such as ocean waves. However, light waves are not entirely like ocean waves, because ocean waves require a material medium (seawater) in which to propagate, whereas light can propagate in vacuum.

Aristotle (384–322 B.C.) censured Empedocles (ca.490–ca.430 B.C.) for having spoken of light as traveling, that is, taking time to go from one place to another. Light for Aristotle was not an effluence of photons which stream from a luminous object with finite speed c. Light for Aristotle was a quality that the medium acquired all at once from the luminous object, just as water may conceivably freeze at all parts simultaneously. The Aristotelian view prevailed for many centuries afterward. A notable exception was the view of Ibn al-Haytham (ca.965–ca.1039), known as Alhazen, who asserted that the movement of light required a finite, though imperceptible, interval of time. Roger Bacon (ca.1220–ca.1292) produced new arguments to defend this view, but still most writers up to the seventeenth century held to the mistaken doctrine of instantaneous transmission.

The fact that the velocity of light is very great, and perhaps infinite, was known from ancient times. Galileo endeavored in 1607 to measure the velocity of light with the aid of lantern signals, but without success, for light traverses earthly distances in extremely short fractions of time. The first measurements that succeeded in estimating the velocity of light were ones that made use of the enormous distances between heavenly bodies. Olaf Roemer (1644–1710) was the first person who showed that the velocity of light c is finite, making use of astonomical observations of the eclipses of the moons of Jupiter. Roemer was a young Danish astronomer living in Paris, and he published only one scientific paper. Though only a page and a half long, it brought him an immortal place in science. This

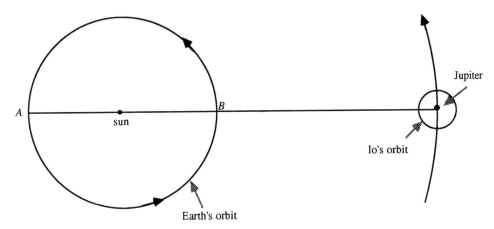

Figure 1.2. Roemer's discovery. The light travel time from Jupiter to the Earth is a minimum when the Earth is at *B* (opposition) and a maximum when the Earth is at *A* (conjunction). The periods between successive eclipses of Io by Jupiter are greater as the Earth travels from *B* to *A*, and the periods are less as it returns to *B*.

paper appeared in the French *Journal des Scavans* on December 7, 1676, and was entitled "Proof of the movement of light."

Roemer showed that light is not propagated instantaneously (Figure 1.2). For a long time, he had observed the rotation of the moons of Jupiter and timed their eclipses as they passed around that huge planet, with the care and accuracy already shown by his countryman, Tycho Brahe (1546–1601). Roemer found the schedules of the moons, that is, how long it takes on the average for each moon to go around Jupiter. However, he noticed that the moons sometimes were ahead of schedule and sometimes behind. In fact, they were ahead of schedule when Jupiter was particularly close to Earth, and they were behind when Jupiter was farther from Earth. The reason for this discrepancy was simple and beautiful. It takes a little while to see the moons of Jupiter because of the time it takes light to travel from Jupiter to the Earth. When Jupiter is closer to the Earth, the time is a little less, and when Jupiter is farther from the Earth, the time is a little more. This is why the moons appear to be a little ahead or a little behind schedule, depending on whether they are closer to or farther from the Earth.

For the moon Io of Jupiter, Roemer found that the eclipse was 11 minutes ahead of schedule when the Earth is closest to Jupiter. Also, he found that the eclipse was 11 minutes behind schedule six months later when the Earth and Jupiter are farthest apart. Thus, there is a 22-minute delay between these two readings. The reason for the delay is the extra time it takes for light from Io to reach Earth, as the light now has to cover the extra distance equal to the diameter of the Earth's orbital path around the sun. Because Roemer was working in the Paris observatory where the diameter of the Earth's orbit was first determined, he presumably knew this diameter. But there is no evidence that he ever troubled to divide this diameter by the time delay of 22 minutes to discover the speed of light. His whole interest was showing that the speed was not infinite.

In Roemer's time, the best known value of the average diameter of the Earth's orbit about the sun was 283,000,000 km. (Today we know it is nearly 300,000,000 km.) A year or two later, Christian Huygens (1629–1695) with full acknowledgment to Roemer, performed the division:

$$c = \frac{283{,}000{,}000}{22(60)} = 214{,}000 \text{ km/s}$$

This great speed was almost unimaginable, and it is not surprising that it took about 50 years for Roemer's demonstration to gain full acceptance. From then on, however, increasingly accurate measurements of the speed of light were obtained from ingenious experiments.

A direct manifestation of the finite speed of light is the phenomenon of the aberration of starlight. This phenomenon was discovered by the English astronomer James Bradley (1693–1762).

The basis of the first terrestrial determination of the velocity of light was performed in 1849 by the French physicist A. H. L. Fizeau (1819–1896). Galileo had placed two observers some distance apart, each observer being equipped with a lamp that could be shut off quickly. The first observer shuts off his lamp, and immediately upon seeing this, the second observer shuts off his lamp. The first observer notes the time that elapses between the instant at which he shuts off his lamp and the instant at which he sees the other observer shut off his lamp. If this observed time interval is proportional to the separation distance of the two observers, then the speed of light is finite. However, a valid result cannot be obtained from Galileo's experiment over the distances available between terrestrial points. The difficulty, aside from the necessity of making measurements of very short time intervals, is that it takes a human observer a comparatively long time to react to a stimulus. A certain finite delay, called the reaction time, elapses between the time when an observer sees the other's lamp being shut off and when the observer shuts off his own lamp. Fizeau overcame this human weakness in the following way. Since the second observer acts merely as a reflector of the shutting off of the light from the first lamp, the second observer can be replaced by a mirror that reflects any light signal without delay. In addition, the measurement of the time interval can be performed by mechanical means in which human reaction time plays no part.

The mechanics of the Fizeau method were improved upon by the French physicist Leon Foucault (1819–1868). By the time Maxwell put forward his electromagnetic theory, a measured speed of close to 300,000 km/s was available to provide the crucial confirmation of his theory.

Maxwell formulated the theory of electromagnetism first in several large articles in the early 1860s and later published it in his two-volume *Treatise*. Maxwell's theory is a field theory in which waves are transmitted at a finite velocity in vacuum, namely, the propagation velocity of the field. The theory provides a definite value for this velocity; namely, the value $1/\sqrt{\varepsilon_0\mu_0}$, where ε_0 and μ_0 are the electric and magnetic constants. Because this value came out close to 300,000 km/s, Maxwell stated that this velocity is the velocity c of light in vacuum. Thus Maxwell showed that light is an electromagnetic wave and also predicted the existence of all other electromagnetic waves.

The speed of light has a fundamental place in modern physics. At present, the most accurate value of the speed of light, as well as of any other electromagnetic wave, traveling in vacuum (or free space) is

$$c = 299{,}792{,}457.4 \pm 1.2 \text{ m/s}$$

or about 186,300 miles/s or about 1 billion feet per second. This is one of the most important constants of physics, for no signal has ever been observed to travel faster. Invariably, in pedagogical discussions, the accurate value of c is rounded off to 300,000 km/s or 186,000 mi/s or 1 gigafoot per second, and we will continue to use that practice.

Light does not propagate instantaneously, although its velocity is indeed tremendous: 300,000 km/s. This huge velocity is hard to conceive, because in our everyday experiences, we deal with far lesser speeds. The speed of a typical space vehicle is a mere 12 km/s. The Earth revolves about the sun, but the speed of the Earth is only 30 km/s. The colossal velocity of light is very extraordinary in itself. However, much more striking is the fact that this velocity is strictly constant. There are essential differences between the propagation of light and the propagation of a material body such as a bullet. The velocity of a bullet depends largely on the design of the rifle from which it is fired and the properties of the gunpowder, while the speed of light is always the same no matter what its source. You can always accelerate or decelerate the motion of a bullet artificially. If you place a box of sand in the path of a bullet, the bullet will lose velocity when it pierces the box, and the bullet will not regain its velocity when it emerges from the box. But this is not so with light. The speed of light in vacuum cannot be changed; a photon (a particle of light) in vacuum cannot be accelerated or decelerated. The speed of light in vacuum is always the same. Let us place a plate of glass in the path of a beam of light. Since the velocity of light in glass is less than in vacuum, the beam travels slower in the glass. However, having passed through the glass, the light regains the speed of 300,000 km/s. Whatever change in velocity the light undergoes in matter, the light propagates with the same velocity (300,000 km/s) as soon as it emerges into vacuum.

In this respect, the propagation of light is more like the wave propagation of sound than of the propagation of a material body such as a bullet. Sound is the vibration of the medium in which it propagates; for example, sound in the air is the vibration of the gas particles making up the air. Therefore, the wave velocity of sound depends upon the properties of the medium and not on the properties of the sound-producing body. Sound velocity cannot be increased or decreased any more than light velocity, even by passing the sound through other bodies. For example, if we place a metal barrier in its path, the sound will change velocity inside the metal, but as soon as it emerges again into the air it will resume its initial velocity.

Both sound and light propagate as waves, and thus they have such properties in common. Place an electric light bulb and an electric bell under the glass hood of an air pump. As we pump out the air from under the hood, the sound of the bell will get weaker and weaker, until it becomes altogether inaudible. The light bulb, on the other hand, will radiate light as usual. This experiment shows that sound propagates in a material medium while light propagates even in a vacuum. Therein lies the essential difference between sound and light propagation. This essential difference is expressed in the fact that the velocity of light in vacuum is absolute (300,000 km/s in all inertial frames), whereas the velocities of other kinds of waves and velocities of material bodies are relative.

MACROWORLD AND MICROWORLD

It is when we deal with the microworld of atoms (small in relation to humans) and the macroworld of astronomy (large in relation to humans) that our concepts of space and time come together. In astronomy, Vega is a nearby star, and yet it is 250 trillion km (2.5×10^{14} km) away. The bulk of the stars in our Milky Way galaxy are about 300 quadrillion (3×10^{17}) km away, whereas the nearest full-sized outside galaxy (the Andromeda galaxy) is 15 quintillion (1.5×10^{19}) km away. We can mathematically accept such macroscopic distances, but we cannot readily visualize them. The trick that brings them into our comprehension is the utilization of the speed of light.

The velocity of light in a vacuum is (as a round number) 300,000 km/s. A light-second is defined as that distance which light travels in 1 second of time, and so is equal to 300,000 km. A more practical unit in astronomy is the light-year. A light-year is the distance which light travels in a year; it is equal to $300,000 \times 365 \times 24 \times 60 \times 60$ km, or 9.5 trillion km. The units light-second and light-year are units representing "time of length," namely, the 1 second or the 1 year that light travels to cover the distances 300,000 kilometers or 9.5 trillion kilometers, respectively.

In this "time-of-length" unit, the star Vega is 27 light-years away, which is a small distance compared to the 32,000 light-years distance of the bulk of the stars in the Milky Way and the 1,500,000 light-years distance of the Andromeda galaxy. The difference between 27, 32,000, and 1,500,000 is easier to visualize than is the difference between 250 trillion, 300 quadrillion, and 15 quintillion. Thus, the light-year is a practical unit when it comes to expressing astronomical distances.

Let us mention another facet. The use of the velocity of light in defining the unit of distance has the virtue of simplifying the connection between space and time. Any message sent by light or any other electromagnetic signal from Vega to Earth will take 27 years. We see that the use of light-years implies that the communication time and the distance is the same number. For example, from Earth to Vega, the time of communication by an electromagnetic signal is $t = 27$ years and the distance is also $x = 27$ light-years. They are numerically equal:

$$t - x = 0 \quad \text{(for a light signal)}$$

However, depending upon the convention used for the directions of the t and x axes, the values of t and/or x might be either positive or negative. Therefore, we write instead

$$t^2 - x^2 = 0 \quad \text{(for a light signal)}$$

The square root of the quantity on the left is called the *time-space interval*; that is,

$$\text{time-space interval} = \sqrt{t^2 - x^2}$$

We therefore conclude that for any light signal, the time-space interval is equal to zero. This is an important concept in the special theory of relativity.

The counterpart to the time-space interval is the *space-time interval* defined as

$$\text{space-time interval} = \sqrt{x^2 - t^2}$$

They are essentially the same thing, except that the quantity under the radical sign in one is the negative of the corresponding quantity in the other. The time-space interval may also be called a *timelike interval*, and the space-time interval may also be called a *spacelike interval*. In this book, as a matter of convention, we will usually make use of the time-space interval instead of the space-time interval. However, in any case, to prevent ambiguity, we always specify which one is being used, and thereby we avoid using the word *interval* by itself.

Another convention makes use of the velocity of light to deal with the tremendously short times which appear in our study of the microworld of atoms. However, instead of using the concept of "time of length," we now use the concept of "length of time." In this regard, we define a light-meter as the time required for light (in a vacuum) to cover a distance of 1 meter. Thus, a light-meter is the length of time for a light signal to travel 1 meter, and is given by

$$1 \text{ light-meter} = \frac{1}{3 \times 10^8} = 3.34 \times 10^{-9} \text{ seconds}$$

so

$$3 \text{ light-meters} = \frac{3}{3 \times 10^8} = 10^{-8} \text{ seconds}$$

To obtain smaller units of time, we can consider the path of light covering smaller and smaller distances. One millimeter is 10^{-3} meters, 1 micron is 10^{-6} meters, and 1 millimicron is 10^{-9} meters. One fermi (the unit used for nuclear distances) is defined as 10^{-15} meters, or one-millionth of a millimicron. Thus we have the following table:

3 light-meters	$= 10^{-8}$ seconds, or a hundred-millionth of a second
3 light-millimeters	$= 10^{-11}$ seconds, or a hundred-billionth of a second
3 light-microns	$= 10^{-14}$ seconds, or a hundred-trillionth of a second
3 light-millimicrons	$= 10^{-17}$ seconds, or a hundred-quadrillionth of a second
3 light-fermis	$= 10^{-23}$ seconds, or a hundred-sextillionth of a second

The fermi was named after Enrico Fermi (1901–1954), and this unit of length is approximately equal to the diameter of the various subatomic particles. A light-fermi is the time required for light to travel from one side of a proton to the other. In other words, the light-fermi is the time required for the fastest known signal to cover the smallest tangible distance. Nuclear reactions that take place in a few light-fermis of time (i.e., on the order of 10^{-23} seconds) are classed as strong interactions. They are the results of the strong nuclear force which can be felt in the most evanescent imaginable time interval.

As we have seen, it is useful in astronomy and modern physics to relate distance and time by means of the velocity of light. If the velocity of light were not a universal constant, this would not be feasible. For example, on the temperature scale, absolute zero is a fixed constant so we can use absolute zero as the benchmark in measuring temperature. For over 100 years, starting with the famous Michelson-Morley experiment in 1887, the proposition that the velocity of light is a fixed constant has been tested experimentally and argued theoretically. Even though it is now universally accepted that its value is the same constant value in all inertial frames of reference, there is no way that we can ever make a direct measurement on some other galaxy or in some other eon. However, until there is some evidence to the contrary, it will be assumed that the velocity (in vacuum) of light and all other electromagnetic waves is the same fixed constant c.

NATURAL UNITS

Originally, the foot and day were considered natural units, because one could measure a distance with his foot and could count the days. However, for people on some other galaxy, these units would be quite arbitrary, so instead, natural units are now defined in terms of fundamental physical constants. One such fundamental physical quantity is the velocity of light in vacuum, which is believed to be the same constant at all places, in all directions, and at all times. In this book, unless otherwise stated, we will now use *natural units based on the velocity of light*. In this section, we want to show the relationship between a quantity measured in natural units and the same quantity measured in conventional units. Therefore, let capital letters denote quantities in SI (Système Internationale) units and the corresponding small letters denote the same quantities in natural units. In SI units, the velocity of light in vacuum is $3 \cdot 10^8$ m/s; in natural units, it is $c = 1$. In SI units, distance is given in meters, which we may denote as X. Also in SI units, time is given in seconds, say, T, and a velocity V is given in meters per second. There are two ways we can go: one is *time of length* (the astronomical way) and the other *length of time* (the subatomic way).

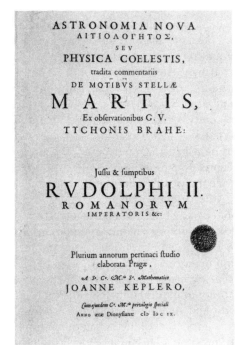

Plate 1.6. Title page of *Astronomia Nova* by Kepler, 1609.

In the time-of-length method, we convert all lengths to times. The quantities in natural units are

$$x = \frac{X}{3 \cdot 10^8}, \quad t = T, \quad v = \frac{V}{3 \cdot 10^8}$$

Here the unit of x is *light-second*, the unit of t is second, and the unit of v is light-second/second. Even though we say light-second, the unit is actually a second, which means that both x and t have the same unit (second), and thus velocity v is a pure number (that is, in natural units velocity is dimensionless).

In the length-of-time method, we convert all times to lengths, so

$$x = X, \quad t = 3 \cdot 10^8 T, \quad v = \frac{V}{3 \cdot 10^8}$$

Thus the unit of x is meter, and the unit of t is *light-meter*. Since both x and t have the same unit (meter), the velocity v is a pure number.

Whichever of the two ways we use, the velocity of light simply will not appear in the mathematical equations explicitly. We must be careful to remember, however, that c is built into the equations with the value $c = 1$.

For mathematical convenience, we thus avoid using the quantity c, the velocity of light. In other words, we measure distance and time in the same unit. It is easy to figure out from the equation $x = ct$ what we do. What distance is a light-second? It is the distance that light travels in 1 second. Thus if we measure all distances and times in the same unit, seconds, then our unit of distance is the light-second, which as a round number is 300,000,000 meters. All our equations come out simpler. The other way that we make distance and time units the same is to measure time in meters. What is a meter of time? A meter of time, or light-meter, is the time it takes for light to go 1 meter, and it is therefore 1/300,000,000 second, or 0.00000000333 second or 3.33 billionths of a second. By using either unit, that is, either the light-second or the light-meter, we put all our equations in a system of units in which c, the velocity of light, is equal to 1. Such units are called natural units, as they make one of the basic parameters of nature, the velocity of light, equal to unity. We will therefore put all our equations in a natural system of units in which $c = 1$. With time and space measured in natural units, all the equations become much simplified.

If we are ever afraid that by using equations with $c = 1$ we shall never be able to convert back to SI or other unnatural units, the answer is quite the opposite. It is much easier to work with and remember equations without the c's in them. It is always easy to put the c's back by looking at the dimensions. For example, in $\sqrt{1 - \beta^2}$, we know that β^2 must be a pure number to be able to subtract it from 1. Thus, if V is velocity in unnatural units, then we have $\beta = V/c$, which is a pure number, and thus

$$\sqrt{1 - \beta^2} = \sqrt{1 - \frac{V^2}{c^2}}$$

It is interesting to note that the foot (the length of the average person's foot) was the original natural unit of distance. However, in terms of modern physics, the gigafoot (defined as 1 billion feet) is the natural unit of distance, because light travels at the speed of $c = 1$ gigafoot per second. For this reason alone, the foot should be retained in preference to the artificial and awkward unit of the meter.

NEWTON'S LAWS OF MOTION

Newton achieved the first great scientific synthesis in his work *Philosophiae Naturalis Principia Mathematica* published in 1687. He presented his concepts as a single connected whole which served as the basis of modern science. Isaac Newton was born in Lincolnshire, England, on Christmas day 1642; his father had already died. Helped by people who recognized his talent, he was able to attend Cambridge University. When the plague struck England in 1665, the universities were closed and he returned to Lincolnshire. In those few years he made some of his most important scientific discoveries and inventions. Newton then returned to Cambridge where he became professor of mathematics. However, it was not until 1687, 22 years after his early work, that all his ideas came together in the *Principia*.
The *Principia* is a difficult book, even today, because Newton insisted on using the methods of Euclidean geometry to derive his dynamical results. Apparently, he had first obtained these results much more simply by the methods of differential and integral calculus, which he had invented for that purpose. The *Principia* is written in three parts. The first part deals with the determination of the motion of a body from the forces acting on it. The second part describes the various forces actually encountered in nature, except of course electricity and magnetism, which were only rudimentally known at that time. The third part analyzes the solar system and the motion of the planets, moons, and comets under the action of the force of gravity. All this and more are treated in the *Principia*, expressed in terms of axioms, lemmas, and theorems in the style as originally put forth by the ancient Greeks in their mathematical expositions.
Newton's three laws of motion are the main results of part one of the *Principia*. The first law is known as the *law of inertia*. It states that a body will continue in a state of rest or of uniform motion in a straight line unless some external force acts on it. This law was already known to Galileo. It was a great step for Galileo to see that a state of uniform motion (i.e., motion with constant velocity) is just as natural for a body as a state of rest. This represented a major innovation, for it was no longer necessary to find an explanation or cause for uniform motion.

Instead, uniform motion was seen to be a natural state of matter. The tendency for a body to maintain its state of rest or its state of uniform motion in a straight line was given the name *inertia* by Galileo. It was for this reason that the first law is called the law of inertia. The harder it is to change the state of motion of a body, the greater the inertia of the body. Inertia is the property of a body to resist any change in its state of motion, such as starting, stopping, or deviating from a straight-line path. The term *mass* is used to refer to the amount of inertia that a body has. The more inertia, the greater the mass.

Newton's second law deals with a change in the state of motion. As a measure of the quality of motion, he used the concept of *momentum*, which is equal to the product of the mass of the body times its velocity. The second law is the *fundamental law of dynamics*. It says that the force applied to a body is equal to the rate of change of its momentum with respect to time. In the case where the mass of a body is constant, we need only deal with the rate of change of the velocity with respect to time. This rate of change is the acceleration. Thus, in this case, Newton's second law says that force is equal to the constant mass times acceleration. In words, this means that for a given force, we can generate a greater acceleration in a body of smaller mass. If the same force is used to push a baby carriage and a car, the baby carriage will experience a greater acceleration than the car.

Newton's third law is known as the *law of action and reaction*. It states that action and reaction are equal and opposite. If one body exerts a force (action) on a second body, then the second body exerts an equal and opposite force (reaction) on the first body.

Let us now return to Newton's second law. If no net force acts on a body (or collection of bodies), then the rate of change of the momentum must be zero. This means that the momentum must be constant. In other words, if there is no net force, the total momentum is unchanged. This constancy of total momentum when no net force acts on a system of bodies is called the *law of conservation of momentum*.

A related concept is that of angular momentum. Suppose we have a body tracing a circular path of radius r. If v is its speed and m is its mass, then the body has momentum mv in the tangential direction. Its angular momentum is mvr. In general, a rotating body possesses angular momentum. Like the law of conservation of momentum, there is also a *law of conservation of angular momentum*, namely, if no net force acts on a system, then its angular momentum remains constant. These two conservation laws are basic to physics.

INERTIAL FRAMES OF REFERENCE

The basic laws of mechanics, Newton's laws, make it possible for us to pick out from among all conceivable reference frames a special class of frames. These special frames are called *inertial frames of reference*, and in them not only the laws

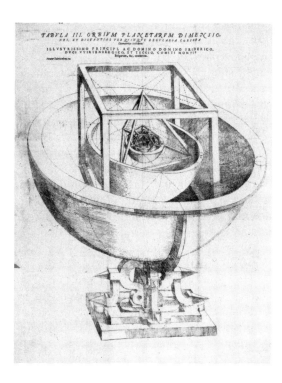

Plate 1.7. An early but unsuccessful attempt by Kepler in 1596 to explain the heliocentric structure of the solar system. The distances between the spherical shells of the orbits of the planets are mathematically determined by an interlocking arrangement of the five regular geometric solids as follows: sun at center, sphere of Mercury, octahedron, sphere of Venus, icosahedron, sphere of Earth, dodecahedron, sphere of Mars, tetrahedron, sphere of Jupiter, cube, sphere of Saturn (outermost).

of mechanics but also all other physical laws look particularly simple. An inertial frame of reference is a frame in which all three laws of Newton are valid. As it turns out there are an infinite number of such frames.

The first law of Newton, the law of inertia, says that a body subjected to no forces moves due to inertia, that is, uniformly and rectilinearly. Let a body be at rest in such an inertial frame of reference. Then, according to the second law of Newton, no forces act on this body. Without touching the body, let us consider the body from the viewpoint of an observer moving relative to the inertial frame with an acceleration a. This observer will note that the body moves relative to him with an acceleration $-a$. If the second law of Newton were valid in his frame, he could say that the body experiences the force $-ma$. But we know from an observer in the inertial frame that there is no force acting on the body. Therefore, the second law of Newton is merely not valid in the reference frame moving relative to the inertial frame with an acceleration. Since the laws of Newton are not valid in all reference frames, Newton had to point out that a certain reference frame was available in which all these laws were valid. The first law of Newton is, in fact, equivalent to this statement. This law postulates that an inertial frame of reference, that is, a reference frame in which the law of inertia is valid, is available. In other words, one can find a reference frame in which a body that interacts with no other bodies moves due to inertia, that is, uniformly and rectilinearly.

All phenomena take place in space and in the course of time. In the special theory of relativity, it is customary to define an *event* as something that occurs at a particular position in space at a particular instant of time. An example of an event is the emission of a light signal from a certain point in space at a certain moment of time. Another example is the presence of a moving particle (a material point) at a given point in space and a given instant of time. When an event is realized, we say that it happened. Any physical phenomenon consists of a sequence of events. Thus a description of an isolated event serves as a basis for the description of any phenomenon, and so we start with the description of a separate event.

First, we must label every point in space. But space is uniform and isotropic, and this implies that all points in space and all directions are equally favored. It should be pointed out at once that here we are dealing with free space, or vacuum. The investigation of phenomena in free space is necessary for the special theory of relativity. Basically, it is sufficient to assume that in the space domain considered, gravitational and electric fields are not very strong, and that there are virtually no material objects present. However, we will single out a certain point by placing on it a material object. Points in space are labeled by means of a coordinate system. The origin of coordinates is taken at the point of the material object. The simplest coordinate system is a Cartesian system. It can be constructed by tracing three mutually perpendicular straight lines, that is, the x, y, z axes. In terms of physics, however, these are not just abstract straight lines. Instead, the coordinate axes are rigid nondeformable solids. Instruments, standards, and other objects of a given reference frame are fixed to them. Thus it should be borne in mind that a physical coordinate system is always a material object.

We now must calibrate the Cartesian coordinate system. From any point M in space, we construct the perpendiculars to the x, y, z axes. In other words we project M into the coordinate axes. We next measure the distances of the point projections from the origin along the respective axes by means of the chosen scale, and thus we obtain the numbers x, y, z (the Cartesian coordinates of M). These distances can be measured by step-by-step transposition of a unit scale along each axis from the origin to the point projection on the axis. Thus through the introduction of Cartesian coordinates, each point in space acquires three numbers, namely, its Cartesian representation (x, y, z).

Second, we must find a way to study motion. Although mechanical motion is the most straightforward type of motion, as say compared to wave motion, its description requires time measurements. Thus we must supplement the coordinate system with clocks. The clocks are needed to register the occurrence of events at various points in space. We must determine how many clocks are needed. In classical mechanics, we tacitly assume that one clock resting in the given coordinate system is enough. But what does this single-clock assumption imply? Suppose the clock is located at the origin of the coordinate system. Events may happen at any points in space, including points far removed from the origin. How does the clock, removed from the place where an event happens, register that event? We analyze this situation as follows. Just at the moment the event occurs, a signal must be sent from the place of occurrence of the event to the clock located at the origin. However, if the velocity of the signal is finite, then the signal will reach the clock at some time instant after the onset of the event. This time lag between the occurrence and the recording of the event depends upon the distance between the point of occurrence and the clock.

In classical mechanics, it is assumed that conceptually there are signals which propagate infinitely fast. Thus in that situation, one clock rigidly fixed to any point in the coordinate system is enough. The onset of an event is registered as follows. At the moment that an event occurs at a given point in space, a signal is sent from that point to the clock, and the time of its arrival is thus the time of the onset of the event. This conclusion follows from the assumption that the velocity of the signal is infinite, so that there is no time lag between the occurrence and registration of the event.

Modern physics, however, claims that all signals are transmitted at a finite speed, and that moreover the ultimate speed is the speed of light in vacuum. Let us now construct a frame of reference which takes this finite signal speed into account. Such a frame is called a *relativistic frame of reference*. It is assumed that the velocity of light in vacuum is the ultimate velocity at which signals are transmitted. This assumption has to be inevitably incorporated into the theory if we want the principle of light to come into effect. We have just seen how a reference frame is constructed in classical mechanics. There we indicated that it was sufficient for each such reference frame to have but one clock, since it was assumed that infinitely fast signals were used. But in the special theory of relativity, it is assumed that the signal velocity is the velocity of light, which, as Roemer discovered, is finite. Thus one clock no longer suffices, because of the finite lag time between the occurrence and the recording of distant events.

To correct this situation, a set of clocks has to be added to the coordinate system which we constructed. In fact, the special theory of relativity requires that clocks be located at every point of space. Of course, this is not needed in practice, but as a matter of principle a clock must be at every point where an event occurs. All the clocks of a given reference frame are motionless relative to it.

Thus it is assumed that it is possible to have at one's disposal as many ideal clocks as is needed. And, in theory, this assumption can be realized. It is known that all atoms of the same kind have the same characteristic oscillation frequency. We can therefore take the atoms themselves as the clocks, and the periods of the atomic oscillations as the time standards. We can deal with length standards in the same way. We can choose the wavelength of a characteristic radiation of a given atom as the unit of length. In fact, that is how the length of the meter was standardized by Michelson at the beginning of the twentieth century.

THE RELATIVITY PRINCIPLE AND THE LIGHT PRINCIPLE

The principle of relativity was first conceived by Galileo. The following formulation was given by Newton in one of his corollaries to the laws of motion:

> The motion of bodies included in a given space are the same among themselves, whether that space is at rest or moves uniformly forward in a straight line.

For example, in a ship traveling at uniform speed, all experiments performed in the ship and all the phenomena observed in the ship will appear the same as if the ship were not moving. As we have seen, this principle is applied in classical mechanics under the assumption that the signal velocity is infinite. However, the situation changed with the introduction of Maxwell's equations of the electromagnetic field, which describe electricity, magnetism, and light in one uniform system. Now the signal velocity is the velocity of light, which although very large, it is not infinite. Moreover, the propagation of light and other electromagnetic waves is quite different from the propagation of other types of waves such as sound waves, water waves, and seismic waves. These other waves propagate in a material substance (air, water, earth) whereas light and other electromagnetic waves can and do propagate in vacuum.

With the success of Maxwell's electromagnetic theory, it became clear that the principle of relativity should be universally applied so as to include the system of Maxwell's equations as well. This means that the finite signal speed (velocity of light) which is an intrinsic part of the Maxwell theory must be applied to mechanics as well. Consequently, it was the equations of mechanics that had to be changed to conform with finite signal speed. Maxwell's theory required no change as it was developed on the basis of finite signal speed.

Of course, in the later part of the nineteenth century much effort was expended to keep the classical laws of mechanics (infinite signal speed) and to relegate Maxwell's equations to some inferior role. However, all these efforts were doomed to failure. In 1904, Poincaré and Paul Langevin (1872–1946) were among the leading scientists invited to attend the World's Fair in St. Louis, Missouri. Michelson was also invited to attend, but at the last minute (September 19) he wrote to the organizers that he was exhausted by continued sleeplessness and would not be able to be present. As we have seen, it was here that Poincaré made the first clear statement of this inclusive principle of relativity, together with the concomitant principle that "no velocity could surpass that of light." (L'état actuel et l'avenir de la Physique mathématique, *International Congress of Arts and Sciences*, St. Louis, Missouri, September 24, 1904.) [The writer's father, Edward Arthur Robinson (1872–1952), was present at this St. Louis World's Fair.] Poincaré regarded these two principles as empirical results derived from experience.

In the following year of 1905, Einstein (Zur Elektrodynamik bewegter Körper, *Annalen der Physik*, Vol. 17, pp. 891–921, 1905) elevated the principle of relativity to the status of a postulate. Postulates are familiar to us from our study of Euclidean geometry, as for example the postulate that a straight line is the shortest distance between two points. A postulate is something assumed without proof as being self-evident or generally accepted. The concomitant principle, which we call the principle of light, was formulated differently by Einstein. We recall that Poincaré had stated this principle by saying that the speed of light is the ultimate signal speed. Einstein in his 1905 paper states instead that "light is always propagated in empty space with a definite velocity c which is independent of the state of motion of the emitting body." (It can be shown that these two formulations of the principle of light are mathematically equivalent.) Einstein also elevated this principle to the status of a postulate.

We can sum up all the requirements of the special theory of relativity in the following two principles:

1. *The principle of relativity.* All the laws of physics are the same in every inertial reference frame. (Galileo's principle applied to all of physics.)
2. *The principle of light.* The ultimate signal speed is the velocity c of light in vacuum (Poincaré's formulation), or the velocity of light is the same value c in every inertial reference frame (Einstein's formulation).

In hindsight it might be said that in the years up to 1904 Poincaré made use of the results of physical experiments ranging from the initial astronomical observations of Roemer to the interferometer experiments of Michelson and Morley. Finally in 1904, he was successful in summarizing all this knowledge by the use of inductive reasoning into the form of these two principles. Einstein, in his initial paper on relativity theory in 1905, started with these two principles and elevated them to postulates. Thus he gave birth to the special theory of relativity as a deductive mathematical discipline. In this way, Einstein was following the path of Euclid, who 3000 years earlier had taken empirical geometrical principles, elevated them to postulates, and created the deductive discipline of Euclidean geometry. Also Einstein was following the example set by Newton who deliberately wrote his *Principia* in the same form as Euclid wrote *The Elements*.

Let us now summarize. The principle of relativity just given goes back to the early work of Galileo, and it was stated clearly by Newton as the basis for his system of mechanics. In classical Newtonian mechanics an infinite signal speed is assumed. With the advent of the electromagnetic theory of Maxwell, it was seen that a finite signal speed is required, and thus in 1904 Poincaré on the basis of inductive reasoning from physical experiments put the principle of relativity together with the concomitant principle of light. In this way it was possible to extend the range of the principle of relativity so it would apply to all the laws of physics, including electromagnetic theory as well as mechanics. Einstein in 1905 elevated these two principles to the status of postulates, and so created the special theory of relativity as a deductive science. As a result, all the other implications of the special theory of relativity follow from these two principles as mathematical results. Also in 1905, Poincaré (Sur la dynamique de l'électron, *C. R. Acad. Sci. Paris*, Vol. 140, pp. 1504–1508, 1905) created an equivalent deductive discipline, one based on the idea of invariance and the resulting space-time symmetry. He devised the mathematical object known as the Poincaré group and specified that all laws of physics must have the space-time symmetry exhibited by this group. The space-time symmetry of the Poincaré group was further developed by Minkowski in 1908 in his work on the unification of space and time, and by Einstein in 1915 in his application of Riemannian geometry to formulate general relativity theory.

RELATIVITY AS A SHEAR IN SPACE-TIME

Let us now consider two inertial reference frames moving relative to each other. The length scales and the clocks of each frame are at rest only with respect to their own reference frame.

In the foregoing section we saw that the special theory of relativity is made up of two principles, namely, the principle of relativity, where Galileo's principle of relativity is extended to cover all phenomena of nature, including Maxwell's electromagnetic theory, and the principle of light, where the ultimate velocity in nature is the velocity of electromagnetic waves in vacuum and this velocity is the same in all inertial frames of reference. But how must a transformation of space and time look to meet both requirements set? Poincaré in 1905 arrived at the answer by constructing the fundamental space-time symmetry existing in nature. He used the powerful mathematical methods of group theory. The result is the Poincaré group. He named the mappings of the homogeneous subgroup of this group the Lorentz transformation. Einstein in 1905 gave another derivation of the Lorentz transformation, and over the following years Einstein and others have given many still different ways of arriving at this fundamental result. In this book, we will give several derivations of the Lorentz transformation because each of them seems to reinforce certain fundamental ideas of relativity theory. Here we lead off with a simple geometric demonstration.

It is the purpose of this section to give a simple demonstration of the required transformation, the Lorentz transform. We have two inertial frames K and K', with K' moving away from K with velocity β (in natural units). For the purposes of this argument, we consider only one spatial dimension, so the space-time coordinates of K are x, t and those of K' are x', t'. Because K' moves away from K with velocity β, it follows that

$$x' = x - \beta t$$

Let the x axis be the horizontal axis (given by $t = 0$) and the t axis be the vertical axis (given by $x = 0$) in a rectangular system. Then the t' axis (given by $x' = 0$) is the line $x - \beta t = 0$ or $x = \beta t$. See Figure 1.3. We have also plotted the line

RELATIVITY AS A SHEAR IN SPACE-TIME

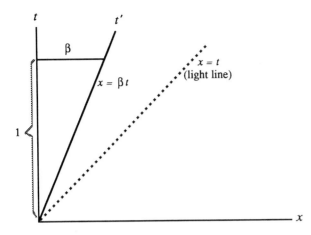

Figure 1.3. The light line bisects the angle between the x and t axes.

$x = t$ in the diagram. We recall that the speed of light in natural units is $c = 1$. Thus the line $x = t$ has slope given by the velocity of light; that is, $c = x/t = 1$. As a result, this line represents a particle (a photon) traveling at the speed of light, so it is called a *light-line*.

Now we must make use of the light principle, namely, the fact that in each frame the speed of light $c = 1$ is the ultimate signal speed. Clearly, this fact is shown by the light-line $x = t$ in the K frame. Similarly it will be shown by the light-line $x' = t'$ in the K' frame. For this to be so, the x' axis must be rotated from the x axis toward the light-line. The amount of rotation must be the same amount (but opposite in direction) as the amount that the t' axis is rotated from the t axis. The rotation of the two axes in opposite directions produces a shearing action. Thus we can describe this mathematical construction as a shear. It follows that the x' axis is given by $t = \beta x$. Since this axis occurs when $t' = 0$, we see that $t' = -\beta x + t$. See Figure 1.4.

We thus have the two tentative equations, which describe the shear symmetry of space x and time t, given by

$$x' = x - \beta t$$

$$t' = -\beta x + t$$

We now apply the relativity principle that tells us that neither frame can be favored, so we must modify these tentative equations by dividing the right-hand sides by the positive square root of the determinant of the coefficients. (This will fix the determinant of the resulting transformation equal to 1.) The required square root is

$$\sqrt{\begin{vmatrix} 1 & -\beta \\ -\beta & 1 \end{vmatrix}} = \sqrt{1 - \beta^2}$$

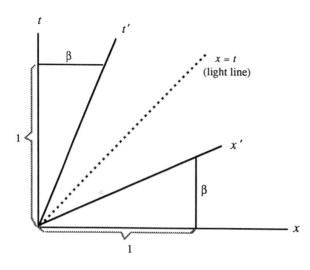

Figure 1.4. The light line also bisects the angle between the x' and t' axes.

If we define $\gamma = 1/\sqrt{1 - \beta^2}$, then we see that the final set of equations is

$$x' = \gamma(x - \beta t)$$

$$t' = \gamma(-\beta x + t)$$

This shear transformation from the x, t coordinates to the x', t' coordinates is the celebrated *Lorentz transform*. The Lorentz transform exhibits the essential feature of the Poincaré group, that is, the *shear symmetry of space-time*.

It is well known that space-space coordinates rotate, for example, when we use true north in one frame and magnetic north in the other frame. In one sentence, *the theory of relativity states that space-time coordinates shear, when we have two inertial frames in relative motion.*

The import of the Lorentz (or shear) transformation is that the time-space interval is an *invariant* (the same in both frames). The invariance of the squared time-space interval is expressed by the equation

$$t'^2 - x'^2 = t^2 - x^2$$

In three spatial coordinates, we have in general

$$t'^2 - x'^2 - y'^2 - z'^2 = t^2 - x^2 - y^2 - z^2$$

Physicists say that two equations are covariant if they have the same mathematical form. For example, the scalar electromagnetic wave equation (with propagation velocity given in the velocity of light) retains the same form when x, y, z, t are transformed to x', y', z', t' by means of the Lorentz transform. Thus we would say that the electromagnetic wave equation is covariant under the Lorentz transformation. Thus Einstein states: "General laws of nature are covariant with respect to Lorentz transformations. This is a definite mathematical condition that the theory of relativity demands as a natural law; and in virtue of this, the theory becomes a valuable heuristic aid in the search for general laws of nature" (Einstein, *Relativity*, 1920, p. 43). Einstein also states: "From the point of view of method, the special principle of relativity is comparable to Carnot's principle of the impossibility of perpetual motion of the second kind; for like the latter it supplies us with a general condition which all natural laws must satisfy" (Einstein, *Nature*, Vol. 106, 1921, p. 783). What Einstein is really saying here is that all the laws of physics are the same in every inertial reference frame and that the speed of light in vacuum is constant and is the ultimate signal speed.

TIME-SPACE INTERVAL BETWEEN EVENTS

As usual we are using natural units for which the velocity of light in vacuum is unity ($c = 1$). Let us conduct an imaginary experiment in which we consider two inertial frames of reference, labeled as K and K'. Both frames have a common axis; that is, x and x' lie along the same line. The frame K' in vacuum moves along this line at the velocity β. At the initial moment $t = t' = 0$, when the origins O and O' coincide, a light flash is triggered. According to the principle of light, the light signal propagates in all directions in the frames K and K' at the same velocity c. Consequently, the wave profile (i.e., the surface of equal phase) looks like a sphere in each of the frames K and K'. The equations of these spheres are

$$x^2 + y^2 + z^2 = t^2$$
$$x'^2 + y'^2 + z'^2 = t'^2$$

We can think of the wave profile as constituting a signal sent in each frame from the origin to an arbitrary point. Thus we deal, in fact, with two events in each frame. Let us consider K. The first event is the initiation of the light signal at $(0, 0, 0)$ and time 0. The second event is the receiving of the light signal at (x, y, z) and time t. The time-space interval τ between these two events is defined as

$$\tau = \sqrt{(t - 0)^2 - (x - 0)^2 - (y - 0)^2 - (z - 0)^2}$$

If we square τ we obtain

$$\tau^2 = t^2 - x^2 - y^2 - z^2$$

But as we have seen, the light profile satisfies $x^2 + y^2 + z^2 = t^2$, and hence we see that $\tau^2 = 0$. In a similar way, we see that

$$\tau'^2 = t'^2 - x'^2 - y'^2 - z'^2$$

is also zero. We have thus established the following result. *The square of the time-space interval between two events, consisting in sending a light signal from one point and its arrival at another point, must be equal to zero in any inertial reference frame*; that is,

$$\tau^2 = 0 \quad \text{and} \quad \tau'^2 = 0$$

The time-space interval between events is basic, and of course it can be defined not only for the sending and arrival of a light signal but for any two events. Let the coordinates of event 1 be defined by the numbers x_1, y_1, z_1, t_1 and the coordinates of event 2 by the numbers x_2, y_2, z_2, t_2. The time-space interval τ_{12} between these events is then defined as

$$\tau_{12} = \sqrt{(t_2 - t_1)^2 - (x_2 - x_1)^2 - (y_2 - y_1)^2 - (z_2 - z_1)^2}$$

THE LORENTZ TRANSFORM AND THE GALILEAN TRANSFORM

Let us now convert the Lorentz transform equations to unnatural units. Let V be the relative velocity between the two frames in meters per second. If c is the speed of light in SI units (300,000,000 m/s), then we have

$$\beta = \frac{V}{c}, \quad \sqrt{1 - \beta^2} = \sqrt{1 - \frac{V^2}{c^2}}$$

Also we change x to X/c, x' to X'/c, t to T, and t' to T', where small letters indicate quantities in natural units and capital letters the corresponding quantities in SI units. Then the Lorentz transformation becomes

$$X' = \gamma(X - VT)$$
$$T' = \gamma\left(-\frac{VX}{c^2} + T\right)$$

with

$$\gamma = 1/\sqrt{1 - \frac{V^2}{c^2}}$$

The Lorentz Transform and the Galilean Transform

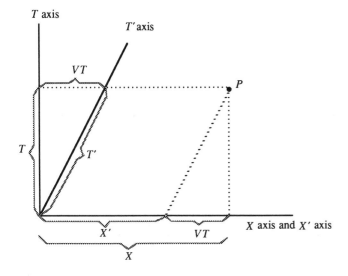

Figure 1.5. The Galilean transformation. The X and X' axes coincide with each other and with the light line. The event P has coordinates (X, T) as well as (X', T'), where $X' = X - VT$ and $T' = T$.

The square of the time-space interval becomes

$$T'^2 - \frac{X'^2}{c^2} = T^2 - \frac{X^2}{c^2}$$

We recall that Galileo performed experiments, and found that as far as terrestrial measurements are concerned the velocity of light is so great it may well be infinite. Thus Galileo would have set $c = \infty$ into the preceding equations. Thus $V/c = 0$ and $V/c^2 = 0$, so they become

$$X' = X - VT$$

$$T' = T$$

This is the Galilean transformation, which applies when the speed of light (the signal speed) is effectively infinite. The Galilean transform is shown in Figure 1.5. With infinite signal speed, one clock is sufficient for all space, and for this reason, the times T and T' in the two inertial frames are identical: $T = T'$. The other equation $X' = X - Vt$ merely states that the two frames are moving away from each other at relative velocity V. In such a situation ($c = \infty$), Newton's definitions of absolute space and absolute time are certainly valid. At the time of Galileo and Newton, the assumption of infinite signal speed was eminently justifiable for all applications to science and technology, and for that matter still is, except in those cases where particles have speeds that become significant in comparison to the speed of light.

CHAPTER 2

SPACE AND TIME

> I could behold
> The antechapel where the statue stood
> Of Newton, with his prism and silent face,
> The marble index of a mind forever
> Voyaging through strange seas of thought alone.
>
> Williams Wordsworth (1770–1850)
> *The Prelude*, III, 59–63

THE SAME AND THE ULTIMATE

Our view of nature for the more than two centuries prior to 1900 was dominated by the ideas of Newtonian physics. The exception to this rule, however, was Maxwell's theory of electricity, magnetism, and light introduced in 1865. Although it was not clearly recognized at the time, Maxwell's theory does not fit very well

Plate 2.1. Sir Isaac Newton (1642–1727)

into Newtonian physics. Maxwell's theory is now counted as part of special relativity; logically, it belongs there. Classical Newtonian physics makes the basic assumption of infinite signal speed. As a result, space and time become two distinct entities, and lengths and time durations are independent of the observer. These assumptions about space and time seem obvious from our everyday experience because, for terrestrial distances and human observers, the speed of light is effectively infinite. The success of Newtonian physics made its assumptions seem well grounded. Relativity takes into account the finite speed of light. It means that space and time are no longer absolute and distinct. In the realm of high-velocity particles, special relativity has been checked by literally billions of observations. Its conceptual revolution unified space and time.

In terms of mathematics, special relativity is quite simple: two inertial frames of reference are related by a space-time shear transformation. It is for this reason we will emphasize the mathematics. However, the physical concept of special relativity is hard. We might say that other types of conceptual revolutions were hard in the past but easily understood today. For example, it was once difficult for people to accept Copernicus and abandon the old geocentric ideas. As we know, there are historical similarities between Copernicus's heliocentric theory and special relativity. In 1887, Michelson and Morley measured how fast light travels from one point to another on Earth. They found that the Earth's motion around the sun does not affect the measured speed of light. This fact is summed up in the light principle: *the speed of light in vacuum, the ultimate signal velocity, is a constant and is the same in all inertial frames.* In a vacuum, light must travel at exactly the speed c; it can propagate neither slower nor faster. We can accept this principle at face value, but with our present understanding, it is beyond human comprehension to explain why the principle is true in terms of some familiar physical model. It may be that in terms of human evolution our brains are not far enough developed ever to understand the nature of light propagation. Here quantum mechanics enters in, and so again we are faced with problems that can be attacked by mathematics but which are beyond human comprehension. For example, we know that quantum mechanics says that light is made up of particles (photons) but light propagates as a wave. This wave-particle duality can be handled by mathematics, but our brains cannot conceive of it in the same sense that we know that eating takes away our hunger. Newton was deeply troubled by such problems. However, he did put forth the theory that gravity represents force at a distance, even though he realized that no physical model in terms of human understanding could explain such a force. In conclusion, we realize that the heliocentric theory of Copernicus is much easier to accept in human terms than the special theory of relativity. The reader therefore should not be troubled if he or she cannot understand some of the applications of the special theory of relativity in human terms. Instead, he or she should rely on the mathematics, even as the captain of a boat in a fog does not see but safely makes his or her way by means of the radio navigation system. We might say that the entire physical world about us, as far as humans can ever comprehend, is nothing more than piecemeal and incomplete systems of mathematical equations, for that is as much as we can ever hope to grasp.

Let us give an analogy or a metaphor. This metaphor is *not* a model of the special theory of relativity, but it may convey to the reader a feeling for the special theory of relativity. Suppose that an automobile going at 60 miles per hour represents the ultimate signal speed. An observer is at point A; she puts one hourglass on a train going 30 miles per hour, and she keeps another hourglass with herself. In 1 hour, her hourglass is empty. The other hourglass becomes empty when it reaches point B, 30 miles from A. As soon as this happens, a messenger drives his car at 60 miles per hour and gives the news to the observer at A. The observer at A would then say that it took 1.5 hours for the hourglass on the train to empty (= 1 hour for the train to arrive at B plus 0.5 hour for the car to drive back to A). Thus, the observer would say that the hourglass on the moving train empties more slowly than does the one that was at rest with herself. With infinite signal speed, both hourglasses would empty in 1 hour. The discrepancy is due to the 60-miles-per-hour signal speed acting as the ultimate speed.

Let us look at another example. At the time of Newton, light was considered to have an infinite speed, $c = \infty$. If Newton was on a stage coach traveling at 5 miles per hour, and if he lit a lamp, he saw the light travel out in the coach at infinite speed. An observer on the ground ahead of the coach saw the same light

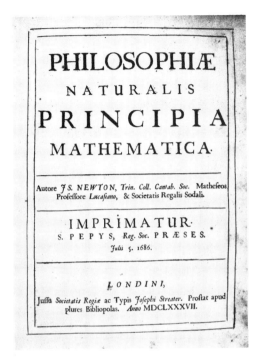

Plate 2.2. Title page of Newton's *Principia*, 1687.

come to her at the speed

$$c + 5 = \infty + 5 = \infty = c$$

That is, the observer on the ground also saw the same light traveling at the same speed, namely, infinity. In this example, the equation $c + 5 = c$ does not bother us, because $c = \infty$.

At the time of Einstein, light was known to have a finite speed, in round numbers $c = 300,000$ km/s. If Einstein was on a spaceship traveling at 100,000 km/s and if he lit a lamp, he saw the light travel out in the ship at 300,000 km/s. An observer on the Earth ahead of the spaceship sees the same light come to her at the speed

$$c + 100,000 = c$$

That is, the observer on the ground also sees the same light traveling at the same speed, namely, 300,000. The equation $c + 100,000 = c$ does bother us, because c is finite ($c = 300,000$). However, the gist of the equation is true. For it to be mathematically true, however, we must redefine the operation of the addition of physical velocities. In other words, we can understand this situation mathematically, but not intuitively.

The light principle states that the velocity of light is the ultimate signal speed, and that an observer in any inertial frame always sees light travel at that ultimate speed, not one iota faster, not one iota slower. Einstein on the spaceship feels himself at rest in the inertial frame of the spaceship; whatever light he sees travels at the speed c = 300,000 km/s. The observer on the Earth feels herself at rest in the inertial frame of the Earth; whatever light she sees travels at the speed c = 300,000. Light in vacuum can travel at no other speed than the ultimate speed. Although the speed of light is not infinite (the ultimate number), it acts as if c = 300,000 were the ultimate number. Only electromagnetic waves act this way; they propagate through vacuum. Other types of waves do not act this way; they propagate through material substances.

Electromagnetic waves (light, microwaves, radio, etc.) occupy a special place in the makeup of the cosmos. We can try to describe them mathematically and with some success; we cannot really understand them. They are beyond human understanding. That is why any description of these things without mathematics is doomed to failure. All the popular nonmathematical accounts of relativity are indeed frustrating.

There is the familiar sense in which motion is relative. If two spaceships are drifting in empty space with their engines cut off, and if they pass each other heading in opposite directions, then it is impossible to say if one, or the other, or both of them are moving. All that is certain is that they are moving relative to each other. Is it really impossible to find out who is moving? We know from experience that no mechanical experiment can tell us who is moving if we are in an inertial frame (a state of uniform translatory motion, that is, no acceleration and no swerving). For instance, if you are going at a steady 60 miles per hour in your automobile, and if you drop a coin, it falls straight to the floor. If you are parked, a coin also falls straight to the floor. This behavior is explained by the Galileo principle of relativity. However, knowing Maxwell's equations, we may ask if there is some tricky experiment using electromagnetic waves that would enable us to tell if our spaceship is moving or not. According to the inclusive principle of relativity, the answer is no. This principle asserts that the laws by which the states of physical systems undergo change are not affected, whether these changes of state be referred to the one or the other of two systems of coordinates in uniform translatory motion. In other words, anyone in any inertial frame can say "I am motionless."

As we have seen, the other principle required for the special theory of relativity is the principle of light (the principle that the speed of light is constant in all inertial frames and that it is the ultimate signal speed). This principle asserts that whenever you measure the speed of a light ray, you will get the same number (300,000 km/s). It does not matter if you are moving toward or away from the light source, and it does not matter if the source is moving toward or away from you. In the foregoing sentence, of course, we can delete the section "It does not matter if you are moving toward or away from the light source, and," since the principle of relativity says that we can always assume we are motionless and thus ascribe all the relative motion between us and the light source to the light source.

As we have said, we cannot understand the light principle (speed of light is constant and ultimate) by the use of any human analogies, but must just accept it. For example, let us try to understand it in terms of sound waves. We can understand sound waves, as they travel through a material substance, the air. Sound waves travel through air at speed $c = 1100$ feet per second. If we suppose that there is no light, can we take this speed of 1100 as representing the ultimate signal speed? Is this not a perfect analogy? We know that the speed at which sound travels through the air has nothing to do with the speed of the source. Suppose you run forward and throw a ball. Then the ball goes faster than it does if you throw it while you are standing still. That is, if you stand still, then

$$\text{speed of ball} = \text{speed of throw}$$

However, if you run, then (approximately)

$$\text{speed of ball} = \text{speed of source} + \text{speed of throw}$$

With this analogy in mind, should not the sound coming from the whistle of a train moving toward you be traveling faster than the sound coming from a train standing still? But, as we have just said, the speed at which sound travels through the air has nothing to do with the speed of the source (the train). A gunshot produces a sound wave whose velocity has nothing to do with the speed of the gun. The ripple caused by a rock thrown out into a lake moves at the same speed as the ripple caused by a rock dropped into the lake. Thus, any wave motion does not depend upon the speed of the source. However, sound waves (and other material waves) do depend upon the speed of the medium through which they travel. Let us now explain.

Inside a moving train, the air is still and traveling with the train at the speed of the train. Outside the train, the air is still with respect to the ground, and so the air outside the train is not traveling with the train but appears to be rushing backward with respect to the train. Thus, we have two different inertial frames: the ground with its exterior air still and the train with its interior air still. These two kinds of air are in motion with respect to each other. Sound outside the train moves in the still air with respect to the ground, and thus travels at the speed of 1100 feet per second with respect to the ground. Sound inside the train moves in the still air with respect to the train, and thus travels at the speed of 1100 feet per second with respect to the train. Thus, there are two sound waves; one traveling in the outside air and the other in the inside air, and the inside air is traveling away from the outside air at the speed of the train. If the speed of the train is V, and if the sound signal is set off inside the train, then the sound reaches a person inside the train at a speed of 1100 feet per second. However, the sound waves inside the train are coming at the person on the ground at a speed of (approximately) $1100 + V$, because the sound is traveling through that inside air at 1100 feet per second, and that inside air is traveling at V feet per second with respect to the ground. Thus, the speed of any particular sound wave depends upon the speed of the medium as well as the speed of the wave in the medium.

Let the speed of water waves be V_0. If a person drops a coin in a dish of water on the train, the speed of the resulting water waves as seen by a person outside the train is approximately $V_0 + V$, whereas if a person drops a coin from a train into a pond outside the train, the speed of the water waves is V_0. The difference, of course, is due to the fact that the water inside the train is moving at a speed V with respect to the water outside the train. Inside each body of water, the waves are moving with the same velocity V_0.

What we have just said about sound and water waves, does it also apply to light waves? The answer is no. In special relativity, it is assumed that light moves through a vacuum, and not through a material medium such as air or water. We cannot say that the vacuum inside the train is moving at a speed of V with respect to the vacuum outside the train. If a light source is set off both inside and outside the train, the person on the ground ahead of the train sees the light outside the train coming at a speed of 300,000 km/s. However, the light waves inside the train are also coming to this person at the same speed. We cannot say, as we did for the sound waves and water waves, that the light waves inside the train are coming toward the outside person at a speed of $300,000 + V$ (i.e., the inside light speed plus the speed of the inside vacuum with respect to the outside vacuum). Thus, the observer always sees light at the same speed, 300,000 km/s.

The reason why the sound versus light analogy fails is that sound waves do not travel at the ultimate velocity, even if we try to make them do it for the purposes of the analogy. To an outside person, the velocity of a sound wave is approximately equal to the velocity of the sound wave in the material plus the speed of the material. However, to the same outside person, the velocity of a light wave is always the velocity of the light wave: 300,000 km/s with respect to himself. We cannot say that there is an inside vacuum and an outside vacuum and that they are moving with some relative velocity V with respect to each other. Thus, for light, we have in effect $c + V = c$ (provided the plus sign is defined in the right way mathematically). We will do so in Chapter 4. The new definition of addition is called the *physical velocity addition formula*.

Let us give another example. Imagine you are sitting on the grass strip in the middle of a cosmic north-south superhighway. Suppose you see photons going north and other photons going south. You measure the speed of the northbound ones; you find each one going exactly at 300,000 km/s, the speed c of light. Each southbound photon is also going 300,000 km/s. Now Einstein comes along the grass strip driving a spaceship going at a steady 100,000 north. You hop on and again measure the same photons, but now with respect to Einstein's spaceship. Before you start measuring, you say to Einstein that we should find the northbound photons passing the spaceship at $300,000 - 100,000 = 200,000$ and the southbound photons passing the ship at 300,000 + 100,000 = 400,000. Einstein says nonsense. Without looking at the road or the grass, you measure the northbound speed and the southbound speed of the photons. Both are still 300,000 km/s. Why? Because light speed is universal. You hop off, and measure again, both ways. On the ground you found a basic law of nature; the speed of light is 300,000 km/s. On the space ship you found exactly the same basic law. The speed of light is the same in all inertial frames of reference, even if the inertial frames are moving relative to each other like the grass and spaceship.

Perhaps you have trouble accepting the result that no matter how fast you go, a photon that passes you always passes you at speed c. Well you should, because this result cannot be understood in terms of human brain capacity. You are on the Earth in an inertial frame. Within your frame, you are always at rest, and a photon always passes you at speed c. This is easier to accept. When you were on the grass, you were at rest. The photons passed you at speed c. When you were on Einstein's spaceship, you were at rest. The photons passed you at speed c. Accept the fact that you, personally, are always at rest, and that photons always pass you at speed c = 300,000 km/s. What applies to you applies to every other inertial frame, and this is the basis of special relativity.

SPACE AND TIME

Space and time are not independent of one another; instead there is an intimate connection between them. The theory of relativity has revealed the nature of this connection. Physical space and physical time have not separate and independent existences, but are selections from something more complex, namely, a blend of space and time that comprises both.

A basketball has any number of different diameters, each pointing in a different direction. It would be inaccurate to single out any one of these as being the sole representative of the height of the ball, to the exclusion of the others. Each one has an equal claim to representing the height, and each can indeed be made the height by turning the basketball the right way up. As long as the ball does not enter into some sort of orientation with other objects, such terms as height, width, or length have little meaning. In the same way, time and space are meaningless in the abstract. However, when the ball is placed on a horizontal floor, one particular diameter immediately becomes the height of the ball. In the same way, when we consider a human observer in the four-dimensional space-time continuum, one particular direction immediately becomes identifiable with her time and the other three directions with her space. We cannot speak of space and time separately in the space-time unity unless we put an observer in it.

Two observers, and in fact any number of observers, in a given inertial frame will all have the same perceptual space and time. However, observers in another inertial frame moving at some constant velocity with respect to the first frame will have a different perceptual space and time. We can say that the nonagreement

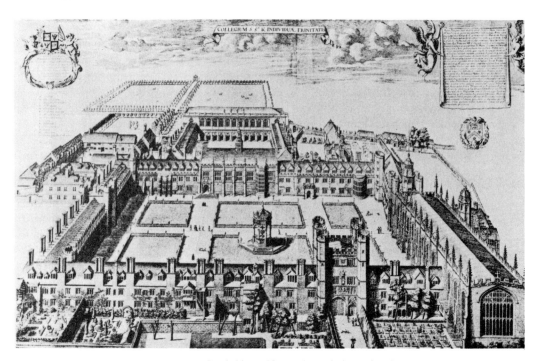

Plate 2.3. Trinity College, Cambridge. Newton's workplace when he published the *Principia*.

is due to the fact that each of the two inertial frames splits up space and time in a different way.

There is a true reference velocity in nature, and it is the ultimate velocity, the speed of light. We perceive that light takes time to travel through space. A particle of light, known as a photon, travels at a uniform finite speed of 300,000 km/s. This means that in 1 second, a photon of light travels 300,000 km. But what does this really mean?

For simplicity, we will use only a two-dimensional space-time continuum, one dimension for distance x and the other for time t. We suppose that we are at rest in our inertial frame. This means that as time passes in our frame, we stay at the same place so we do not move any spatial distance x. In other words, when we are at rest, we are moving through time only. Moreover, we are moving through time at a fixed rate which can neither be increased nor decreased.

Now let us look at a photon. The photon is moving away from us at a velocity of 300,000 km/s. This is the ultimate spatial velocity in the universe. The photon according to us is moving through space at a fixed rate that can neither be increased nor decreased. The photon represents the reference by which the space-time entity is split into space and time. This is how it is done.

The photon is the antithesis to us, the observer. At rest in our inertial frame, we pass only through time. The photon has no rest mass, but is pure kinetic energy. It travels through vacuum (the absence of mass) so it can have no inertial frame. Thus in its own perception the photon must travel only through space, not through time. In other words, the photon diverts all of its space-time motion to space. It is going through space as fast as possible, so there is nothing left for traveling through time. Time does not pass for the photon. It does not age. It is at rest with respect to time. (Recall that the time-space interval between two events, consisting of the sending and the receiving of a photon, must be zero in any inertial frame.)

Let us summarize. At rest with respect to space in our inertial frame, we are traveling through time as fast as possible. At rest with respect to time in vacuum, the photon is passing through space as fast as possible. These are the two limits between which all other motion with respect to us will lie.

Light operates outside of time. If a person could attach herself to a photon and ride on a beam of light, she would discover that time would stop. Light brings information about the time instant shown by a clock. This information carried by light does not change during the transmission. If a photon leaves the clock at high noon, then for that photon, it will always be high noon. It does not matter how long it takes for the photon to reach the destination in our reckoning of time, for when that photon arrives, the photon is still saying that the time is high noon. To the photon, time does not flow. Its starting time is the same as its arrival time. Light does not age. It is ageless. The timelessness of light is an essential feature of relativity theory.

Let us now return to our basketball analogy. We consider only one spatial coordinate x and the time coordinate t. The basketball then is a circle. Let observer A be in one inertial frame and observer B be in another inertial frame moving with velocity β with respect to the first. The ball is placed on a horizontal floor (Figure 2.1).

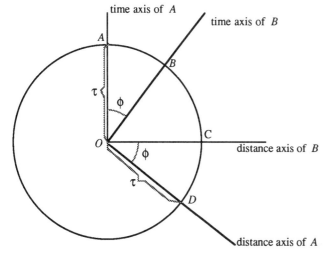

Figure 2.1. Circle of radius τ. The angle ϕ is defined by the equation $\beta = \sin \phi$. The inertial frame of A is defined by its time and distance axes, and similarly for B.

Let the vertical radius OA represent the time axis of A, and let the horizontal radius OC represent the distance axis of B. The angle ϕ is defined by the relation

$$\beta = \sin \phi$$

The angle ϕ so defined as the inverse sine of the velocity has a special name, the *Gudermannian angle*. Define radius OB as the radius swung at angle ϕ in a clockwise direction from vertical radius OA. Likewise define radius OD as the radius swung at angle ϕ in a clockwise direction from horizontal radius OC. By inspection of Figure 2.1, we see that radii OB and OC represent a shearing of radii OA and OD through an angle ϕ. Therefore, since OA is the time axis of A, it follows that OB (i.e., the shear of OA) is the time axis of B. Likewise, since OC is the distance axis of B, it follows that OD (i.e., the antishear of OC) is the distance axis of A. Thus we have established the space and time axes of A, and their sheared counterpart, namely, the space and time axes of B.

The construction in Figure 2.1 is called *Gudermannian space-time diagram*, and it has the following special feature: *radius OB is perpendicular to radius OD*. This follows because radius OA is perpendicular to radius OC (as by definition they are the vertical and horizontal axes), and radius OB and radius OD represent a clockwise rotation of the vertical and horizontal axes by an angle ϕ. Thus the Gudermannian diagram represents both a shear and a rotation. The shear exhibits the space-time symmetry of the Poincaré group (Lorentz transform) while the rotation allows us to use familiar Euclidean geometry (more specifically, the Pythagorean theorem).

Let the circle have radius equal to τ. Then radius OA represents τ seconds in the lifetime of A. Because A lies on the time axis of A, it follows that A is at rest in its frame, which means that A does not move during time period τ. Similarly, OB represents τ seconds in the lifetime of B. Because B lies on the time axis of B, it follows that B is at rest in its frame, which means that B does not move during time period τ. We thus have

$$OA = A\text{'s perception of } A\text{'s time} = \tau$$
$$OO = A\text{'s perception of } A\text{'s distance} = 0$$
$$OB = B\text{'s perception of } B\text{'s time} = \tau$$
$$OO = B\text{'s perception of } B\text{'s distance} = 0$$

Now let us look at cross-perceptions. Let us look at A's perception of B. This means that we must project point B onto the axes of A. The projecting lines must be parallel to the axes of A. In Figure 2.2, draw the line BE parallel to OD (i.e., the distance axis of A). Then BE gives the distance x of B in A's frame. The line OE lies on the time axis of A, so OE is the time t of B in A's frame.

Space and Time

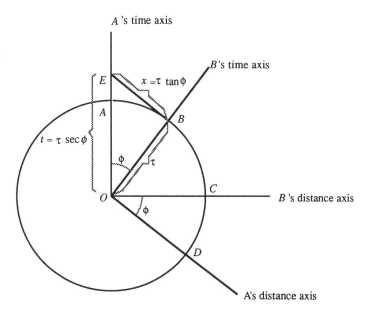

Figure 2.2. Gudermannian space-time diagram.

Thus we have

$$OE = A\text{'s perception of }B\text{'s time} = t$$
$$BE = A\text{'s perception of }B\text{'s distance} = x$$

The observer A perceives B moving through space at constant speed β. A's perception of the *speed of B through space* is

$$\beta = \frac{A\text{'s perception of }B\text{'s distance}}{A\text{'s perception of }B\text{'s time}} = \frac{x}{t} \qquad (2.1)$$

This is the relative speed of the two frames, and from Figure 2.2 we see that

$$\beta = \frac{x}{t} = \frac{BE}{OE} = \sin \phi$$

This equation merely confirms the Gudermannian definition previously employed. But now let us define an unconventional quantity, namely, A's perception of the *speed of B through time*, defined as

$$\alpha = \frac{B\text{'s perception of }B\text{'s time}}{A\text{'s perception of }B\text{'s time}} = \frac{\tau}{t} \qquad (2.2)$$

From Figure 2.2 we see that

$$\alpha = \frac{\tau}{t} = \frac{OB}{OE} = \cos \phi$$

Thus according to A, the space-time of B which is represented by the radius OB is split up by means of a time-speed α and a distance-speed β, where

$$\alpha^2 + \beta^2 = 1$$

Figure 2.2 has the following interpretation. A perceives B moving through space, so B's speed through time is accordingly reduced. In time period t (in A's frame), body B moves a distance $x = t \sin \phi$ (in A's frame), and so B ages a reduced time $\tau = t \cos \phi$ (in B's frame). In other words, B extrinsically travels through $t \sin \phi$ light-seconds of space and, in so doing, intrinsically ages $t \cos \phi$ seconds of time. Meanwhile, A at rest in its inertial frame ages the full t seconds while it watches this happen. These components satisfy the Pythagorean theorem

$$x^2 + \tau^2 = t^2$$

or

$$(t \sin \phi)^2 + (t \cos \phi)^2 = t^2$$

If we arrange the quantities of B's frame on the left and A's frame on the right, we have

$$\tau^2 = t^2 - x^2$$

which demonstrates the invariance of the square of the time-space interval.

Finally, let us define the quantity γ as the reciprocal of α. Thus, in terms of the Gudermannian angle ϕ, we have the *Greek trio* of alpha, beta, gamma:

$$\alpha = \cos \phi, \quad \beta = \sin \phi, \quad \gamma = \sec \phi$$

These three constants appear repeatedly throughout the book, and we will reserve the symbols α, β, γ, and ϕ only for their usage in this sense.

When $\phi = 90°$, B becomes a photon. Suppose $t = 1$. The photon travels through $\sin 90° = 1$ light-second of space. The photon ages $\cos 90° = 0$; that is, the photon does not age at all. A ages 1 second as it watches this happen. The photon has the maximum speed through space ($\beta = c = 1$), the velocity of light. Its velocity through time ($\alpha = 0$) is zero. The observer A is stationary in space; the photon is stationary in time.

Consider one second in the lifetime of A. Because A is at rest in its frame, A does not move during this time period. At the other extreme, consider the distance of one light-second in the itinerary of a photon. Because time is at rest for a photon, the photon does not age during this distance span. Observer A and the photon represent the two extremes. Observer A does not move but only ages ($\phi = 0$). The photon does not age but only moves ($\phi = 90°$).

Let τ seconds of B's time according to B transpire. Let t be A's perception of this amount τ of B's time. Then by equation (2.2)

$$t = A\text{'s perception of } B\text{'s time} = \frac{B\text{'s perception of } B\text{'s time}}{\alpha} = \frac{\tau}{\alpha} \quad (2.3)$$

Then by equations (2.1) and (2.3) we have

$x = A$'s perception of B's distance

$$= \beta (A\text{'s perception of } B\text{'s time}) = \frac{\beta\tau}{\alpha} \quad (2.4)$$

In terms of the Gudermannian angle ϕ, the equations (2.3) and (2.4) are

$$t = \frac{\tau}{\cos \phi} = \tau \sec \phi \quad (2.5)$$

$$x = \tau \frac{\sin \phi}{\cos \phi} = \tau \tan \phi \quad (2.6)$$

Let us now summarize. In Figure 2.2 the circle has radius τ. The radius OB makes an angle ϕ with the vertical radius OA. The tangent BE meets radius OA produced at E. Because OB is equal to τ, it follows that OE is equal to t, and BE is equal to x. We can make the following interpretation. A perceives B moving through space. Meanwhile B regards itself as at rest, say, during a time period τ. However, A's perception is that B actually aged an amount $t = OE = \tau \sec \phi$ and moved a distance $x = BE = \tau \tan \phi$. The time-space interval according to B (distance 0, time τ) is τ. The time-space interval according to A (distance x, time t) is $\sqrt{t^2 - x^2}$. We can verify that these two time-space intervals are equal by trigonometry. Note that the required equality

$$\tau = \sqrt{t^2 - x^2}$$

follows from the trigonometric identity

$$\sqrt{\tau^2 \sec^2 \phi - \tau^2 \tan^2 \phi} = \tau \sqrt{\sec^2 \phi - \tan^2 \phi} = \tau$$

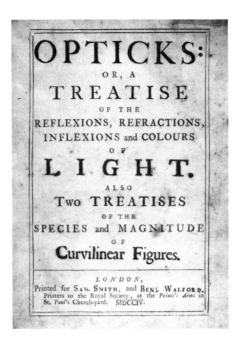

Plate 2.4. Title page of Newton's *Opticks*, 1704.

THE FABLE OF THE TWO ESKIMOS

Explorers in the arctic regions during the early part of this century came upon a group of Eskimos who lived on or at least very near to the North Pole. They discovered an amazing thing. Apparently these Eskimos were very old. In fact, they seemed to live forever. Realizing the deprivations these native people suffered by living in such a desolate and cold region, several of the Eskimo babies were resettled in more temperate climates, some even in tropical regions. One child who was placed at a location on the Earth's equator lived a certain time span, which we will call τ units of time. Those who were placed at locations above the equator but below the North Pole lived longer than this equatorial life span. In fact, the farther north a child was placed, the longer was its life span. This puzzled the medical profession for many years, but recently it was discovered that the North Pole Eskimos are extremely sensitive to the effects of the ultraviolet radiation in the sun's rays. At the equator, the sun is directly overhead, so the sun's rays hit the Earth there broadside, that is, at normal incidence. At the North Pole, the sun is on the horizon, so the sun's rays hit the Earth there end-on, that is, at tangential incidence. In other words, at the equator the normal component of the sun's rays is 1, and the tangential component is 0. At the Pole, the normal component is 0, and the tangential component is 1. It is the normal component, that is, the overhead component, that carries the lethal dose. Thus, at the North Pole an Eskimo is free and clear, whereas at the equator she gets hit head on

THE FABLE OF THE TWO ESKIMOS

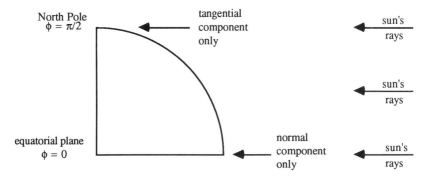

Figure 2.3. Eskimos are adversely affected only by the normal component.

(Figure 2.3). The equator is at latitude $\phi = 0$, and the North Pole is at latitude $\phi = 90°$.

Now let us look at an intermediate latitude ϕ. The effect there will lie between the two extremes. Let the sun's rays hit a point at this latitude. The normal component will be in the direction of the Earth's radius, whereas the tangential component will be at right angles to this radius (Figure 2.4). The normal component is $\cos \phi$, and the tangential component is $\sin \phi$.

Now came the intensive medical research. Great amounts of statistical data were analyzed on the most powerful computers. A simple relationship finally emerged. The life span t of an Eskimo is proportional to the reciprocal of the normal component; that is,

$$t = \frac{\tau}{\cos \phi}, \quad \text{or} \quad t = \tau \sec \phi$$

where the constant τ is the proportionality constant. In other words, the smaller the amount of sunlight that hits her from above the longer she lives.

Next we need a measure of how far the Eskimo is from the equator. To use natural units, we let the radius of the Earth be equal to the same constant τ. Then the arc distance from the point at latitude ϕ to the equator is just equal to $\tau\phi$,

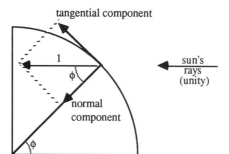

Figure 2.4. Normal component at latitude ϕ is $\cos \phi$.

58   SPACE AND TIME   CHAPTER 2

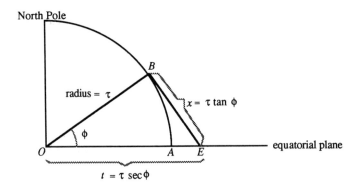

Figure 2.5. Lifetime t and distance x defined for an Eskimo at latitude ϕ.

where the angle ϕ is in radians. However, instead of this arc length, it is more convenient to use the straight-line length of the tangent line from the point to the plane of the equator. This distance is $x = \tau \tan \phi$. In Figure 2.5, line segment OB is the radius (equal to τ) at latitude ϕ, line segment BE is $x = \tau \tan \phi$, and line segment OE is $t = \tau \sec \phi$.

We have now given the hard facts. Now we must recount the *fable of the two Eskimos*, call them A and B. Eskimo A lives at the equator, which is at latitude $\phi = 0$. He computes his distance and lifetime as

$$x_A = \tau \tan 0 = 0 \qquad t_A = \tau \sec 0 = \tau$$

That is, he is zero distance from the equator, and he has the (minimum) lifetime of τ. Eskimo B lives north of the equator, at latitude ϕ. Her lifetime and distance are

$$x_B = \tau \tan \phi, \quad t_B = \tau \sec \phi$$

Both A and B agree on these measurements, and these measurements are consistent.

Now comes the scheme of the old man (Figure 2.6). He moves Eskimo B to a point on the equator at central angle ϕ from Eskimo A who is already on the equator. Thus the fable is about two Eskimos, not at the North Pole, but both on the equator at two separate points. But there is more. The old man inserts a magnet in the compass of each Eskimo so that each of their compasses give false readings. Using his false compass Eskimo A still thinks Eskimo B is north of him at latitude ϕ. Eskimo A thus computes his space-time coordinates and B's space-time coordinates exactly as given. But those he computed for B are wrong, because B is actually on the equator, and not at north latitude ϕ as he mistakenly believes.

Meanwhile, Eskimo B sees the sun directly overhead, so she realizes she is

THE FABLE OF THE TWO ESKIMOS

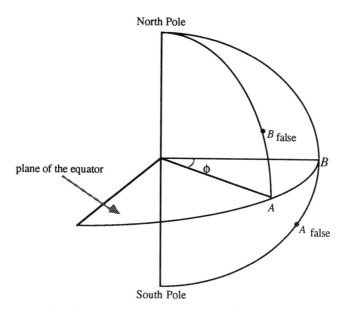

Figure 2.6. The old man puts A and B on the equator separated by angle ϕ. However, the old man tricks A into thinking that B is at B_{false} with latitude ϕ and tricks B into thinking that A is at A_{false} with latitude $-\phi$. Thus each of A and B obtains wrong (or relativistic) measurements of the other's space and time, despite the fact that each obtains the correct (or proper) measurements of his or her own space and time.

on the equator $\phi = 0$. Thus she computes her space-time coordinates as

$$x'_B = \tau \tan 0 = 0, \qquad t'_B = \tau \sec 0 = \tau$$

However, her false compass makes her believe that Eskimo A is at latitude ϕ south, that is, at $-\phi$. She therefore computes his space-time coordinates as

$$x'_A = \tau \tan (-\phi) = -\tau \tan \phi, \qquad t'_A = \tau \sec (-\phi) = \tau \sec \phi$$

But those computed for A are wrong, because A is actually on the equator and not at south latitude ϕ as B mistakenly believes.

The tale of the two Eskimos can be summarized in this way. The old man puts both Eskimos on the equator, but by trickery with their direction-finders, confuses them so that each thinks that the other is not on the equator. When each Eskimo computes his or her (proper) time and distance, the results are correct. But when each Eskimo computes the other's time and distance, the results are wrong. However, the square of the time-space interval is the same for every pair of time-space measurements; that is, the following equalities are true

$$t_A^2 - x_A^2 = t_B^2 - x_B^2 = t_A'^2 - x_A'^2 = t_B'^2 - x_B'^2$$

because the following equalities are trigonometric identities:

$$\tau^2 - 0^2 = \tau^2 \sec^2 \phi - \tau^2 \tan^2 \phi = \tau^2 \sec^2 \phi - (-\tau \tan \phi)^2 = \tau^2 - 0^2$$

This fable is the special theory of relativity, no more, no less. Albert Einstein summed it all up by saying

> Raffiniert ist der Herrgott aber boshaft ist er nicht.

When asked what he meant by that, he answered,

> Die Natur verbirgt ihr Geheimnis durch die Erhabenheit ihres Wesens, aber nicht durch List.

In terms of the Eskimo fable, we can make this statement. The old man is subtle (because he changed the compass directions), but he is not malicious (because he kept the time-space interval invariant). When asked what is meant by that, the following answer might be given. Nature hides her secrets because of her essential loftiness (because a deep understanding of space-time is kept from the Eskimos) but not by means of ruse (because there is a perfect space-time symmetry existing in nature).

The critical idea in this fable is due to Sir Joseph Larmor in 1901 when he originated the concept of the *dilation of time*, namely, the lifetime τ of an Eskimo at latitude 0 is perceived to increase to the dilated value $\gamma\tau$ at latitude ϕ. The gamma or the *dilation factor* is given by $\gamma = \sec \phi$. In this fable, we recognize that the latitude ϕ is identical to the Gudermannian angle ϕ that we introduced in the previous section.

As we have just demonstrated, there is a perfect equanimity in the perspectives of space-time as seen by A and B. Each has the same proper lifetime, but sees the other's lifetime dilated by the gamma factor. This is the symmetry of space-time. However, when we come to applying these principles, we often introduce some extraneous nonsymmetry, and it is for this reason that some seeming paradoxes crop up. The introduced asymmetry is that we are now going to favor Eskimo A by looking at everything from his perspective.

Eskimo A is on the equator and knows that his proper lifetime is τ. He is also able to measure the distance x to Eskimo B. Thus our hypothesis is that Eskimo A has two given quantities, namely, x and τ, and from these he will compute all other quantities.

Because $x = \tau \tan \phi$, Eskimo A can use this equation to compute the latitude

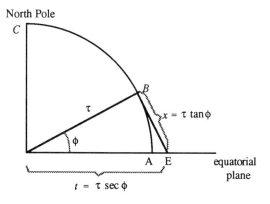

Figure 2.7. Given τ and x, Eskimo A computes the dilated lifetime t of Eskimo B.

ϕ of Eskimo B. He can then compute the dilated lifetime of Eskimo B by the formula $t = \tau \sec \phi$. See Figure 2.7.

Distance over time equals velocity. Eskimo A has determined that Eskimo B has distance x and time t, so he can immediately compute the velocity β as

$$\beta = \frac{x}{t} = \frac{\tau \tan \phi}{\tau \sec \phi} = \sin \phi$$

So far, so good. But now comes the asymmetry. Because both A and B are Eskimos, Eskimo A decides to put himself in B's shoes (or, better, in B's Eskimo boots). Eskimo A says that since his lifetime is τ, the lifetime of B must also be τ, not t. Since distance is velocity times time, Eskimo A therefore reasons that the distance to Eskimo B must be $\lambda = \beta\tau$, instead of the measured value $x = \beta t$. Since $\beta = \sin \phi$, Eskimo A computes the distance to Eskimo B as (Figure 2.8)

$$\lambda = \tau \sin \phi$$

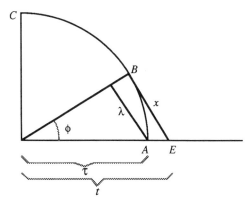

Figure 2.8. By placing himself in B's boots, A denies that B's lifetime is the dilated value t and makes it the proper value τ instead. To be consistent, A must also deny that B's distance is x and make it λ instead. Since $\tau = \alpha t$, it follows that $\lambda = \alpha x$; that is, λ is the contracted value of x. This is FitzGerald length contraction.

The computed distance λ is less than the measured distance $x = \tau \tan \phi$, because

$$\lambda = \tau \sin \phi = \frac{x}{\tan \phi} \sin \phi = x \cos \phi$$

and $\cos \phi$ is less than unity. Define the alpha factor as $\alpha = \cos \phi$. Thus the measured distance x is contracted to the value $\lambda = \alpha x$.

This critical idea in the fable is due to George FitzGerald when he originated the concept of the *contraction of distance*, namely, the distance x of an Eskimo at latitude ϕ is perceived to shrink to the contracted value αx. The alpha or the *contraction factor* is given by $\alpha = \cos \phi$. In relativity theory, we see that the Greek trio $\alpha = \cos \phi$, $\beta = \sin \phi$, and $\gamma = \sec \phi$ are, respectively, the contraction factor, the velocity, and the dilation factor. All three are simple trigonometric functions of the latitude (or Gudermannian angle) ϕ.

Let us now summarize in the form of the table:

	A's Perspective in His Own Boots	A's Perspective in B's Boots
Distance to B	x (given)	λ (computed as αx)
Lifetime of B	t (computed as $\gamma \tau$)	τ (given)

The table indicates a seeming contradiction, namely, that the same distance has two different values, x and λ. But this is accepted as physical fact, as the saga of the elementary particle, the muon, will illustrate. This story is told in the next section. The key quantity is the velocity:

$$\beta = \frac{x}{t} = \frac{(\lambda/\alpha)}{\gamma \tau} = \frac{\lambda}{\tau}$$

In terms of β the preceding table becomes

	A's Perspective in His Own Boots	A's Perspective in B's Boots
Velocity	$\beta = x/t$	$\beta = \lambda/\tau$
Distance	x (given)	$\lambda = \beta \tau$
Time	$t = x/\beta$	τ (given)

This table will appear in the next section.

THE FABLE OF THE MUON AND THE MOUNTAIN

The first of the short-lifetime particles found by physicists was the muon in 1938. The muon is a particle much like the electron, but about 200 times heavier. Pions and a set of strange particles were discovered in the 1940s among the products of cosmic-ray collisions. These particles are even more short-lived than the muons. With the building of large accelerators (or atom smashers), more and more particles were discovered. Today well over 100 of them have been identified. However, most of these particles are not elementary, but are made up of simpler entities called quarks. Yet the muon, the first to be discovered, seems to be not made up of quarks and so is truly elementary. The muon plays no essential role in the atomic nucleus.

We close this chapter with one of the most important physical confirmations of the special theory of relativity. We call it the fable of the muon and the mountain. It can be summed up by the saying: *if the mountain will not come to the muon, then the muon must go to the mountain.* This seems fair, but the muon must travel a distance of 19.9 light-μs, whereas the mountain needs only travel a distance of 1.99 light-μs. (Note: μs is the abbreviation for microsecond, or one-millionth of a second.) Yet each journey is the same, and they each go at the same speed. Why? Because time is not absolute. (In this example, we use approximate numbers for the physical variables to make the arithmetic easy.)

Plate 2.5. The Tower of London, Newton's workplace as Master of the Mint when he published the *Opticks*.

From cosmic rays hitting the Earth from outer space, muons are created by collisions high up in the atmosphere. These muons are very energetic particles, some moving with 199/200 of the velocity of the light. These muons are known to be unstable, with a lifetime of about $\tau = 2 \times 10^{-6}$ sec (2 μs) when they are measured at rest in a laboratory on Earth. Yet when they are created by collisions in the atmosphere, they travel at speed $\beta = 199/200$, and they cover a distance of 19.9 light-μs on the trip down to the Earth before they decay. It seems puzzling that they last for only $\tau = 2$ μs in the laboratory, but they can travel a distance of $x = 19.9$ light-μs in the atmosphere.

Let us look at a muon created high up in the atmosphere. The Earth represents the mountain in the fable. If we consider the mountain at rest, then this muon goes to the mountain at a speed of $\beta = 199/200$. On the other hand, if we consider this muon to be at rest, then the mountain comes to the muon at the same speed $\beta = 199/200$. We summarize:

	Muon Goes to Mountain (or the Flying Muon)	Mountain Comes to Muon (or the Flying Mountain)
Velocity	$\beta = 199/200$	$-\beta = -199/200$
Distance	$x = 19.9$ light-μs	$\lambda = \beta\tau = (199/200)(2) = 1.99$ light-μs
Time	$t = x/\beta = 19.9/(199/200) = 20$ μs	$\tau = 2$ μs

In the table, the velocity β, the distance x, and the time τ are the physical observations that we can measure. The time t and the distance λ are then computed.

When the muon goes to the mountain, the observed distance is $x = 19.9$ light-μs, but when the mountain comes to the muon the computed distance has decreased, or contracted, to the value $\lambda = 1.99$ light-μs. This computational result is called the FitzGerald contraction of distance.

When the mountain comes to the muon, the observed travel time is $\tau = 2$ μs, but when the muon goes to the mountain, the computed travel time has increased, or dilated, to $t = 20$ μs. This computational result is called the Larmor dilation of time.

The possibility of length contraction and time dilation represented a great stumbling block in physics in the late nineteenth century. Relativity removed this puzzle. We are looking at the flying muon's internal lifetime clock. To the muon, the period of this clock is $\tau = 2$ microseconds, but to us observing the clock of the moving muon, that period is dilated to 20 μs. That is, the flying muon's lifetime is computed to be 20 μs. This computation is based on the fact that it hurls through the atmosphere a distance we measure on Earth as 19.9 light-μs. However, from the muon's own point of view (i.e., the case of the flying mountain), its lifetime is a normal 2 μs, but the thickness $x = 19.9$ of our atmosphere that rushes past it is contracted to 1.99 light-μs. Thus the flying mountain makes a shrunk trip of $\lambda = 1.99$ in the stationary muon's lifetime of $\tau = 2$ μs, whereas the stationary mountain sees the flying muon make a trip of $x = 19.9$ in its flying lifetime of $t = 20$ μs. From either point of view, the relative velocity between the Earth frame and the muon frame is the same, namely,

$$\beta = \frac{\lambda}{\tau} = \frac{1.99}{2} = \frac{199}{200}, \qquad \beta = \frac{x}{t} = \frac{19.9}{20} = \frac{199}{200}$$

At this point, the reader might say, "Stop! This is utter nonsense! How can the same journey be a distance of both 19.9 light-μs and 1.99 light-μs?" We recall that special relativity is based on the relativity principle and the light principle, which are perfectly symmetrical in application. Why then should we get such an unsymmetrical result of 19.9 versus 1.99? This result shows a clear favoritism for the flying mountain, which only has to cover the short distance of 1.99, versus the flying muon, which must cover the long distance of 19.9.

The answer is that there is a third principle, always unexpressed, that invariably and necessarily crops up in applications. This is the *recognition principle*, which says that physical measurements are always confined to certain recognized inertial frames, and the resulting favoritism introduces an asymmetry such as we have observed in the muon-mountain fable. The mountain is the favored reference frame because the atmosphere is attached to the mountain, not to the muon.

Now let us explain the reason for the unsymmetrical result. The journey is through the atmosphere, and the atmosphere belongs to the mountain. Already the reader can see that the muon is going to get the bad deal. When the flying muon goes to the mountain, it must travel through the atmosphere which is at rest, even as the mountain is at rest. However, when the flying mountain comes to the muon, the atmosphere moves together with the mountain. This asymmetry is summarized in the table:

	Muon Goes to Mountain (or the Flying Muon)	Mountain Comes to Muon (or the Flying Mountain)
Muon	Moving	At rest
Atmosphere (attached to mountain)	At rest (rest length = 19.9)	Moving (moving length = 1.99)
Mountain	At rest	Moving

As the table shows, the fable is one of the mountain and the atmosphere versus the muon, that is, two against one. This causes the unsymmetrical result of 19.9 versus 1.99 as the length of the same journey. If we reversed the situation by attaching the atmosphere to the muon, then we would get the opposite result, namely, the flying muon would have the short distance of 1.99 to travel and the flying mountain would have the long distance of 19.9.

The fable of the muon and the mountain illustrates that the same phenomenon can be considered from two points of view. In the case of the mountain coming to the muon, the muon (call it A) is at rest with a lifetime of $\tau = 2$ μs. In the case of the muon going to the mountain, the muon (call it B) is moving with velocity $\beta = 199/200$ with respect to the mountain. Because $\beta = \sin \phi$, the angle ϕ is $\sin^{-1} \beta$. The rest muon A disintegrates in 2 μs. It travels nowhere. The moving muon B takes a longer time to disintegrate. This longer or dilated time is

$$t = \tau \sec \phi = 20 \text{ μs}$$

During this time, moving muon B travels the distance

$$x = \beta t = \tau \tan \phi = 19.9 \text{ light-μs}$$

This phenomenon in routinely observed in high-energy physics. Its explanation is as follows. Stationary muon A is not moving through space ($\beta = 0$), so it is traveling through time as fast as possible ($\alpha = 1$). It therefore ages rapidly and dies in $\tau/\alpha = 2/1 = 2$ μs. Moving muon B is traveling through space at a speed $\beta = 199/200$, and so it is traveling through time at the reduced speed of $\alpha = \sqrt{1 - \beta^2} = 0.1$. Its aging is inversely proportional to this α speed, so we see it die in $t = \tau/\alpha = 2/(0.1) = 20$ μs. (A trip at a reduced speed takes longer; this is a trip at reduced speed through time.) The constant $\gamma = 1/\alpha = 10$ is the time dilation factor. In time $t = 20$ μs, the muon travels a distance $x = \beta t = 19.9$ light-μs.

A photon (or light particle) is an extreme case. It travels through space at the maximum possible rate ($\beta = 1$) so it travels through time at a zero rate ($\alpha = 0$). Thus it does not age.

Finally, let us look at the world from the point of view of the moving muon B. In its frame, it is at rest, and the mountain is moving to muon B at a speed β. In the 2-μs lifetime ($\tau = 2$) of muon B (at rest in its frame), the mountain comes to the muon a distance

$$\lambda = \beta \tau = 1.99 \text{ light-μs}$$

But as we have seen, this same journey measured when the muon goes to the mountain is the distance

$$x = \beta t = 19.9 \text{ light-μs}$$

Thus the distance of 19.9 (muon goes to mountain) shrinks or contracts to a mere distance of 1.99 (mountain comes to muon). What is the contraction factor? Solving the two foregoing equations, we have

$$\beta = \frac{\lambda}{\tau} = \frac{x}{t}$$

so

$$\lambda = \frac{\tau}{t} x$$

The Fable of the Muon and the Mountain

But $t = \tau/\alpha$, so

$$\lambda = \alpha x$$

The shrinking factor is therefore equal to α, the speed through time. Thus the constant α is the FitzGerald length contraction factor.

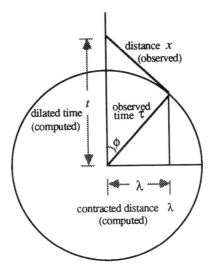

Figure 2.9. Observed and computed space and time coordinates.

$\beta = \sin \phi$ = relative velocity of muon and mountain (observed)

τ = lifetime of muon perceived by muon at rest (observed)

$x = \beta t$ = distance covered by moving muon perceived by mountain at rest (observed)

$t = \gamma \tau$ = lifetime of moving muon perceived by mountain at rest (computed)

$\lambda = \beta \tau$ = distance covered by moving mountain perceived by muon at rest (computed)

In summary, the essential features of the fable of the muon and the mountain are incorporated in Figure 2.9 and in the following table:

Constant	Interpretation (1)	Interpretation (2)
$\alpha = \cos \phi$	Speed through time	Length contraction factor
$\beta = \sin \phi$	Speed through space	
$\gamma = 1/\alpha$		Time dilation factor

In the muon-mountain fable, there are three variables x, τ, and β that can be observed. However, only two of these three variables are independent, as the third can be computed from the other two. For example, if we measure x and τ, then we can compute β in two steps. First, compute ϕ by the formula $\tan \phi = x/\tau$, and then compute β as $\beta = \sin \phi$. In other words,

$$\beta = \sin \tan^{-1} (x/\tau)$$

Now suppose that we instead measure τ and β. Then compute ϕ by the formula $\beta = \sin \phi$, and compute x by the formula $x = \tau \tan \phi$; that is,

$$x = \tau \tan \sin^{-1} \beta$$

Finally, suppose that we measure x and β. Compute ϕ by the formula $\beta = \sin \phi$ and then compute τ by the formula $\tau = x \cot \phi$; that is,

$$\tau = x \cot \sin^{-1} \beta$$

The fact that we can measure three quantities, but only two are independent, means that we can compute the third one, and then compare the computed value with the measured value. The degree of agreement gives a test of the special theory of relativity. All such tests have to be statistical, because the required measurements cannot be made on a given individual muon, neither in theory nor practice.

Let us now try to summarize the fable of the muon and the mountain. As we have seen, when the muon goes to the mountain, it must cover a distance of 19.9 light-μs (which is 5.97 km), whereas the mountain making the same journey must cover a distance of only 1.99 light-μs (which is 0.597 km). But how can the same journey be both 5.97 km and 0.597 km? The problem is that we are looking at this journey in a much too materialistic way. Instead, we must look on it as a journey of devotion. The muon is going to devote its entire lifetime to the journey. When the length of any journey represents a lifetime, it cannot be fairly measured in kilometers; instead, it is a journey through life and must be so treated.

The muon knows that its own lifetime is a short 2 μs, and it is traveling at a speed of 199/200. Thus, the journey of one lifetime is the short distance of 2(199/200) = 1.99 light-μs. The mountain perceives the muon out there flying at a great speed through space and therefore at a reduced speed through time. Therefore, the mountain perceives that the lifetime of the muon is greater than it really is. In other words, the mountain believes that the lifetime of a muon is greater than 2 μs. The mountain believes that the lifetime of the muon is the long value of 20 μs. In this long lifetime, the muon can travel the long distance 20(199/200) = 19.9 light-μs.

Here we have called the lifetime of 2 μs and the corresponding distance of 1.99 light-μs *short* and the 20-μs lifetime and the 19.9 light-μs distance *long*. This terminology exposes the nature of the fable. But the correct terminology is to call the 2-μs lifetime and the 19.9 light-μs distance the given quantities; then the computed 20 μs lifetime is said to be dilated, and the computed 1.99 light-μs distance is said to be contracted.

CHAPTER 3

MAXWELL'S EQUATIONS AND RELATIVITY

Hail holy light, offspring of Heaven first born
Or of the Eternal Coeternal beam
May I express thee unblamed? Since God is light,
And never but in unapproached light
Dwelt from Eternity, dwelt then in thee,
Bright effluence of bright essence increate.

John Milton (1608–1674)
Paradise Lost, III, 1-6

Plate 3.1. James Clerk Maxwell (1831–1879)

KEPLER'S EQUATION

Johannes Kepler (1571–1630) worked with the great naked-eye astronomer Tycho Brahe (1546–1601) for the 18 months preceding Tycho's death. Tycho had given Kepler his painstakingly gathered observations, and Kepler kept control of that data when he moved to Prague. In a letter, Kepler said "I confess that when Tycho died, I quickly took advantage of the absence, or lack of circumspection, of the heirs, by taking the observations under my care, or perhaps usurping them."

The year 1609 saw the publication of Kepler's book *Astronomia Nova* in which he gave his first two laws of planetary motion. However, Kepler did something unusual in science writing, because in his book Kepler describes every blind alley, detour, trap, and pitfall that he encountered in arriving at these laws. It is good that he did this, because it establishes that these laws did not come out of the air but were the result of hard work. In a way, this is the most interesting aspect of scientific research. As we know, the theory of relativity also followed a difficult path in its development, involving many obstacles and obstructions.

In his preface, Kepler writes,

> What matters to me is not merely to impart to the reader what I have to say, but above all to convey to him the reasons, subterfuges, and lucky hazards which led me to my discoveries. When Christopher Columbus, Magellan, and the Portuguese relate how they went astray on their journeys, we not only forgive them, but would regret to miss their narration because without it the whole grand entertainment would be lost.

Kepler's Equation

Many of the concepts introduced by Kepler, together with the names he used to describe them, have persisted to this day. The term "true anomaly" is used to describe the angle from periapsis to the orbiting object measured in the direction of motion. Another term called "eccentric anomaly" was introduced by Kepler in connection with elliptical orbits. In Figure 3.1, we see an ellipse with the periapsis labeled. The point P represents the orbiting object. The *true anomaly* is the angle ϕ'_c. Also shown is an auxiliary circle circumscribed about the ellipse. A line perpendicular to the major axis has been extended through P to where it intersects the auxiliary circle. The angle ϕ_c is called the *eccentric anomaly*.

Let the *eccentricity* of the ellipse be β. Then Kepler's equation for relating the eccentric anomaly to the true anomaly is

$$\cos \phi'_c = \frac{\cos \phi_c - \beta}{1 - \beta \cos \phi_c}$$

For the purposes of this book, it is more convenient to use the complementary angles

$$\phi = \frac{\pi}{2} - \phi_c, \qquad \phi' = \frac{\pi}{2} - \phi'_c$$

Then Kepler's formula becomes

$$\sin \phi' = \frac{\sin \phi - \beta}{1 - \beta \sin \phi}$$

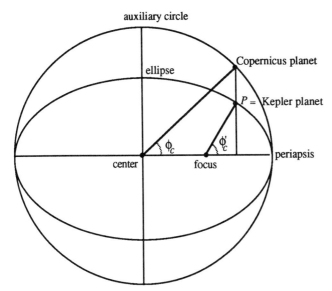

Figure 3.1. The true anomaly ϕ'_c of the Kepler planet P and the eccentric anomaly ϕ_c of the Copernicus planet.

If we define

$$v = \sin \phi$$
$$v' = \sin \phi'$$

then *Kepler's equation* takes the form

$$v' = \frac{v - \beta}{1 - v\beta}$$

This equation will find use in the next section.

OLYMPIC METAPHOR OF RELATIVITY

Each athlete in the Olympics tries for the perfect ten. In preparation before the actual competition starts, each player gauges her competitors' abilities relative to herself. She is required to mark a person better than herself with a positive grade and a person worse than herself with a negative grade. Thus, on her scale of ability, she places herself in the center, at the point 0. She would place a perfect athlete at the point 1, which represents the highest grade possible, the grade of a "perfect ten" athlete. At the other extreme is the point -1, representing the worst possible case. She would place all competing athletes at various points between these two extremes on her scale. This scale is called the athlete's *frame of reference*. Each competing athlete creates her own frame of reference, so there is one for each athlete.

For simplicity we consider three athletes: A, B, C. According to her feelings, A makes out the scale

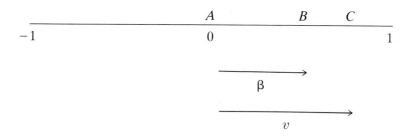

This scale shows that according to the feelings of A, athlete B rates a value β, and C rates a value v. The values β and v are chosen relative to the fact that there exists a "perfect ten" to whom the rating 1 is given. This is the *perfective aspect* of relativity theory.

Next, let us look at B. According to his feelings, he makes out the scale

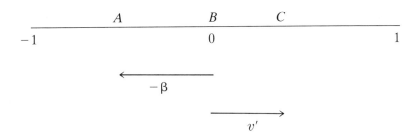

This scale shows that according to the judgment of B, athlete A rates a value of $-\beta$, and C rates a value of v'.

Let us first look at the relationship between A and B. Athlete A rates B as β, and athlete B rates A as $-\beta$. This relationship is as we would expect. There are only the two of them involved, and they can make this clear-cut decision.

But now let us consider the third party C. Athlete A rates C as v, and B rates C as v'. What is the relationship of these two ratings? Since A rates B as β, and A rates C as v, athlete A might conclude that B rates C as $v - \beta$. The quantity $v - \beta$ is called an *apparent rating*, because A makes this rating in her frame and not in B's frame. But the *true rating* of C by B is the one given in B's own frame, namely, v'. Can we set the apparent rating $v - \beta$ equal to the true rating v'? That is, can we conclude that

$$v' = v - \beta \tag{3.1}$$

Now let us look at B. Since B rates A as $-\beta$ and B rates C as v', athlete B would conclude that A rates C as $-(-\beta) + v'$. That is, the apparent rating made in B's frame is $v' + \beta$. Can we conclude that this apparent rating is equal to the true rating v of C by A made in A's own frame? That is, can we conclude that

$$v = v' + \beta \tag{3.2}$$

Let us look at the two conclusions (3.1) and (3.2). They are certainly consistent. One can be obtained from the other by merely transposing the β. If we could treat a physical phenomenon in this way, then it would give a system of measurement that is at once simple and beautiful.

We have already treated the perfective aspect of relativity theory, the standard of athletic perfection that gets a 1 in every athlete's frame of reference, the mark of a perfect ten. Now we come to the other aspect of relativity theory, the *perspective* aspect. Here the word perspective is used in the sense of a subjective evaluation of relative significance, that is, a point of view. First, we are willing to concede that if A likes B by β, then B likes A by $-\beta$. But because two people have an honest difference in their points of view, we cannot concede the conclusions (3.1) and (3.2) with respect to a third party C. If A likes B by β, and A likes C by v, then A cannot conclude that B likes C by $v - \beta$. In other words, the feelings of A cannot be directly transcribed into the feelings of B. Likewise, if B likes A by $-\beta$, and B likes C by v', then B cannot conclude that A likes C by $v' + \beta$.

When it comes to the feelings of two people, extreme care must be used to describe their individual relationships with a third person. For example, a child might dislike an uncle by 0.4 (i.e., the child might like the uncle by -0.4) and like an aunt by 0.6, but she cannot conclude that the uncle likes the aunt by 1.0 (a perfect score). And what if the child dislikes the uncle by 0.5 and likes the aunt by 0.8? She cannot conclude that the uncle likes the aunt by $0.5 + 0.8 = 1.3$, because that value is above the ultimate possible score of 1.0. Each person has her own point of view, but it does not readily transcribe into the point of view of someone else. This is the most difficult point of relativity theory, but one faced every day in human relationships.

The final formulation of relativity theory is based upon a dictum made by William of Ockham who lived from about 1285 to 1349. This dictum is known as *Ockham's razor*, because it gives a cutting edge to science. Ockham's razor is a rule in science and philosophy stating that formulas should not be aggrandized needlessly. This rule is interpreted to mean that the simplest of two or more competing formulas is preferable. There can be any number of formulas that relate the feelings of A and the feelings of B with respect to a third party C. Ockham's razor says that we must take the simplest possible formula. Anything more complicated is thrown out. Newton wrote "Nature is pleased with simplicity." Henry David Thoreau expressed the same idea when he said, "Simplicity, simplicity, simplicity."

According to the perspective aspect, each person must be treated in an equivalent way in any formula expressing his or her point of view. In other words, each person's formula must look exactly like every other person's formula. Let us explain. The known facts are

$$A\text{'s feelings} \begin{cases} A \text{ likes } B \text{ by } \beta \\ A \text{ likes } C \text{ by } v \end{cases}$$

$$B\text{'s feelings} \begin{cases} B \text{ likes } A \text{ by } \beta' \text{ (which is equal to } -\beta) \\ B \text{ likes } C \text{ by } v' \end{cases}$$

First, let us find the feelings of B from those of A. The formula would have the form

$$v' = f(\beta, v) \qquad (3.3)$$

Here we are using the mathematical notation of a *function f*. The equation says that v' is a function of β and v. In other words f stands for the formula. We put β and v into the formula and we get out the required v', as seen in the diagram:

$$\begin{array}{c} \beta \\ \downarrow \\ v \rightarrow \boxed{f} \rightarrow v' \end{array}$$

The theory of relativity requires that A and B must have the same perspective. What is good for the goose is good for the gander. The same formula f must work to find the feelings of A from those of B. Thus we have

$$v = f(\beta', v')$$

The same formula f is used, but now we put in β' and v' and get out the required v, as seen in the diagram:

$$\begin{array}{c} \beta' \\ \downarrow \\ v' \rightarrow \boxed{f} \rightarrow v \end{array}$$

At this point it is good to make note of a certain mathematical convention used in relativity theory. We use certain variables for the feelings of A, for example, β, v. The corresponding variables for the feelings of B make use of the same symbols, but have primes attached to them, for example β', v'. Thus we call the frame of reference of A the *unprimed system* and the frame of reference of B the *primed system*. These two frames are related by the fact that

$$\beta' = -\beta$$

Thus the foregoing equation may be written as

$$v = f(-\beta, v') \tag{3.4}$$

which is

$$\begin{array}{c} -\beta \\ \downarrow \\ v' \rightarrow \boxed{f} \rightarrow v \end{array}$$

Equations (3.3) and (3.4) express the perspective aspect of relativity theory, which can be expressed as the following rule. In equation (3.3), interchange primes on the variables (i.e., change v to v' and change v' to v) and change β to $-$ β. The result is equation (3.4).

Likewise, take equation (3.4) and change v, v', β to v', v, $-$ β. The result is equation (3.3). Equation (3.3) is called the *direct formula*, and equation (3.4) is called the *reflexive formula*. They say that A and B have the same point of view expressed in terms of their own perspectives.

As we know, relativity theory has two aspects: perfectivity and perspectivity. For emphasis, at this point let us repeat the perfective argument. In anyone's frame, the value 1 is the ultimate value, and $-$ 1 is the lowest possible value. No value may lie outside these limits. If someone is a perfect ten, then this person must appear as a perfect ten for everyone else. This is the perfective quality of relativity theory. A perfect ten appears as a perfect ten to all concerned. If this were not so, then the person would not be a perfect ten. Suppose C is a perfect ten. Then A's frame would be

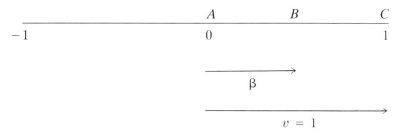

and B's frame would be

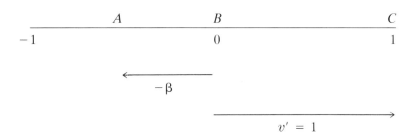

Equation (3.1), namely $v' = v - \beta$, does not work because it gives $1 = 1 - \beta$, which is not true. As a result, we must reject equation (3.1) as being a suitable candidate for the required formula (3.3). However, we can readily find the required formula to compute v' (i.e., B's like of C) from v (i.e., A's like of C). By Ockham's razor, or by Thoreau's philosophy, the simplest possible formula that preserves both the required perfective and perspective aspects must be used. The greatest scientific advance in the early part of the seventeenth century was Kepler's discovery that the planets travel in elliptical orbits about the sun. As we have seen in the first section, *Kepler's equation* is

$$v' = \frac{v - \beta}{1 - v\beta} \qquad (3.5)$$

What could be more simple than saying that Kepler had in fact found the right formula for relativity theory. This fact would have delighted Henry David Thoreau when he said "Simplicity, simplicity, simplicity." Kepler's equation (3.5) is the formula that gives B's feelings (primed) in terms of A's feelings (unprimed).

First, let us confirm the perfective aspect. Let C be a perfect ten. Thus A likes C by $v = 1$. The formula gives

$$v' = \frac{1 - \beta}{1 - \beta} = 1$$

so B also likes C by 1. Thus the perfective aspect holds for the formula.

Next let us confirm the perspective aspect. This means that the formula that gives A's feelings in terms of B's feelings should have the same form as equation (3.5). Thus we take equation (3.5) and change v' to v, change v to v', and change β to $-\beta$. The result, called the *reflexive equation*, is

$$v = \frac{v' + \beta}{1 + v'\beta} \qquad (3.6)$$

This is the required equation of the same form to give A's feelings v (unprimed) from knowledge of B's feelings v' (primed). Put $v' = 1$ in this equation. The result is

$$v = \frac{1 + \beta}{1 + \beta} = 1$$

which confirms that this equation is also perfective.

The final test is that we must show that the direct equation (3.5) and the reflexive equation (3.6) are consistent. We do this as follows. Solve equation (3.5) for v. The result, called the *inverse equation*, turns out to be the same as equation (3.6), the reflexive equation. The fact that the inverse equation is the same as the reflexive equation establishes the required consistency. This is shown by means of the diagram

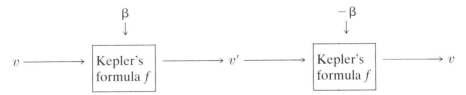

Remember our example when a child likes her uncle by $\beta = -0.4$ and likes her aunt by $v = 0.6$. Then, according to the theory of relativity, we can definitely and precisely conclude that the uncle likes the aunt by the result given by Kepler's equation (3.5), namely,

$$v' = \frac{0.6 - (-0.4)}{1 - 0.6(-0.4)} = 0.806$$

Suppose instead the child likes her uncle by $\beta = -0.5$ and likes her aunt by $v = 0.8$. Then according to the theory of relativity the uncle likes the aunt by

$$v' = \frac{0.8 - (-0.5)}{1 - (0.8)(-0.5)} = 0.929$$

Not even Dr. Sigmund Freud (1856–1939) had a formula like this by which he could compute the feelings of his patients. Kepler's formula is the magic of the special theory of relativity. In psychology, there is no formula, and no mathematics is required. In physics, there is a formula, and as a result mathematics is required.

Let us summarize the special theory of relativity. Each person has his or her own frame of reference, which is at once perfective and perspective. According to Ockham's razor, the simplest possible formula consistent with perfectivity and perspectivity must be used to relate any two different frames. The result is the formula given by Kepler's equation (3.5). This is all there is to special relativity. All the necessary mathematics is encompassed in the formula devised by Kepler nearly four centuries ago. We work out the mathematics in the next section.

MATHEMATICAL MODEL OF RELATIVITY

In the preceding section, we presented the essential mathematical structure of the special theory of relativity. It is based on a person and his idea of perfection and perspective. This person creates his own frame of reference by placing himself at 0, perfection at 1, and negative perfection at -1. He places all other people between these two limits. If person A places person B at β, then necessarily person B places person A at $-\beta$. In the evaluation of a third party, the placement must be perspective. Perfectivity and perspectivity, under Ockham's rule, gives Kepler's equation, which relates person A's like of person C to person B's like of person C. If person A likes C by the amount v, then Kepler's equation (3.5), which is

$$v' = \frac{v - \beta}{1 - v\beta}$$

gives v', the amount by which person B likes person C. The quantity v' represents

MATHEMATICAL MODEL OF RELATIVITY

the actual or physical evaluation of C by B, according to B's frame. The quantity $v - \beta$ represents the apparent evaluation of C by B according to A's frame. The apparent is not the same as the actual.

In the special theory of relativity, the quantities β, v' and v are velocities. That is, each is the ratio of a distance over a time. Each person has his own frame, with his own measurement of distance and of time. Let x and t represent distance and time, respectively, for person A, and x' and t' for person B. In particular, let

x = distance from A to C, according to A

t = time from A to C, according to A

x' = distance from B to C, according to B

t' = time from B to C, according to B

Then it follows that

$$v = \frac{x}{t}, \quad v' = \frac{x'}{t'}$$

With this notation, Kepler's equation (3.5) becomes

$$\frac{x'}{t'} = \frac{(x/t) - \beta}{1 - (x/t)\beta}$$

Multiply both the top and bottom of the right-hand side by t. Then

$$\frac{x'}{t'} = \frac{x - \beta t}{t - \beta x}$$

This equation says that x' is to $x - \beta t$ as t' is to $t - \beta x$. Denote the proportionality constant by γ. Then we have

$$x' = \gamma(x - \beta t)$$
$$t' = \gamma(t - \beta x) \tag{3.7}$$

These are known as the *Lorentz equations*.

Similarly, equation (3.6) becomes

$$\frac{x}{t} = \frac{(x'/t') + \beta}{1 + (x'/t')\beta}$$

which is

$$\frac{x}{t} = \frac{x' + \beta t'}{t' + \beta x'}$$

Because of the perspectivity between A and B, the same constant of proportionality is required. Thus we have

$$x = \gamma(x' + \beta t')$$
$$t = \gamma(t' + \beta x') \tag{3.8}$$

These are known as the *reflexive Lorentz equations*. Substitute x' and t' given by (3.7) into the first equation of (3.8). We obtain

$$x = \gamma[\gamma(x - \beta t) + \beta\gamma(t - \beta x)]$$

which is

$$x = \gamma^2 x (1 - \beta^2)$$

so

$$\gamma^2 = \frac{1}{1 - \beta^2}$$

The positive square root

$$\gamma = \frac{1}{\sqrt{1 - \beta^2}}$$

gives the required constant in terms of β. Since the magnitude of β cannot exceed

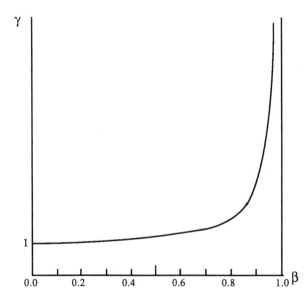

Figure 3.2. Dilation factor γ plotted as a function of velocity β.

1, it follows that the magnitude of γ cannot be less than 1. The constant γ is called the *dilation factor*. A plot of γ versus β is shown in Figure 3.2.

In this section and the preceding one, we have constructed a mathematical model of special relativity and derived the Lorentz equations. But up to this point we have not specified the physical linkage. That is the subject matter of the remainder of this chapter.

WAVE MOTION

In baseball, the first player in the batting order is chosen to be a person who can get on base one way or another. He does not have to be a slugger. Following him in the batting order come the sluggers, because their long blasts can drive the runner home.

We have chosen the Olympic metaphor of relativity as the lead-off batter in this book. This metaphor describes the special theory of relativity. It contains the essential mathematics that is required for this important subject. An understanding of this material means that you are on first base. The rest of the book is spent trying to get you to home plate. Like a baseball game, there is a lot of repetition but almost always there are some inspiring moments. And the game is not over until the end of the last inning.

The repetition in this book is done on purpose. We try to look at the same relativistic formulas over and over again from different points of view. We feel that understanding comes from being conversant with the subject. We are trying to drive home a few runs, but in addition we expect the reader to go to other relativity books for further information, just as you go to other ball games. No one book will do. In this book, we concentrate on the mathematics of relativity theory, and purposely neglect the philosophy of the subject.

Living on Spaceship Earth, humans are dependent upon wave motion to bring them knowledge of the vast universe surrounding them. The wave-particle duality says that light, although made up of small particles called photons, travels as a wave. As we will see in the next chapter, the wave motion of light makes it possible for us to utilize the Doppler effect to uncover much about the structure and composition of the universe.

Walt Whitman writes[7]

In cabin'd ships at sea,
The boundless blue on every side expanding,
With whistling winds and music of the waves, the large imperious waves,
Or some lone bark buoy'd on the dense marine,
Where joyous full of faith, spreading white sails,

---

[7] Whitman, Walt. "In Cabin'd Ships at Sea." From *The Complete Poems and Prose of Walt Whitman*, Ferguson Bros. & Co. Printers, Philadelphia, 1888.

> She cleaves the ether mid the sparkle and the foam of day,
> > or under many a star at night,
> By sailors young and old haply will I, a reminiscence of the land, be read,
> In full rapport at last.
>
> Here are our thoughts, voyager's thoughts,
> Here not the land, firm land, alone appears, may then by them be said
> The sky o'erarches here, we feel the undulating deck beneath our feet,
> We feel the long pulsation, ebb and flow of endless motion,
> The tones of unseen mystery, the vague and vast suggestions
> > of the briny world, the light-flowing syllables,
> The perfume, the faint creaking of the cordage, the melancholy rhythm,
> The boundless vista and the horizon far and dim are here,
> And this is ocean's poem.

Far below the lofty heights of the poets are the applied physicists who work with wave motion in the form of observational data. Geophysics is such a discipline. A geophysicist must probe the unknown depths of the Earth using the necessarily incomplete data that she can glean from the surface. She is denied the precise experiments that the experimental physicist can design and control, and she is also denied the luxury of elaborate theories because of the sheer complexity of the underground strata. Furthermore, she is faced with the same types of problems as those faced in relativity theory. Instead of light waves, she uses seismic waves to probe the unknown depths. She sends out signals which are returned to her by reflection from underground interfaces separating different rock strata. Surprisingly, the reflecting quality of the subsurface geologic interfaces (i.e., the ratio of imperfections on the reflecting surface to the wavelength of the seismic waves) is about the same as the reflecting quality of a good optical mirror (i.e., the ratio of the imperfections on the mirror to the wavelength of light). Just as space-time diagrams make up the essential data in relativity theory, so do space-time plots represent the basic seismic data collected in geophysics. The analysis of these seismic records involves the untangling of all the multiple reflections, interference patterns, and diffractions which are well known in the theory of light. To make any sense out of the tremendous amount of data involved, the geophysicist must use the same mathematical methods as do all other branches of physics that involve wave motion. This unifying mathematics falls under the general heading *wave equation*.

THE WAVE EQUATION

The wave equation begins with the search for perfection undertaken by Pythagoras (ca.582–ca.497 B.C.). His was the first great insight into the mathematical analysis of waves. He discovered that the pitch of a sound from a plucked string depends upon the string's length and that harmonious sounds are given off by strings whose

The Wave Equation

lengths are in the ratio of whole numbers. However, significant additional progress was impossible until the invention of calculus, more than 2000 years later. Newton's invention permitted the English mathematician Brook Taylor (1685–1731) to make the first productive attempt at the quantification of wave motion. Taylor is best known, of course, for his discovery of the Taylor series, familiar to every calculus student.

Consider the shape of a stretched string. At each point, the string has a certain slope. The slope of this slope represents the *curvature* of the string at any moment. Now consider the motion of any particular point on the string. This point has a *velocity* and an *acceleration* (the time rate of change of velocity). When the string is straight, it is in equilibrium position, so there is no net transverse force acting on any point on the string. However, when the string is curved, the tension on the string exerts a restoring force which attempts to move it back to its equilibrium position. The more the curvature the greater this restoring force. Taylor noted this and carried out the following reasoning:

Given: Curvature is proportional to force.   (Taylor's discovery)
            Force is proportional to acceleration.   (Newton's law)
Conclusion: Curvature is proportional to acceleration.   (wave equation)

This conclusion when translated into mathematical symbols is the scalar wave equation. Taylor could not fully develop the properties of this equation because calculus was still in its infancy. However, the most important fact about this equation is the constant of proportionality. Taylor found that this vital constant is equal to the ratio of the linear density of the string divided by the tension in the string. Now comes one of the most startling discoveries in the history of physics.

Linear density has dimensions of mass per unit length. Tension has the dimension of force. Thus the wave-equation constant has dimensions

$$\text{constant} = \frac{\text{density}}{\text{tension}} = \frac{(\text{mass/length})}{\text{force}}$$

But force is equal to mass times acceleration (Newton's second law). Since acceleration is equal to length divided by time squared, force has dimensions (mass)(length)/[(time)(time)]. Thus the constant has dimensions

$$\text{constant} = \frac{(\text{mass/length})}{(\text{mass})(\text{length})/[(\text{time})(\text{time})]}$$

$$= \frac{(\text{time})(\text{time})}{(\text{length})(\text{length})}$$

$$= \frac{1}{(\text{velocity})(\text{velocity})}$$

The last equality follows from the fact that velocity is equal to length/time. Thus the constant has the dimensions of the reciprocal of the square of a velocity. Solving for this velocity, we obtain

$$\text{velocity} = \frac{1}{\sqrt{\text{constant}}} = \sqrt{\frac{\text{tension}}{\text{density}}}$$

No equation in physics proved more important than did this one of Brook Taylor. The implications are astounding. This equation gives the velocity of traveling material waves on the string in terms of the physical makeup of the string (i.e., its tension and density). Thus Taylor found that wave velocity depends upon the physics. This is the type of linkage that makes physics a science. There is no such linkage in economics where wave motion predominates (in the form of business cycles, etc.). There are many incomplete linkages found to date in medicine (where EKG and EEG waves are vital signs). But wave motion in geophysics has physical linkage; so also in optics, acoustics, electronics, relativity theory, quantum mechanics, and every other branch of physics involving wave motion. Brook Taylor, although overshadowed by his great contemporary Sir Isaac Newton, was the one who first linked wave motion to physical properties in the form of the wave equation. Today, we are still following his footsteps.

Since stress is force per unit area, Taylor's discovery can also be stated as

$$\text{curvature} = -\kappa \times \text{stress}$$

where $-\kappa$ is the constant of proportionality. The most far-reaching application of this idea was made by Einstein in 1915 in the theory of general relativity. In special relativity, space-time is flat, but in general relativity space-time is curved. Einstein reasoned that matter influences the curvature of space-time, and curvature of space-time influences the motion of matter. In this way Einstein extended Taylor's result to apply, not just to a string, but to all of space-time, and he wrote

$$\text{curvature of space-time} = -\kappa \times \text{stress of matter and energy}$$

Expressed in the compact notation of tensors, this becomes *Einstein's field equation* of general relativity, namely,

$$R_k - \frac{1}{2} g_k R = -\kappa T_k$$

Ironically, the mathematician David Hilbert (1862–1943), whom Einstein had consulted on the problem of general relativity, became interested, and actually arrived at the field equation first, presenting it to the Royal Academy of Sciences in Göttingen in 1915. Five days later Einstein also presented it, the end product for which he had been searching for so many years, to the Prussian Academy in Berlin. (See J. Mehra, *Einstein, Hilbert, and the Theory of Gravitation*, Reidel, Boston, 1974.)

MAXWELL'S EQUATIONS

The history of physics from Newton to Maxwell provides a narration that certainly fulfills Kepler's dream of a journey whose story is more entertaining than the voyages of discovery of Columbus, Magellan, and the Portuguese. But we must pass over all this and go directly to James Clerk Maxwell (1831–1879).

Maxwell's Equations:

$\nabla \cdot E = \dfrac{\rho}{\varepsilon_0}$	This is Gauss's law. It relates electric field E to electric charge density ρ.
$\nabla \cdot B = 0$	This relates magnetic field B to "magnetic charge or monopoles" (which does not exist, so the right hand side is always zero).
$\nabla \times E = -\dfrac{\partial B}{\partial t}$	This is Faraday's law. It says that the electric field is produced by a changing magnetic field.
$\nabla \times B = \mu_0 j + \varepsilon_0 \mu_0 \dfrac{\partial E}{\partial t}$	This is Ampere's law with the addition of Maxwell's displacement current. It says that a magnetic field is produced by an electric current density j (Oersted) and/or by a changing electric field in the form of the displacement current $\partial E/\partial t$ (Maxwell).

An explanation of symbols in Maxwell's equations is as follows:

E = electric field	ε_0 is an electric constant, the permittivity.
B = magnetic field	μ_0 is a magnetic constant, the permeability.
ρ = electric charge density	
j = electric current density	$c = \dfrac{1}{\sqrt{\varepsilon_0 \mu_0}}$ is the speed of light
$\dfrac{\partial E}{\partial t}$ = displacement current	

Figure 3.3. Maxwell's equations, or the unification of electricity, magnetism, and light.

Maxwell graduated from Cambridge University in 1854 and started his research on electricity and magnetism. In time, Maxwell drew together almost everything that was known about electricity and magnetism into a set of four equations, now known as *Maxwell's equations* (Figure 3.3). The first two equations relate the way the electric and magnetic fields extend through space to the presence and distribution of their sources (charges and monopoles). The first equation, Gauss's law, shows how electrical charges produce an electric field E. It says that the divergence $\nabla \cdot \boldsymbol{E}$ of the electric field is the electric charge density ρ divided by the constant ε_0. This equation sums up the experimental results of Charles Augustin de Coulomb (1736–1806). The second equation uses the fact that there is no monopole; that is, there is no magnetic analog to the electric charge. Thus the right-hand side of the second equation is zero. This equation says that the divergence $\nabla \cdot \boldsymbol{B}$ of the magnetic field is zero. The third equation is Faraday's induction law, which states that a time-varying magnetic field is always accompanied by an electric field. Michael Faraday (1791–1867) in 1831 discovered that a changing magnetic field can produce an electric field and generate an electric current. The third equation says that the rotation $\nabla \times \boldsymbol{E}$ of the electric field is equal to the negative rate of change $\partial \boldsymbol{B}/\partial t$ of the magnetic field. Faraday's discovery made possible the invention of electric motors, which change electricity into mechanical motion, and electric generators, which convert motion into electricity. The last equation is a modification of Ampère's law, which describes the magnetic field set up by a current. This fact that a current (that is, moving electrical charges) can generate magnetism was the discovery of Hans Christian Oersted (1777–1851) in 1819. However, the equation as given by André Marie Ampère (1775–1836) did not seem to have the required symmetry that Maxwell wished, so Maxwell boldly amended the equation by adding on another term $\partial \boldsymbol{E}/\partial t$. In effect, Maxwell argued that if in Faraday's law a varying magnetic field created an electric field, then in Ampère's law a varying electric field should create a magnetic field. Thus, in addition to the fact that currents produce magnetic fields, as Oersted had realized, Maxwell realized that changing electric fields $\partial \boldsymbol{E}/\partial t$ would also produce them. To Maxwell, the symmetry was irresistible. The fourth equation says that the rotation $\nabla \times \boldsymbol{B}$ of the magnetic field is equal to the sum of two terms. The first term is the magnetic constant μ_0 times the current density $\boldsymbol{j}$. The second term is $\mu_0 \varepsilon_0$ times the rate of change $\partial \boldsymbol{E}/\partial t$ of the electric field. This is the crucial contribution of Maxwell, and the key quantity $\partial \boldsymbol{E}/\partial t$ is called the *displacement current*. The resulting four Maxwell equations are to electromagnetic theory as Newton's laws are to mechanics.

On December 8, 1864, Maxwell delivered his paper "A Dynamical Theory of the Electromagnetic Field" to the Royal Society. Combining his four equations, he arrived at the wave equation. The theoretical implication was utterly amazing. Electric and magnetic fields in a vacuum could support waves, which we call *electromagnetic waves*. The crucial parameter in the wave equation is the velocity c of the wave, in this case the electromagnetic wave. Maxwell's result is that the *velocity* of electromagnetic waves in vacuum is given by the equation

$$c = \frac{1}{\sqrt{\mu_0 \varepsilon_0}}$$

The result is the direct descendant of Taylor's equation for the velocity of material waves in a string. We have already discussed Taylor's linkage. Maxwell's linkage is given by the physical quantities μ_0, the (magnetic) permeability of free space, and ε_0, the (electric) permittivity of free space. These two quantities, μ_0 and ε_0, are fundamental constants of nature and can be directly evaluated by magnetic and electric measurements which do not involve wave motion. Their values are

$$\mu_0 = 4\pi \cdot 10^{-7}, \quad \varepsilon_0 = 8.85415 \cdot 10^{-12}$$

which give

$$c = 2.9979 \cdot 10^8 \text{ m/s}$$

Spectacular! Maxwell from his equations was able to show that the speed of propagation of electromagnetic waves turned out to be just about 300,000 km/s, almost exactly equal to the measured speed of light. "The agreement of the results," he reported, "seems to show that light is an electromagnetic disturbance propagated through the field according to electromagnetic laws."

Maxwell did not live to see the triumph of his theoretical results. Maxwell died of cancer in 1879 at the age of 48, nine years before Heinrich Hertz confirmed his theory by generating electromagnetic waves in the laboratory. Electromagnetic waves cover a broad spectrum, and they include light as well as all the other such waves (radio, television, microwaves, X rays, etc.) that we utilize everyday. All electromagnetic waves are governed by the same wave equation, and all have the same velocity c given by Maxwell's result $c = 1/\sqrt{\mu_0 \varepsilon_0}$.

The implications of Maxwell's electromagnetic theory started a revolution in physics. Would the velocity of light measured inside a moving vehicle be the same as the velocity of light on the ground? If not, then $c = 1/\sqrt{\mu_0 \varepsilon_0}$ would be different in almost every physical context, and μ_0 and ε_0 could not be universal constants. Maxwell opened Pandora's box, and in response he suggested two experiments to confirm or deny the constancy of the velocity of light. One was an astronomical experiment to measure the one-way velocity of light making use of the eclipses of the moons of Jupiter. The other was an Earth-bound experiment to measure the two-way velocity of light by means of mirrors. This experiment would give possible variations in the velocity of light on a double journey between two fixed mirrors. The instrumental difficulties of this experiment served as a challenge to the young Albert A. Michelson, who then invented his famous interferometer to carry out the job.

Our assurance of the constancy of the speed of light is based on a series of refined experiments beginning with the classic one performed by Albert A. Michelson (1852–1931) and his friend E. W. Morley (1838–1923) in 1887. They used the Earth itself as the moving vehicle (that is, as a moving inertial frame). The Earth moves at a speed of about 30 km/s in orbit about the sun. Michelson and Morley compared the round-trip speed of light along the line of the Earth's motion with the round-trip speed of light perpendicular to this line. They repeated this experiment at different times of the year so that the Earth would be moving in different directions with respect to the fixed stars. No effect of the motion of the

Earth on the relative speed of light in the two perpendicular directions was ever observed. To the accuracy of their experiment, they concluded that the measured speed of light in the two perpendicular directions was always the same. More refined experiments since then have continually reduced this uncertainty. As a result, this fact is now accepted universally: the *speed of light (in vacuum) is the same constant in every direction in every inertial frame*, and the numerical value is theoretically that given by Maxwell, $c = 1/\sqrt{\mu_0 \varepsilon_0} = 2.9979 \times 10^5$ km/s. Moreover, *this speed is the ultimate speed; no energy-carrying wave or physical entity can go faster than light.*

PHYSICAL THEORY OF RELATIVITY

Material waves are waves caused by the vibrations of a physical material or medium. The traveling waves on a stretched string result from the transverse vibrations of the string particles, so these waves are material waves. Also, sound waves are material waves as they result from the vibrations of air molecules. However, electromagnetic waves (including light) are the only waves observed in nature that are not material waves. Electromagnetic waves require no medium for their propagation, as they can and do propagate in vacuum. (Gravity waves theoretically predicted also propagate at the speed of light in vacuum, but they have not yet been observed in nature. Also, neutrinos travel at the speed of light.) In the following discussion, we will use the simple term "light" as the representative of all the electromagnetic waves. Because light consists of the visible electromagnetic waves, it was the only type of electromagnetic wave known up to the time of Maxwell and Hertz. Today, we are familiar with many other types such as radio waves, microwaves, and X rays, but light remains the one closest to our senses.

We now want to link the mathematical model of relativity as given earlier with physics. In the physical world, we must find the required perspectivity and perfectivity qualities. In regard to perfection in the physical world, there can be no doubt:

Let there be light: and there was light.

According to all human experience, light must be the physical entity that plays the perfective role in relativity theory. Light has no rest mass. It is pure energy of motion.

To have an inertial frame of reference, there must be inertia, and to have inertia, there must be mass. A material medium has mass, so therefore one can always fix an inertial reference frame to a material medium. By definition, the mass defining this inertial reference frame is at rest. In our Olympic metaphor an ordinary athlete can always define a reference frame by placing herself at rest (i.e., at the zero position).

Let us consider another material medium moving with velocity β relative to

Plate 3.2. Thomas Alva Edison (1847–1931). Inventor of the light.

the first one. Because this one has mass, we can fix an inertial reference frame to it. The presence of mass in each of these reference frames lets us distinguish one from the other. In the Olympic metaphor, a second athlete can define his reference frame by placing himself at rest (i.e., at the zero position). We distinguish the two athletes by means of their relative rating β.

As we have seen, the presence of a medium with mass always distinguishes one reference frame from all others. But vacuum has no mass. Therefore it is impossible to find a reference frame for vacuum. There cannot be any reference frame in which vacuum is at rest. Consider the plight of the perfect ten in the Olympic metaphor. The perfect ten cannot create a scale of reference for himself. He cannot place himself at rest (i.e., at the zero position) because he is perfect, and there can be no values above him. A perfect ten can have no frame of reference. He can never be at rest.

Consequently, all inertial reference frames are equivalent relative to vacuum. As a result, the velocity of electromagnetic waves (in vacuum) must be the same, $c = 1/\sqrt{\mu_0 \varepsilon_0}$ (in SI units) or $c = 1$ (in natural units), in all frames of reference. In the Olympic metaphor, the perfect ten athlete must be placed at one on every other athlete's scale. Thus light gives us both the perfective and the perspective aspects required. Therefore, light is the link that makes the theory of relativity a physical theory.

We can sum up as follows. Light being perfect can have only one velocity, namely, $c = 1/\sqrt{\mu_0 \varepsilon_0}$ (in SI units) or $c = 1$ (in natural units). This constancy of velocity holds in all inertial frames of reference. Light cannot travel faster than this, and light can never travel slower than this. Light can only be perfect, no more and no less. Other things must travel slower than light. At this point, let us say that in one spatial dimension (x axis only) light can only travel to the right (represented by $c = +1$ on the scale) or to the left (represented by $c = -1$ on the scale). Thus the points of perfection and negative perfection are each occupied by light, but traveling in opposite directions.

As we have seen, the perspective and perfective qualities, used in conjunction with Ockham's razor, gave us the Lorentz transformation of coordinates. The validity of this transformation must be checked out in terms of Maxwell's equations. It can be shown that Maxwell's equations do not alter their appearance under the Lorentz transformation. It therefore follows that electromagnetic phenomena are described the same way in different inertial frames of reference. In other words, electromagnetic phenomena are perspective. It also can be shown that wave equation retains the same form under Lorentz transformation, with the wave velocity c remaining invariant (perfective).

Finally, let us remark that the linkage of the mathematical model of relativity (i.e., the Olympic metaphor) to the physical world depends upon our idea of perfection. Suppose that in another galaxy a snail represented perfection. Then there the physical theory of relativity would be based upon the velocity of a snail, not the velocity of light. In our experience, we cannot base the physical theory of relativity on the velocity of sound, simply because sound is not a perfect ten. There are many better athletes than sound, who can run faster. For example, our supersonic airplanes can win in a race with sound waves.

There is a hypothetical particle called the tachyon that has been postulated to travel faster than light. But if such a particle existed, light would not be perfect. If this were so, we might have to modify Maxwell's equations. Without Maxwell's equations, we cannot do physics as we know it. In fact, we are so addicted to Maxwell's equations that we will do anything to protect them. There can be no tachyon. Light is perfect.

Let us now summarize this section by means of a balance sheet. On the left-hand side, one has the mathematical model, a purely abstract construction. On the right-hand side, we have the physical reality, our concept of nature.

Mathematical Model of Special Relativity	Physical Theory of Special Relativity
Variable t	Physical time t
Variable x	Physical distance x
Perspectivity	Principal of relativity
Perfectivity	Principal of light

Remember, we have used Ockham's razor (simplicity), which means that the mathematical model was chosen as the simplest one possible based on the two qualities of perspectivity and perfectivity. Throughout the book we will be discussing the two corresponding physical principles. As shown in the table, they are, respectively, the principle of relativity and the principle of constant and ultimate velocity of light. This second principle is more concisely called the light principle. We have purposely used different names on the left from those on the right to keep the mathematics separate from the physics.

Appendix: Derivation of Kepler's Equation

Let an ellipse called the *Kepler ellipse* have major semiaxis a and minor semiaxis b. Then c is defined as $c = \sqrt{a^2 - b^2}$ and the eccentricity β is defined as $\beta = c/a$. In Figure 3.4, the first quadrant is shown. We have

$$OA = a, \quad OB = b, \quad OC = c = a\beta$$

The point C is a *focus*. Let P be a arbitrary point on the ellipse. Then the radius vector is defined to be $r = CP$. The angle ϕ' that r makes with the vertical is called the *Kepler angle*.

Define point D such that $OD = a/\beta$. Then the vertical line DH through D is called the *directrix*. Define the perpendicular distance from P to the directrix as $PQ = d$. The ellipse is defined as the locus of point P such that the ratio CP to PQ is equal to the *eccentricity* β; that is,

$$\beta = \frac{r}{d}$$

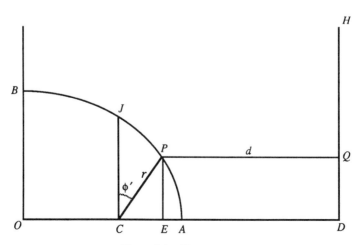

Figure 3.4. Kepler ellipse.

To use this equation, we must express d in terms of r. First, we need CD, which is

$$CD = OD - OC$$
$$= \frac{a}{\beta} - a\beta = \frac{a}{\beta}(1 - \beta^2)$$

Thus the required expression for d is

$$d = ED = CD - CE$$
$$= \frac{a}{\beta}(1 - \beta^2) - r \sin \phi'$$

Therefore, $r = \beta d$ is

$$r = a(1 - \beta^2) - r\beta \sin \phi'$$

We now solve for r. We have

$$r(1 + \beta \sin \phi') = a(1 - \beta^2)$$

so

$$r = \frac{a(1 - \beta^2)}{1 + \beta \sin \phi'}$$

This is the *polar equation for the ellipse*.

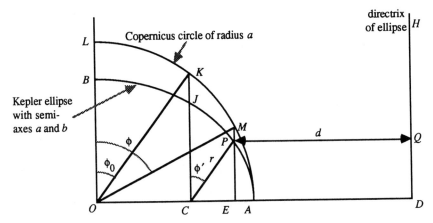

Figure 3.5. Copernicus circle and Kepler ellipse: $OL = OK = OM = OA = a$, $OB = b$, $OC = c = a\beta$

Let us now draw the circumscribed circle of radius a. See Figure 3.5. This is called the *Copernicus circle*. Vertical line *CJ* is extended to cut the circle at *K*. Angle ϕ_0 defined as angle *LOK* is called the *eccentric angle*. We have

$$\sin \phi_0 = \frac{OC}{OK} = \frac{c}{a} = \beta$$

That is, the eccentricity is equal to the sine of the eccentric angle. Vertical line *EP* is extended to cut the circle at *M*. Angle ϕ defined as angle *LOM* is called the *Copernicus angle*.

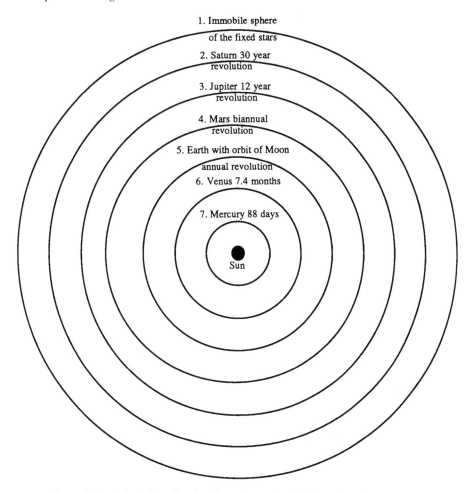

Figure 3.6. A depiction of a plate from Copernicus' *De Revolutionibus* showing the Copernican system, with the sun in the center. The Latin labels are translated into English. (Copernicus had 9 months for Venus and 80 days for Mercury, but they are actually 7.4 months and 88 days, respectively.)

Now we are ready to derive Kepler's formula. We have

$$CE = OE - OC$$

or

$$r \sin \phi' = a \sin \phi - a \sin \phi_0$$

We now substitute into this equation the previously defined expression for r. We obtain

$$a(1 - \beta^2) \sin \phi' = (1 + \beta \sin \phi')(a \sin \phi - a \sin \phi_0)$$

The constant a cancels out. We now solve for $\sin \phi'$. The result is *Kepler's formula*

$$\sin \phi' = \frac{\sin \phi - \sin \phi_0}{1 - \sin \phi \sin \phi_0}$$

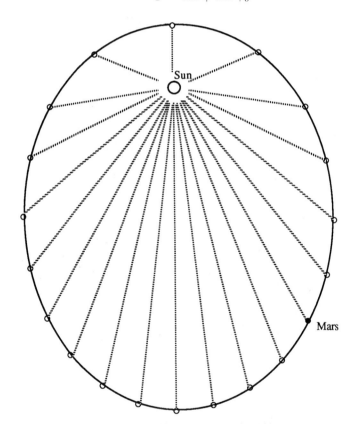

Figure 3.7. Kepler's discovery of the elliptic orbit of Mars. The dotted lines and positions marked on the ellipse indicate equal areas and equal intervals of time on the orbit.

which expresses the sine of the Kepler angle ϕ' in terms of the sines of the eccentric angle ϕ_0 and the Copernicus angle ϕ. If we define $\beta = \sin \phi_0$, $v = \sin \phi$, and $v' = \sin \phi'$, then the *Kepler formula* is

$$v' = \frac{v - \beta}{1 - v\beta}$$

The mean radius from the focus to the ellipse is defined as the average of the smallest distance from the focus to the ellipse and the largest distance from the focus to the ellipse. That is,

$$\text{mean radius} = \frac{(a - c) + (a + c)}{2} = a$$

Thus the Copernicus circle is the circle whose radius a is equal to the mean radius of the Kepler ellipse (Figures 3.6 and 3.7).

CHAPTER 4

SYMMETRY OF THE POINCARÉ GROUP

To see a World in a Grain of Sand
And a Heaven in a Wild Flower
Hold Infinity in the palm of your hand
And Eternity in an hour.

William Blake (1757–1827)
Auguries of Innocence

POINCARÉ SYMMETRY OF SPACE-TIME

Scholars define folklore as the body of traditional customs, beliefs, tales, songs, and the like that are transmitted by word of mouth from one generation of a small society to the next. The wisdom, imagination, and spirit of the common people

Plate 4.1. Henri Poincaré (1854–1912)

is preserved in these fables. Science is also concerned with folklore, but now the folklore does not deal with fables such as the one about John Smith and Pocahontas but instead with stories about subatomic particles and distant galaxies. Modern elementary particles are extensions and elaborations of the *atomos* of the ancient Greeks. In the centralized sun of Copernicus, William Harvey (1578–1657) found a model for the heart as the center of the circulation of the blood.

In India, there is an old fable about four blind people who in their travels come upon an elephant in the forest. None of them has ever encountered an elephant before. The first touches the trunk and cries out that it is a snake. The second catches the tail and says it is a rope. The third finds the elephant's leg and concludes it is the trunk of a tree. The fourth clutches the ear and is convinced it is the great leaf of a tree. This fable makes us realize that we can find ourselves in a similar predicament in science. We are not able to recognize the entirety of space-time, but only see a limited and distorted image of the true reality. Each inertial observer sees a different picture called his local or proper picture. His local picture gives a good perception of reality within his inertial frame, but the picture becomes more and more distorted for frames moving at greater and greater relative speeds. In terms of the fable, we can say that when we are close to the elephant's leg, the elephant does indeed seem like the trunk of a tree. However, this local perception of the elephant becomes a gross distortion as we move farther away from his leg and other features become prominent.

Our perception of space and time is always from a local viewpoint. Space and time in another inertial frame does not give the same perception. To each his own. Everyone who has heard about relativity knows the term *space-time* and knows that the correct thing is to use this expression and not *space and time*. The term space-time conveys the idea that time is bound up with space, and our perception of space and time as individual entities is conditioned by the actual inertial frame in which we are at rest.

To understand the space-time concept, let us first examine the conventional perception of space-time. An event occurs, say, the conquest of Mt. Everest by mountain climbers. Suppose we wish to describe where and when this event occurred. We have to mention four quantities, say, the latitude and longitude, the height above the ground, and the time. The first three give the position in space, while the fourth gives the time. However, the first three may be assigned in a different way and still give the same position in space. For example, we might take Cartesian coordinates with the origin at the center of the Earth. Still other specifications of the position could be obtained from Cartesian systems rotated or translated with respect to the given system. There are an infinite number of coordinate systems, each of which can uniquely fix the given position in space. All are equally legitimate. The choice of which one is merely one of convenience. When we say that space has three dimensions, we mean that three quantities are needed to specify a point in space. However, the method of assigning these quantities can be chosen in any number of different ways. Thus we are accustomed to the conventional wisdom that the three spatial axes required for describing three-dimensional space are quite arbitrary.

Because time is one dimensional, conventional thought said that the only arbitrary elements in the reckoning of time were the unit and the point of origin. The unit could be seconds, minutes, hours, days, or years. The origin could be the founding of the Roman republic, or the birth of Christ. Although the time unit and origin are arbitrary, conventional wisdom held that the time axis was fixed and unique. The net result was that the method of fixing position in space and the method of fixing a point in time were considered wholly independent of each other. It is for this reason that people used to regard space and time each as absolute and distinct.

The theory of relativity shows that there is no longer distinct entities of absolute space and absolute time which can be applied without ambiguity to any part of the universe. Instead there are only the various proper distances and proper times which apply uniquely only within their respective inertial frames. Proper distance and proper time for one frame agree quite well with the corresponding proper quantities in another frame if and only if the two frames are not in rapid motion with respect to each other. In fact, the proper distances and proper times agree exactly only when the two inertial frames are at rest relative to each other (in which case the two frames blend together as one frame). For rapid relative motion, the disagreement is great.

Space and time as we know are perceptions or manifestations of an underlying reality. Each inertial frame has its own proper (or local) space and time variables, and they are related to each other through this underlying reality. The relationship is based on the fundamental symmetry of the space-time domain. This symmetry has a precise mathematical formulation based upon group theory. It is the symmetry of the Poincaré group. To understand this symmetry, we will introduce a metaphor in the form of a fable. We call it the fable of the king's messengers.

THE FABLE OF THE KING'S MESSENGERS

A great and good king had a large army fighting on the frontier. It was important for him to know the army was advancing or retreating, but he could not rely on the reports from the commanding general. As a result, he sent out two messengers, separated by a time period t_0. Each messenger had the order to ride to the army, turn around, and return to the king. The king would measure the time period t_2 between the two messengers on their return.

The king used this decision rule.

1. If the receiving period t_2 is equal to the sending period t_0, then the army is stationary.
2. If the receiving period t_2 is less than the sending period t_0, then the army is retreating.
3. If the receiving period t_2 is greater than the sending period t_0, then the army is advancing.

To arrive at this decision rule, the king used the following reasoning. If the army is retreating, then the second messenger will find the army at a closer distance than will the first messenger. As a result, the second messenger will return sooner than otherwise, and therefore the receiving period will be less. On the other hand, if the army is advancing, the second messenger will have to ride farther, so he will return later than otherwise. As a result, the receiving period will be greater.

Now we must impose the Poincaré symmetry onto this fable. There are two conditions. The *first condition* says that what is good for the goose is good for the gander. Thus, the general could send out two messengers, each to ride to the king and return, and the same decision rule would apply. If the army were retreating, the receiving period would be less than the sending period. If the army were advancing, the receiving period would be greater than the sending period. The *second condition* says that all the messengers ride at the ultimate speed in nature. No traveler is faster than these messengers, and all other travelers are slower. All the messengers travel at the same ultimate speed, which we take to be $c = 1$.

To see how things work in this fable, consider the king A and the army B. In Figure 4.1, we have the space-time diagram for the inertial frame of the king A. The time axis (vertical line on left) represents the so-called *world line* of A, that is, the line that tells us where to find A in the space-time diagram at any specified time. Since we are measuring time t in the inertial frame of A, it follows that A is at rest so A is always at the same place at any given time. The line B is the *world line* of B. If we assume that the army is not moving, this line is also

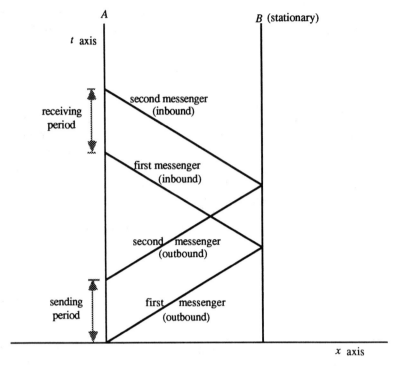

Figure 4.1. If B is stationary with respect to A, then the period received by A is the same as the period sent by A.

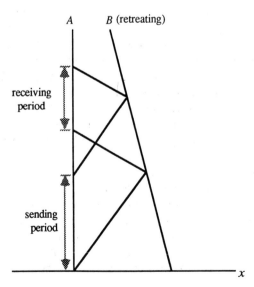

Figure 4.2. If B is coming toward A, then the period received by A is less than the period sent by A.

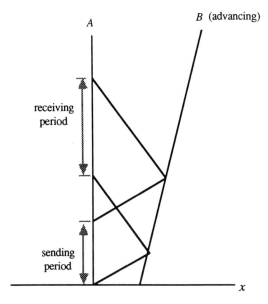

Figure 4.3. If B is going away from A, then the period received by A is greater than the period sent by A.

vertical. The 45° lines show the space-time paths of the two messengers. Because the army B is stationary (a vertical world line), the receiving period is equal to the sending period.

If we assume that the army is retreating, then the world line of B slopes to the left, as shown in Figure 4.2. Now we see that the receiving period is less than the sending period. Finally, if we assume that the army is advancing, then the world line of B slopes to the right, as shown in Figure 4.3. Here the receiving period is greater than the sending period.

RELATIVISTIC DOPPLER FACTOR

Let us now analyze the case of the advancing army. For simplicity, we assume that the king A and the army B are at the same place when the first messenger is sent out. Of course, this messenger never goes anywhere. We take this place and time as the origin. See Figure 4.4. The second messenger leaves A at time t_0, turns at B, and returns to A at time t_2. The total time elapsed for the journey is $t_2 - t_0$, so according to A the second messenger traveled a total distance $c(t_2 - t_0)$, half of it toward B and the other half back from B. Accordingly A sets the distance of B at the instant when the second messenger arrives at B as x_1, where

$$x_1 = \frac{c(t_2 - t_0)}{2}$$

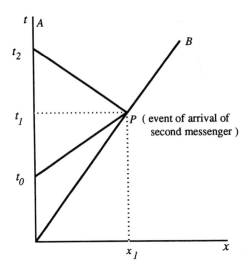

Figure 4.4. The period sent by A is t_0 and the period received by A is t_2. According to A, the second messenger arrives at B at distance x_1 and time t_1.

(This point of arrival is shown at P on B's world line.) By A's reckoning, the arrival at P happened at the half-way instant t_1, namely,

$$t_1 = \frac{t_2 + t_0}{2}$$

The king A therefore concludes that the army B is moving away from him with a velocity

$$\beta = \frac{x_1}{t_1} = \frac{(t_2 - t_0)c/2}{(t_2 + t_0)/2} = \frac{t_2 - t_0}{t_2 + t_0}$$

(where we have set $c = 1$).

When A and B are together at the origin, each of their clocks show time equal to 0. We know that the time recorded by A's clock at P is $(t_0 + t_2)/2$. The question is: What is the time recorded by B's clock at P? To seek the answer to this question, we appeal to the symmetry of the Poincaré group. Notice that after the messenger left A at t_0, it reached B at P. Suppose that the time recorded by B is kt_0, where k is a constant factor. Similarly, if A sent the messenger at $2t_0$, then B would receive the messenger at $2kt_0$, and so on. By the Poincaré symmetry, the same factor k must operate between B and A. That is, if B sends a messenger to A at time τ_1, as measured by B's clock, he must reach A at time $k\tau_1$ as measured by A's clock. Set $\tau_1 = kt_0$ (i.e., the time at P by B's clock). Thus when the messenger from B (sent at P) comes to A, the time by A's clock is $k\tau_1 = k(kt_0) = k^2 t_0$. Because this time is the return time t_2, we have

$$t_2 = k^2 t_0 \quad \text{or} \quad k^2 = \frac{t_2}{t_0}$$

Relativistic Doppler Factor

From the foregoing, the equation for the velocity β is

$$\beta = \frac{(t_2/t_0) - 1}{(t_2/t_0) + 1}$$

Replacing t_2/t_0 by k^2 we have

$$\beta = \frac{k^2 - 1}{k^2 + 1}$$

Solving for k, we obtain

$$k = \sqrt{\frac{1 + \beta}{1 - \beta}}$$

This is an important equation. It gives the k factor in terms of the velocity β of B.

We recall that we assumed that the messenger reached P at time $\tau_1 = kt_0$, as recorded by B's clock; that is,

$$\tau_1 = t_0 \sqrt{\frac{1 + \beta}{1 - \beta}}$$

How does this value of τ measured by B compare with the time t_1 measured for the same event P by A? Time t_1 is (since $t_2 = k^2 t_0$)

$$t_1 = \frac{t_2 + t_0}{2} = \frac{k^2 t_0 + t_0}{2} = t_0 \frac{k^2 + 1}{2}$$

But

$$k^2 + 1 = \frac{1 + \beta}{1 - \beta} + 1 = \frac{2}{1 - \beta}$$

so

$$t_1 = \frac{t_0}{1 - \beta}$$

The event P has two times associated with it, namely, time t_1 by A's clock and time τ_1 by B's clock. We have just found an expression for each of these times in terms of t_0. We can eliminate t_0 by substituting $t_0 = t_1(1 - \beta)$ into the equation for τ_1. We have

$$\tau_1 = t_0\sqrt{\frac{1+\beta}{1-\beta}} = t_1(1-\beta)\sqrt{\frac{1+\beta}{1-\beta}}$$

which gives

$$\tau_1 = \sqrt{1-\beta^2}\, t_1$$

or

$$t_1 = \frac{\tau_1}{\sqrt{1-\beta^2}}$$

This last equation says that the time t_1 as measured by A of a moving event P is dilated when compared to the time τ_1 as measured by B at rest with the event. The dilation factor is

$$\gamma = \frac{1}{\sqrt{1-\beta^2}}$$

We have thus obtained the phenomenon of *time dilation* from the fable of the king's messengers. It is worth emphasizing that we made use of Poincaré symmetry, namely, the symmetry between the inertial frames of A and B and the constancy of the speed of the messengers ($c = 1$).

The fable of the king's messengers displays the Poincaré symmetry of spacetime in terms of the factor

$$k = \sqrt{\frac{1+\beta}{1-\beta}}$$

which in relativity theory is known as the *relativistic Doppler factor*. We shall always define k in this way, that is, with the plus sign upstairs and the minus sign downstairs, and where we allow the intrinsic sign of β to be either positive or negative.

DISTANCE AND TIME

Protagoras, a Sophist philosopher of ancient Greece in the fifth century B.C., declared "Man is the measure of all things." The search for suitable units of measure has gone on for millennia. Two of the most important units deal with *distance* (or *length*) and with *time* (or *duration*). A unit of distance represents the separation between two points in space, and with this unit, we can measure how far it is from here to there. A unit of time represents the separation between two points in time, and with this unit we can measure how long it is from now to then.

The present International System of Units, or Système Internationale d'Unitès (SI), grew out of the metric system introduced in France in 1791. The unit of length is the meter (abbreviated m), and the unit of time is the second (abbreviated s). As we know, velocity (or speed) is a quantity representing distance divided by time. Thus the unit of velocity would be meters per second (m/s). The kilometer (km) is 1000 m, so often velocities are expressed as km/s.

Let us consider the measuring procedures for basic physical quantities. The most important physical measurements are those of distance (meters) and time (seconds). We are accustomed to measuring the length of an object when the object is motionless relative to us. If an object is at rest, it is sufficient to transpose a meter stick along the given length the necessary number of times. This is how we do it in everyday life when measuring, say, the length of a parked train.

It is also easy to measure the time between two events occurring at the same point. We suppose that a clock is located at this point. We register the moment when the first event occurs and the moment when the second event occurs. The difference of these two readings of the clock gives the time between the events. It is in this way the duration of a baseball game is ascertained.

But now comes a more difficult problem. How do we measure the length of a train moving relative to us? Suppose that a train is moving past us at great velocity and we want to determine its length. It is not easy for one person to do this. She has to record simultaneously the position of the front of the train (the locomotive) and the rear of the train (the caboose) relative to certain motionless points on the ground. But as soon as she notes the position of the locomotive and begins to turn her head, the caboose will have gone ahead. Consequently, special care must be taken to mark simultaneously the positions of the locomotive and the caboose. Having marked these simultaneous positions on the ground, we can readily measure the distance by conventional means for measuring motionless objects.

We thus see that if we wish to measure an object that is moving relative to ourselves, we must arrange for the ends of the object to be observed simultaneously. We therefore need a set of stationary clocks at different places which are synchronized. Failure to achieve synchronicity has dire consequences, because with nonsynchronized clocks, we would record the positions of the front and back ends at different times. As a result, we would get an incorrect measure of the length of the object.

Another difficult problem is the task of measuring the time between events occurring at different points in space. Suppose that we want to measure the time it takes a runner to run a 100-meter race. The events in this race are represented by the runner's start and finish. And there is only one clock. The starter's gun serves as a signal to start the race and to actuate a clock located at the finish.

Sound propagates in the air at 340 meters per second, so it takes 100/340 = 0.294 seconds to actuate the clock. Thus, the runner starts running 0.294 seconds before the official located at the finish starts timing the event. This is not very essential because the speed of the runner (about 10 meters per second) is small compared to the velocity of sound (340 meters per second). But from this example, we perceive that the determination of the time duration between two events taking place at two different spatial points is a problem that requires attention.

In the special theory of relativity, it is assumed that each inertial frame has its own coordinate system and motionless clocks are located at all its spatial points, wherever needed. These clocks are synchronized within that reference frame, so that equal readings of the clocks at all the spatial points correspond to the same moment of time within the entire frame.

The special theory of relativity shows that length and time are intrinsically related. The connecting link is a velocity, and a particular velocity. It is the velocity of light, abbreviated c. The problem that special relativity addresses is the definitions of distance and time. We must be more clear in our definitions. Specifically, we must say that a unit of distance is the separation of two events in space, both events being observed at the same instant of time. In other words, we must observe both spatial points simultaneously to measure the distance between them. Similarly we must say that a unit of time represents the separation of two events in time, both events being at the same place in space. In other words, we must observe both time points at the same spatial point to measure the time between them. To use these definitions, we must be able to be clear on what we mean by occurring at the same time (simultaneous) and by occurring at the same place.

PHYSICAL VELOCITY AND APPARENT VELOCITY

Let us now examine why distance and time are fundamental. When an object is moving from one place to another, we say that it is in motion. If it stops moving, we say that it is at rest. The distance that an object moves is obviously an important thing to know when describing its motion. Equally important is the time it takes to move that distance. Distance and time are related to another important property of motion, *speed*. Speed refers to how fast an object is moving. The speed of an object is defined simply as the distance x the object travels divided by the time t it takes to travel this distance: speed = distance/time = x/t. The definition of speed as the distance divided by time is actually the definition of average speed. Instantaneous speed, on the other hand, is the speed an object has at any instant of time. The speedometer of a car indicates instantaneous speed. If you were to drive your car 50 miles in one hour, the average speed would be 50 miles per hour, but during that hour the instantaneous speed may have varied between 40 and 60 miles per hour, depending upon the traffic. If an object moves at the same speed for a period of time with no speeding up or slowing down, we say that it has constant speed (or uniform speed). In this case, the average speed and the instantaneous speed are the same, and either is referred to as simply the speed.

When specifying the motion of an object, not only the speed but the direction of motion is important. The term *velocity* is used to signify both the speed and direction of the moving object. Speed then refers to the magnitude of this velocity. Note that constant speed is not always the same as constant velocity. For example, an automobile moving at a constant speed of 40 miles per hour as it rounds a curve does not have a constant velocity, because its direction is changing at each instant. An object has a constant velocity (or uniform velocity) only if its speed and its direction do not change.

In special relativity, we consider the case of uniform motion in a straight line, so instantaneous velocity and average velocity are the same. As it is well known, calculus was invented by Sir Isaac Newton so that he could handle, among other things, the instantaneous velocity of bodies in nonuniform and curvilinear motion. However, because special relativity theory is concerned with uniform rectilinear motions, calculus is not required, and so we can write this book using only pre-calculus mathematics. (Note: A rectilinear motion is a motion in a straight line. "Rectilinear" is an old term going back to Newton.)

One of the first big stumbling blocks that one encounters in high school is the so-called word problem in first-year algebra. A typical such problem is this one:

> On the river bank we observe the current flowing downstream at 10 km per hour and a boat traveling downstream at 25 km per hour. How fast is the boat going with respect to the current?

Little does one realize that this problem is so difficult that it finally took Einstein to solve it. The solution of this problem requires the special theory of relativity.

Let us try to analyze this problem. There are three elements in the problem. We call the river bank A, the current B, and the boat C. Take the positive direction downstream. The problem says

$$B \text{ moves from } A \text{ at a velocity of } 10 \text{ km/hour}$$

$$C \text{ moves from } A \text{ at a velocity of } 25 \text{ km/hour}$$

and asks

$$C \text{ moves from } B \text{ at what velocity?}$$

Let us seek a solution in the following way. At time 0 let A, B, C be at the same place. One hour later A is still at the same place, but B has moved 10 km downstream from A. Also C has moved 25 km downstream from A. See Figure

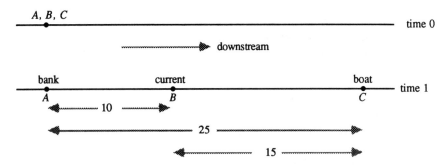

Figure 4.5. At time 0, points A, B, and C are at the same location, whereas at time 1, points B and C have moved away from A distances of 10 and 25, respectively.

4.5. Using this line of reasoning, it follows that C is $25 - 10$, or 15, km from B at the end of 1 hour. Therefore, C has traveled from B a distance of 15 km in 1 hour, so the boat is going at 15 km per hour with respect to the current.

Is this answer correct? No. It is wrong. Why is it wrong? It is wrong because light (in vacuum) travels at the ultimate physical velocity (in round numbers 300,000 km/s), and this velocity of light is the same in all inertial frames. All we can ask at this point is: What does that have to do with the price of tea in China, or with anything else?

The answer is that the foregoing line of reasoning can be used to obtain a velocity greater than that of light. That is why the reasoning must be rejected. To see this, change the numbers to

B moves from A at a velocity of $-100,000$ km/s

C moves from A at a velocity of $250,000$ km/s

as seen in Figure 4.6. Then

C moves from A at a velocity of $100,000 + 250,000$ km/s or $350,000$ km/s

This velocity exceeds the velocity of light.

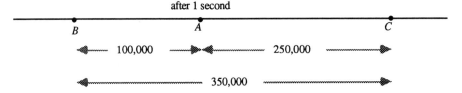

Figure 4.6. At time 0, points A, B, and C were at the same location. At time 1 (shown in the diagram), point B has moved to the left of A, and point C has moved to the right of A, so that B and C are separated by 350,000 km in the given frame.

When it comes to combining two velocities, each measured in a different inertial frame, we must be more careful. The required expression is Kepler's formula (given in Chapter 3; Kepler's formula is derived from the Lorentz transform in Appendix 1 of this chapter). *Kepler's formula* is

$$v' = \frac{v - \beta}{1 - v\beta} \tag{4.1}$$

where the variables have the following interpretation:

β = physical velocity at which B moves from A $\}$ in A's frame
v = physical velocity at which C moves from A

v' = physical velocity at which C moves from B $\}$ in B's frame

Used in this context, Kepler's formula is known as the *physical velocity addition formula*. Equation (4.1) gives v' in terms of v and β.

If in equation (4.1) we change v, v', β to v', v, $-\beta$, respectively, we obtain the *reflexive equation*

$$v = \frac{v' + \beta}{1 + v'\beta} \tag{4.2}$$

The reflexive equation is the same as the inverse equation. This equation gives v in terms of v' and β.

In equation (4.1) the quantities are expressed in natural units (with $c = 1$). To convert to unnatural units, let $v' = V'/c$, $v = V/c$, and $\beta = V_0/c$. Then Kepler's formula (4.1) becomes

$$V' = \frac{V - V_0}{1 - (VV_0/c^2)} \tag{4.1a}$$

Returning now to the boat problem, we have $V_0 = 10$ km/hour and $V = 25$ km/hour. The value of c is 300,000 (3600) km/hour, since there are 3600 seconds in 1 hour. The denominator of equation (4.1a) becomes

$$1 - \frac{VV_0}{c^2} = 1 - 2.14 \times 10^{-16}$$

which is so close to one that to a very close degree of approximation we can take it to be one. Thus equation (4.1a) reduces very closely to

$$V' = V - V_0 = 25 - 10 = 15$$

which is the high school answer, so the high school answer is correct for all intents and purposes. However, in the second problem

$$\beta = -100{,}000/300{,}000 = -1/3$$
$$v = 250{,}000/300{,}000 = 5/6$$

so in natural units

$$v' = \frac{1/3 + 5/6}{1 + (1/3)(5/6)} = 0.9130434 \text{ light-sec/sec}$$

or in SI units

$$V' = 273{,}913 \text{ km/s}$$

Thus the computed velocity is less than the velocity of light, as required.

Recall the light principle: "No velocity can exceed the velocity of light." This statement refers to a physical velocity, that is, the velocity of a particle or a wave that carries energy. Thus, this statement can be reworded as "The velocity of any energy-carrying entity can never exceed the velocity of light." As we know, a photon is a particle of light that carries the light energy; the photon travels at the velocity of light, and it can never travel at a greater velocity or at a lesser velocity.

However, in everyday life, there is movement which is not the movement of an energy-carrying particle or wave. Let us consider a single frame K in which two objects are going away from the origin with speeds β and v, respectively. See Figure 4.7.

As the diagram shows, object B travels a distance β to the right in one unit of time, whereas object C travels a distance v to the right in the same amount of

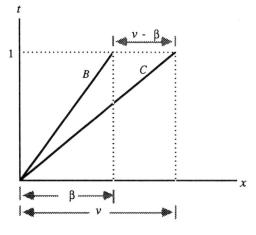

Figure 4.7. In one unit of time in the given frame, B moves a distance β to the right and C moves a distance v to the right, so B and C are separated by a distance $v - \beta$.

time. Thus the distance between the two objects increases by $v - \beta$ in one unit of time, so the apparent velocity of the two objects in frame K is

$$v'' = v - \beta \tag{4.3}$$

This law in which no change in reference frame occurs is called the *apparent velocity addition formula*. That is, no transformations of coordinates are required. In general, the apparent velocity addition formula can be established as follows. In time t, object B moves a distance $x_0 = \beta t$ to the right and object C a distance of $x = vt$ to the right. Thus, the relative distance moved by the two objects as seen by an observer within the frame is $x'' = x - x_0$, so the apparent velocity of the two objects within the frame is

$$\frac{x''}{t} = \frac{x - x_0}{t} = \frac{x}{t} - \frac{x_0}{t}$$

which is the same as equation (4.3).

The reflexive equation as well as the inverse equation to equation (4.3) is

$$v = v'' + \beta \tag{4.4}$$

In summary, the physical velocity addition formulas (4.1) and (4.2) govern velocities between frames, and the apparent velocity addition formulas (4.3) and (4.4) govern velocities within a frame. The interplay between these two addition formulas must be understood to work out some of the famous examples of relativity theory.

Let us consider an example. Suppose we have two objects, one a galaxy moving to the left with speed 0.5 so $\beta = -0.5$, and the other a photon moving to the right with speed $v = 1$. Then in the given frame, the apparent velocity between the two objects is

$$v'' = v - \beta = 1 - (-0.5) = 1.5$$

Thus within a reference frame, a relative velocity can be greater than 1; in this case, it is 1.5. However, as we know, a velocity such as this is not the *physical* (or *group*) velocity of an energy-carrying entity, but it is the *apparent* (or *phase*) velocity between two objects. Such apparent velocities in a given reference frame can have any value and, in particular, can have values greater than the velocity of light.

As another example, suppose we have a galaxy moving to the right with speed $\beta = 0.5$ and a photon moving to the right with speed $v = 1$. Then, in the given frame, the apparent velocity between the two objects is

$$v'' = v - \beta = 1 - 0.5 = 0.5$$

In these two examples, we have combined the speed of light with the speed of a galaxy and have come up with a resultant speed in each case not equal to the speed of light. However, these computed speeds are not physical speeds, but apparent speeds. The physical speed of the photon must always be unity. Let us check this. In the first example ($\beta = -0.5$, $v = 1$), the physical velocity of the photon with respect to the galaxy is

$$v' = \frac{1 - (-0.5)}{1 - (1)(-0.5)} = 1$$

whereas in the second example ($\beta = 0.5$, $v = 1$), it is

$$v' = \frac{1 - 0.5}{1 - 0.5} = 1$$

Both are one, as is required.

TWO-WAY VELOCITY OF LIGHT

Suppose we want to figure out how fast a horse can run. First, we use a straight rack of length x. We need observer A at the beginning and observer B at the end. Both observers must have their watches synchronized; that is, each watch must read the same time as the other. Let observer A see the horse start at time T_1, and observer B see the horse finish at time T_2. We conclude, therefore, that the horse took one-way time $T_2 - T_1$ to run the distance x. The speed of the horse is therefore $x/(T_2 - T_1)$. Such a velocity is called a *two-clock velocity*, because we need a clock here and another clock there to time the horse over the straight course. We may also describe this velocity as a *one-way velocity*, since it pertains to a single direction of motion.

Next, suppose we want to figure out how fast a horse can run around an oval race track. Now let the distance be given by $2X$. Because the start and finish are at the same point, the time is measured by a single clock at this common point. The horse starts at time T_1 on this clock and finishes at time T_3. We call $2X/(T_3 - T_1)$ a *one-clock velocity* or a *round-trip velocity*. A round-trip velocity applies to anything that comes back to its starting point by a closed path. The round-trip velocity is found by dividing the total length of the path traversed by the time interval between departure and return. This time interval is measured on the single local clock at the common point of start and finish.

An important special case of the round-trip velocity is the *two-way velocity*. We encounter this sort of velocity in the case of a ball that bounces back when thrown against a wall. It is the type of velocity measured in seismic exploration, where the path is made up of the direct journey of a seismic wave from a point on the Earth's surface to a subsurface reflecting horizon plus the return journey to the same point on the Earth's surface. The two-way velocity is found by measuring the total distance, from the start to the point of reflection and back, by the time of flight. This two-way time is measured on a single clock.

Let us now think about a 50-meter swimming event. If a 50-meter pool were available, the event could be a one-way journey equal to the length of the pool. A swimmer would clock a certain one-way velocity. On the other hand, if a 25-meter pool were available, the event would need to be a two-way journey, there and back. A swimmer would clock a two-way velocity. It is well known that a swimmer gains time by making a turn, because he can get a good push with his feet from the solid wall of the pool. Thus a swimmer's two-way velocity in a short pool is always greater than his one-way velocity in a long pool. On the other hand, if the swimmer had to make a turn in the water at a rope crossing the midpoint of a 50-meter pool, then he would lose a lot of time in the turn, and his two-way velocity would be much smaller than his one-way velocity. Also the exact distance the swimmer covers would be in question, for as we know a rope can stretch and bend.

In seismic exploration, it is assumed that the reflection process is instantaneous. It is agreed that it takes no time whatsoever for a seismic wave to make a 180° turn at a reflecting interface. It is also often assumed that the reflecting interface is perfect, which in fact it is not, as all reflecting surfaces contain many defects. The two-way velocity is exactly equal to the one-way velocity, no more, no less. This is quite unlike the case in a swimming race where a turn can either speed up or slow down the racer depending upon whether it is at a wall or a rope.

With this background, let us look at the question of measuring the velocity of light. The velocity of light is one of the most important parameters of nature, and a great deal of experimental effort has been spent in trying to determine it. What sort of velocity is it which has been determined when we say $c = 299,792.4574$ km/s? The answer is that it is a two-way velocity, that is, a one-clock velocity. All the experiments performed to measure the velocity of light have been one-clock experiments, and all assume that the process of making a 180° turn by reflection at the mirror is instantaneous and that the mirror surface is perfect. The process of reflection involves the absorption of the incoming photon by the mirror and the subsequent emission of another photon going in the opposite direction. It is assumed that this physical process neither takes any time nor changes any distance whatsoever. The analysis of this process is in the realm of quantum mechanics and is beyond the scope of relativity theory.

The question now comes up whether it is possible to measure the velocity of light without mirrors, that is, to measure a one-way velocity. The answer is that it is perfectly possible to measure such a one-way travel time. However, then we would be faced with the problem of synchronizing the two clocks required, and the synchronization depends upon the unknown velocity. As a result, P. W. Bridgman (*A Sophisticate's Primer of Relativity*, p. 44) concludes that we cannot isolate any information about the properties of light from a one-way experiment. Thus in principle, we can neither measure the one-way velocity of light nor by a one-way experiment synchronize two clocks located at two different places.

There is a true story of an Englishman (Robert Elston) in Sweden. No employer would give him a job unless he had a work permit. On the other hand, the police would not give him a work permit unless he had a job. This situation went on indefinitely, although he met all the requirements of both the employer and the police. It is the same story for light. We cannot measure the one-way velocity of light unless we have two synchronized clocks, one at the source and the other at the receiver. On the other hand, we cannot synchronize the two clocks required for a one-way experiment unless we know the one-way velocity of light.

Let us now discuss these two problems, velocity measurement and clock synchronization. Here we will make use of unnatural (SI) units. First, consider the problem of synchronization. How do we synchronize two clocks located at different points in space? For convenience, we will make use of one-dimensional space X. Let the first clock, called the source clock, be located at point X_1. Let another clock, called the receiver one, be located at point X_2. This particular clock does not differ in any detail from the first clock. It is assumed that the coordinates X_1 and X_2, and therefore the separation distance $X_2 - X_1$, are known, with no clocks being necessary to determine these distances. Both these clocks are in the same inertial reference frame. Let us now send a signal from the source clock to the receiver clock. Our purpose is to synchronize the receiver clock with the source clock. At an arbitrarily chosen moment T_1 on the source clock, a signal is sent to the receiver clock. The signal travels in a straight line at a constant speed V along the known path of length $X_2 - X_1$. The arrival of the signal at the receiver clock is registered by an observer or by means of a device. At the instant of signal arrival, the reading of the receiver clock is set to the value

$$T_2 = T_1 + \frac{X_2 - X_1}{V} \tag{4.5}$$

This setting puts the receiver clock in synchronization with the source clock. In equation (4.5), we recognize the quantity $(X_2 - X_1)/V$ as the one-way signal travel time from the source point to the receiver point. Therefore, equation (4.5) says that the time T_2 of receipt is set equal to the sum of the time T_1 of the source plus the one-way travel time. Equation (4.5) clearly shows that to compute T_2 we must know the one-way signal velocity V. Thus we have shown that we cannot synchronize the two clocks unless we know the one-way signal velocity V.

Now let us discuss the problem of velocity measurement. How is the velocity of a particle or a signal found? Let the signal move along the X axis. To obtain its velocity, we must know the position X_1 of the signal at the instant T_1 and also its position X_2 at the instant T_2. Provided that the motion is uniform, the one-way velocity is equal to

$$V = \frac{X_2 - X_1}{T_2 - T_1} \qquad (4.6)$$

Time T_1 is the time registered by the source clock at point X_1 when the signal starts from that point. Time T_2 is the time registered by another clock (the receiver clock) at point X_2 when the signal arrives at that point. Both clocks keep accurate time. In addition, one must be sure that the receiver clock shows time T_1 at the same instant as when the source clock shows time T_1. Only in this case would the determination of velocity V by equation (4.6) be valid. This means that the clocks must be synchronized to determine one-way velocity. Thus we have shown that we cannot determine one-way signal velocity V unless we can synchronize the two clocks.

In this discussion, we have not specified the nature of the signal. In principle, we can use any signal: a snail, pony express, or light. It is seen that equation (4.5) and equation (4.6) are in fact the same equation. We thus have one equation, either (4.5) or (4.6), and two unknowns, V (required for one-way clock synchronization) and T_2 (required for one-way velocity measurement). Therefore, there is no direct way to find either of these two unknowns by a forward method, and one would have to resort to various approximate trial-and-error iterative methods and other indirect means.

Not being able to measure the one-way velocity of any signal, including light, by any direct method, we are generally content with measuring the two-way velocity which is based on mirrors. Although generally speaking one can use any signal, it is convenient to choose a light signal in vacuum for the purpose of clock synchronization, because a light signal propagates at the same velocity in all inertial frames. Let us now describe a two-way synchronization procedure. The coordinates X_1 and X_2 of the source and receiver clocks, respectively, are known. At an arbitrarily chosen moment, T_1, a light signal is sent from the source clock to the receiver clock. The light signal travels in vacuum along a straight path and hits the receiver at time T_2. A mirror at the receiver reflects the light signal and sends it back to its source. Let T_3 be the time instant that the source clock registers the returned signal. The receiver clock can now be synchronized with the source clock as follows. The instant T_2 of reflection as registered by the clock at the mirror (the receiver clock) is set equal to

$$T_2 = T_1 + \frac{T_3 - T_1}{2} \qquad (4.7)$$

This setting puts the receiver clock in synchronization with the source clock. In equation (4.7), we recognize the quantity $T_3 - T_1$ as the two-way travel time from source to receiver and back to source. Therefore, equation (4.7) says that the time T_2 is set equal to the sum of time T_1 of the source plus one-half the two-way travel time. Compare this equation with equation (4.5). Whereas velocity V appears in equation (4.5), it does not appear in equation (4.7). Therefore we can carry out two-way synchronization with light signals without knowing the velocity of light.

All terrestrial experiments to determine the velocity of light have been two-way experiments. That is, the velocity of light is determined by a beam traveling along a closed path, from source to receiver and then by reflection back to source. In this way only one clock, the one at the source, is required. The equation for the velocity is

$$c = \frac{2(X_2 - X_1)}{T_3 - T_1} \tag{4.8}$$

Here $2(X_2 - X_1)$ is the two-way distance, and $T_3 - T_1$ is the two-way time. For this reason, the velocity c determined by (4.8) is called the two-way velocity. Equation (4.8) makes a critical assumption, one that in principle can never be verified experimentally, and so it is taken only on faith. The assumption is that the velocity of light propagating "there" is identical to the velocity of light propagating "back." It is impossible to ascertain experimentally whether this assumption is true or false, because as we have seen, we cannot measure a one-way velocity. To verify the assumption, we would have to measure the one-way velocity "there" and also measure the one-way velocity "back," and we cannot do this. If this assumption were wrong, then you would have to kiss relativity theory goodbye. The theory of relativity, however, proceeds from the assumption that the velocity of light in vacuum is the same in all directions. The justification is that the totality of experimental data does not contradict either this assumption or any of the consequences of the theory of relativity.

How does a magician do his tricks? The classic answer is: He does them with mirrors. Upon what is the theory of relativity based? The classic answer is: It is based upon tricks done with mirrors. The classical mirror trick is Maxwell's suggested experiment, carried out by Michelson and Morley in 1887, the subject matter of the next chapter.

THE ELECTROMAGNETIC SPECTRUM

The principle of light says that an electromagnetic (EM) wave travels at the ultimate velocity c which is the same in all inertial frames. This means that if we measure the speed of an EM wave in one inertial frame and also measure its speed in another inertial frame, we will obtain the same value c. Thus, there is no way that we can distinguish the two inertial frames upon the basis of the speed of an EM wave.

Any information that an EM wave can carry is therefore not contained in its velocity.

What does contain the information? The answer is that an EM wave carries information in terms of its frequency content. Radio signals are EM waves that carry all sorts of information, from broadcasting stations and also from stars and galaxies. Television signals are also EM waves. Infrared EM waves carry the radiant energy from the sun, which keeps us warm and stimulates plant growth. Light is made up of visible EM waves, the ones that make possible all you know by seeing. Ultraviolet light, which bees can see, also is radiated by the sun. Although we are not aware of them, we are constantly exposed to the EM waves known as X rays, which are produced in certain stars, including the sun.

All EM waves can mathematically be placed on a frequency scale. The result is called the *electromagnetic spectrum*. We can compare it with the chart of the keyboard of a piano. The keys on the left side produce low notes; these sounds have low frequencies or equivalently long wavelengths. (The wavelength is the distance between two crests of a wave. For example, in a rolling sea, there is a long distance from the top of one swell to the top of the next swell, so such waves have long wavelengths.) When we move to the right on the piano, striking notes as we go, the pitch becomes higher; these sounds have high frequencies or equivalently short wavelengths. (For example, in a choppy sea, there is a short distance from one crest to the next, so choppy sea waves have short wavelengths.)

The information represented on the EM spectrum has the same format as the information represented on the piano sound spectrum just described. On the left are the low-frequency waves (radio, television, microwaves). As we proceed to the right, we pass the infrared, visible, and ultraviolet parts of the spectrum. On the right are the high-frequency waves (X rays, gamma rays). However, this arrangement by frequency is about the only way that the electromagnetic spectrum is similar to the sound spectrum. Sound is not part of the electromagnetic spectrum. Sound is made up of material waves that travel about 340 meters per second in air. Electromagnetic waves do not require any material substance for their propagation, and they travel at the ultimate speed of $c = 300{,}000$ km/s in vacuum.

Our theoretical knowledge about the electromagnetic spectrum has come since its mathematical discovery in 1864 by James Clerk Maxwell. Our way of life depends upon that knowledge. It has given us a new awareness, from the submicroscopic world to the universe of stars and galaxies. Before Maxwell, reality was made up of everything that humans could touch, feel, smell, see, and hear. Now we know that at best this represents one-millionth, or one-billionth, of reality. Nearly everything that is going to affect our tomorrows will require instruments working in ranges of reality that are nonhumanly sensible.

Following our usual practice, we will use the generic term "light" to designate all EM waves, including both the visible part of the electromagnetic spectrum (light as we know it) and the much greater invisible part (all the other EM waves). Nearly all books on relativity theory follow this practice, most likely because the word "light" has a poetic quality, whereas the term "EM wave" seems cold and scientific. Since red visible light has lower frequency than blue visible light, the generic terms "red" and "blue" are used as synonyms for "low frequency" and "high frequency," respectively. Thus the red end of the spectrum would include the radio waves, and the blue end the gamma waves. A shift to the red means a shift to lower frequencies (whatever the original frequency was), and similarly a shift to the blue means a shift to higher frequencies.

Credit for the empirical discovery of the electromagnetic spectrum goes to the diversified genius of Sir Isaac Newton, who passed the sun's light through a prism which broke the light up into a spectrum of colors. This was the man-made counterpart of nature's spectrum, the beautiful rainbow. But the great interest in spectral analysis made its appearance only a little more than a century ago. The prominent German chemist Robert Wilhelm Bunsen (1811–1899) repeated Newton's experiment of the glass prism. Only Bunsen did not use the sun's rays as Newton did. In Bunsen's experiment, the role of pure sunlight was replaced by the burning of an old rag that had been soaked in a salt (sodium chloride) solution. The exquisite rainbow of Newton did not appear. The spectrum, which Bunsen saw, only exhibited a few narrow lines, nothing more. One of these lines was a bright yellow.

The role of the glass prism consists only in sorting the incident rays of light into their constituent frequencies (the process known as dispersion). The Newton rainbow is the extended continuous band of the solar spectrum; it indicates that all the frequencies of visible light are present in pure sunlight. The lonely yellow line, which appears when the light source is the burning salt-soaked rag, indicates that the spectrum of table salt contains a single prominent frequency. Further experiments showed that this yellow line belonged to the element sodium. No matter what the substance in which sodium appears, that element makes its whereabouts known by its bright yellow spectral line. As time went on, it was found that every chemical element has its own characteristic line spectrum. Furthermore, the line spectrum of a given element is always the same, no matter in what compound or substance the element is found. Thus, the line spectrum identifies the element, and in this way we can determine what elements are in substances, from the distant stars to microscopic objects.

The successes of spectral analysis were colossal. However, the spectral theory of the elements could not be explained by classical physics. As we know, quantum physics was born starting with the work of Planck in 1900. Spectral theory was explained in 1925 and 1926 by the quantum mechanics developed by Werner Heisenberg (1901–1976) and Erwin Schrödinger (1887–1961).

Electromagnetic waves are a stream of elementary particles called photons. The energy of a photon is proportional to its frequency. The constant of proportionality is known as Planck's constant h, which is equal to the energy of a photon with the frequency of 1 Hz. The numerical value of Planck's constant is

$$h = 6.626 \times 10^{-34} \text{ joules}$$

This constant was introduced in 1900 by Max Planck (1858–1947) in his momentous work which inaugurated the discipline of quantum mechanics in physics. The relationship between the energy E of a photon and its frequency f is given by *Planck's equation*

$$E = hf \qquad (4.9)$$

For example, a photon of red light ($f = 4.5 \times 10^{14}$) has energy

$$E = 6.626 \times 10^{-34} \times 4.5 \times 10^{14} = 2.98 \times 10^{-19} \text{ joules}$$

Because the energy of electromagnetic radiation is proportional to frequency, it follows that X rays and gamma rays have high energy. The photon of a gamma ray with frequency $f = 10^{22}$ has energy 6.6×10^{-12} joules, sufficient energy to break apart delicate molecules found in the human body.

THE DOPPLER EFFECT

The most important tool used in extracting information from "light" (i.e., EM waves) makes use of the Doppler effect. The Doppler effect is exhibited by all types of wave motion. The effect was discovered in 1842 by Christian Johann Doppler, an Austrian physicist, who was born in Salzburg, November 29, 1803, and who died in Venice, Italy (then under Austria), March 17, 1853. Doppler's critics tested his mathematical results by an experiment in Holland. For a couple of days, they had a locomotive pull a flat car back and forth at different speeds. On the flat car were trumpeters, sounding various notes. On the ground, musicians with a good perception of absolute pitch recorded the note as the train approached and as it receded. When the train approached, these musicians found that the pitch increased. When the train went away, they found that the pitch decreased. Doppler's equations were confirmed.

Let us now consider a light source and a receiver, which move relative to each other. Because no material medium is needed for light propagation, it follows that it is only the relative speed β of the source and the receiver that is essential to our argument. All our conclusions will pertain to vacuum.

In our discussion, we will assume that the source emits a monochromatic light wave, that is, a sinusoidal wave characterized by a single fixed frequency f_0. (More complicated sources would have to be treated by means of the mathematical techniques of Fourier analysis.) If we stay at a fixed point and watch the light wave pass, then the time duration that we measure between two adjacent crests of the light wave is called the period t_0. The period is equal to the reciprocal of the frequency; that is,

$$t_0 = \frac{1}{f_0} \qquad (4.10)$$

For purposes of exposition, it is simpler to consider the light wave as a sequence of short pulses, each pulse separated in time from the next by a time duration t_0. Each pulse can be presented by a photon, so we can think of a light wave of frequency f_0 as a train of photons passing a fixed point, where each photon is separated in time from the next by the period t_0. When we speak of frequency, we stay at a fixed point and watch the train of photons pass. As soon as one photon passes, we have to wait a time t_0 (the period) before the next one passes. Now, instead, let us take a snapshot of the entire train at all spatial points, but at one fixed time. The distance between two adjacent photons is called the wavelength λ_0. Each photon travels a distance of 1 wavelength λ_0 in a time duration of 1 period t_0. The velocity of light is therefore equal to

$$c = \frac{\lambda_0}{t_0}$$

In natural units, the velocity of light is 1 ($c = 1$), so in natural units the wavelength equals the period ($\lambda_0 = t_0$).

Up to this point, we have seen that we can characterize a monochromatic light source in a given inertial frame as a stream of photons, separated in time at a fixed point by the period t_0, and separated in space at a fixed time by the wavelength $\lambda_0 = ct_0$. If the receiver is at rest at a fixed point in this inertial frame, then the receiver will register the wavetrain in the same form as it is sent. That is, both source and receiver will register the same period t_0 (and accordingly the same wavelength $\lambda_0 = ct_0$).

One of the main difficulties encountered in following arguments in special relativity is that in working with two different inertial frames, we keep switching back and forth between frames. To reduce that difficulty here, we are going to carry out as much of the mathematics as we can in one inertial frame. Then at the last minute, but with plenty of warning, we will switch to the other frame. When we work in one inertial frame, we can use common sense. Specifically, in a fixed inertial frame, light does not have to appear to travel at the given value c, but can appear to travel at various apparent velocities. It is only when we switch to the different perception of another inertial frame that common sense goes out the window, and we are in the realm of relativity theory.

Let us confine ourselves to a given inertial frame K. First, we consider the case when the receiver recedes from the source at speed β. We assume that β is a positive number. The source emits a light wave of period t_0 and wavelength ct_0. Suppose that at time $t = 0$ the receiver is 1 wavelength from the source. At that moment $t = 0$ the source emits a photon. That is, at time 0 the photon is at $x = 0$ and the receiver is at the point $x = ct_0$. Here we have a race in which the receiver is trying to run away from the photon, like a slow runner trying to run away from someone faster. How far does this photon have to travel to catch up with the receiver? Let the catch-up time be denoted by t_1. In this time the photon traveling at speed c covers a distance equal to ct_1. That is, at time t_1, the photon

THE DOPPLER EFFECT

catches the receiver, so both the photon and receiver are at the point $x = ct_1$. See Figure 4.8. The length ct_1 of the photon's journey (i.e., the catch-up distance) is equal to the sum of

1. The distance from the source to the point where the receiver was at time 0 (i.e., the head-start distance), plus
2. The distance that the receiver travels from time 0 to time t_1 (i.e., the get-away distance).

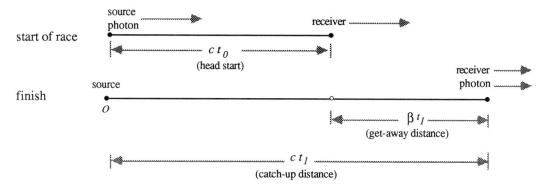

Figure 4.8. Race to get away.

We recall that the head-start distance is equal to the proper wavelength ct_0. The get-away distance is computed as follows. The time duration of the receiver's journey is the same as the time duration t_1 of the photon's journey because the race starts at the same time 0, and ends at time t_1, when the photon catches the receiver. Since the receiver travels at speed β, the length of its journey is βt_1. That is, the receiver's get-away distance is βt_1. Therefore,

$$ct_1 = ct_0 + \beta t_1$$

which gives the travel time as

$$t_1 = \frac{ct_0}{c - \beta} \qquad (4.11)$$

This equation can also be derived by introducing a *phantom racer* as follows. The photon starts at the origin $x = 0$. The receiver has a head start of 1 wavelength $x = ct_0$. The photon travels at the high speed c, whereas the receiver travels at the low speed β. Both are in a race which ends when the photon catches the receiver. The question is: What is the time duration t_1 of the race? To answer this question, we can substitute an equivalent race for the given race. The *apparent speed* between photon and receiver is $c - \beta$. The *phantom racer* goes at the apparent speed, and the race consists of covering the head-start distance ct_0. Thus the time duration t_1 of the race is equal to distance ct_0 divided by speed $c - \beta$. The result is the same as the equation (4.11), as we wanted to show.

Next we consider the case when the receiver approaches the source at speed β. Again we assume that β is a positive number. At time 0 the race starts; the photon is at $x = 0$ and the receiver is at $x = ct_0$. In the first case, we had a race in which the receiver was trying to get away from the photon. In this case, we have a race in which the receiver is trying to come together with the photon. This is the same as the case of two lost friends who, upon seeing each other at a distance, rush toward each other, one slow and the other fast, until they meet. Let the meeting time be denoted by t_2. In this time the photon traveling at speed c covers a distance ct_2. That is, at time t_2, the photon meets the receiver, so both the photon and receiver are at the point $x = ct_2$. See Figure 4.9. The length ct_2 of the photon's journey (i.e., the meeting distance) is equal to the difference of

1. The distance from the source to the point where the receiver was at time 0 (i.e., the separation distance), minus
2. The distance that the receiver travels from time 0 to time t_2 (i.e., the come-together distance).

The separation distance is the proper wavelength ct_0, and the come-together distance is βt_2. Thus

$$ct_2 = ct_0 - \beta t_2$$

which gives the travel time as

$$t_2 = \frac{ct_0}{c + \beta} \qquad (4.12)$$

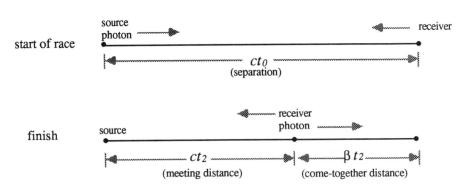

Figure 4.9. Race to meet.

The Doppler Effect

Let us now summarize. In natural units, $c = 1$, so equation (4.11) becomes

$$t_1 = \frac{t_0}{1 - \beta} \tag{4.13}$$

and equation (4.12) becomes

$$t_2 = \frac{t_0}{1 + \beta} \tag{4.14}$$

In each case, the source emits a light wave of period t_0. In the first case, the receiver moves away from the source at speed β. Then equation (4.13) gives the period t_1 registered by the receiver. In the second case, the receiver moves toward the source at speed β. Then equation (4.14) gives the period t_2 registered by the receiver. By inspection, we see that the right-hand sides of equations (4.13) and (4.14) have the same form, where one can be obtained from the other by replacing β by $-\beta$. In the discussion up to this point, we have considered only one frame of reference, the frame in which the source is stationary and the receiver is moving.

We now consider *another inertial frame* in which the source is moving and the receiver is stationary. Call this frame K' with primed coordinates. The relative speed of frames K and K' is given by β. Let τ_1 be the period of the received light wave in frame K', where τ_1 corresponds to the period t_1 in frame K. Likewise, let τ_2 correspond to t_2. Because the receiver is at rest in frame K', it follows by definition that τ_1 and τ_2 are proper times. In frame K, the receiver is moving, so t_1 and t_2 are dilated times. Thus the relationships between t_1 and τ_1, and t_2 and τ_2 are given by

$$t_1 = \gamma \tau_1, \qquad t_2 = \gamma \tau_2$$

where γ is the dilation factor $\gamma = 1/\sqrt{1 - \beta^2}$. As a result, equation (4.13) becomes

$$\tau_1 = \frac{t_0}{\gamma(1 - \beta)} \tag{4.15}$$

and equation (4.14) becomes

$$\tau_2 = \frac{t_0}{\gamma(1 + \beta)} \tag{4.16}$$

Substituting the expression for γ into equations (4.15) and (4.16) they become, respectively,

$$\tau_1 = \sqrt{\frac{1 + \beta}{1 - \beta}} \, t_0 \quad \text{or} \quad \tau_1 = k t_0 \tag{4.17}$$

$$\tau_2 = \sqrt{\frac{1-\beta}{1+\beta}}\, t_0 \quad \text{or} \quad \tau_2 = k^{-1} t_0 \qquad (4.18)$$

Here

$$k = \sqrt{\frac{1+\beta}{1-\beta}}$$

is the *relativistic Doppler factor*, which was introduced earlier. In this section, we have assumed that β is intrinsically positive, so $k > 1$ and $k^{-1} < 1$. Equation (4.17) says that when the source and receiver move away from each other, the received period τ_1 is greater than the transmitted period t_0. This is the same result that we found using the fable of the king's messengers. Similarly, equation (4.18) says that when the source and receiver move toward each other, the received period τ_2 is less than the transmitted period t_0.

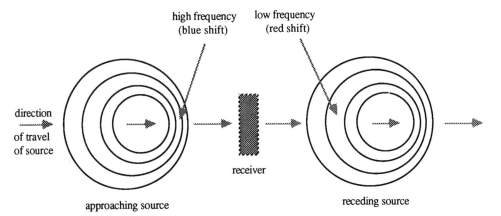

Figure 4.10. The Doppler effect. The frequency of an approaching source is high because all the waves are crowded together. Conversely, the frequency is much lower as the source recedes because the waves are spread out.

We are now ready to write down the final result. See Figure 4.10. We recall that the frequency is defined as the reciprocal of the period. That is $f_0 = 1/t_0$, $\nu_1 = 1/\tau_1$, and $\nu_2 = 1/\tau_2$. Thus by taking the reciprocal of each of equations (4.17) and (4.18), we obtain, respectively,

$$\nu_1 = \sqrt{\frac{1-\beta}{1+\beta}}\, f_0 \qquad (4.19)$$

and

$$\nu_2 = \sqrt{\frac{1+\beta}{1-\beta}} f_0 \tag{4.20}$$

In both equations (4.19) and (4.20), f_0 is the source frequency measured in frame K in which the source is at rest. Equation (4.19) gives the received frequency ν_1 in frame K' in which the receiver is at rest, for the case when frame K' moves away from frame K with speed β. Equation (4.20) gives the received frequency ν_2 in frame K' in which the receiver is at rest, for the case when frame K' moves toward frame K with speed β.

We recall that the element sodium exhibits a strong yellow line in its spectrum. This is called the D line in the line spectrum of sodium; it is much brighter than the other lines in the spectrum. The frequency of the D line is $f_0 = 5.09 \times 10^{14}$ Hz. The energy of a photon of this frequency is

$$E = 6.626 \times 10^{-34} \times 5.09 \times 10^{14} = 3.37 \times 10^{-19} \text{ joules}$$

Suppose that sodium is present in a distant galaxy and that the galaxy is receding from Earth at a speed of $0.1c$. The electromagnetic waves of the yellow sodium light are thus expanded between the galaxy and Earth, so on Earth the sodium D line appears to have frequency given by equation (4.19); that is,

$$\nu_1 = \sqrt{\frac{1-0.1}{1+0.1}} 5 \times 10^{14} = 4.52 \times 10^{14}$$

The color associated with frequency ν_1 is orange-red instead of the bright yellow of f_0.

On the other hand, suppose the galaxy is approaching Earth at a speed of $0.1c$. The electromagnetic waves of the yellow sodium light are thus compressed, so on Earth the sodium D line appears to have frequency given by equation (4.20); that is,

$$\nu_2 = \sqrt{\frac{1+0.1}{1-0.1}} 5 \times 10^{14} = 5.53 \times 10^{14}$$

The color associated with frequency ν_2 is greenish instead of the bright yellow of f_0.

The shift of frequency due to the Doppler effect tells us about the radial speed of a distant galaxy. If the shift is toward the red, the galaxy is receding from Earth. If the shift is toward the blue, the galaxy is approaching Earth. When the large Palomar telescope was put into service in 1923, the astronomer Edwin Hubble (1889–1953) discovered that the color of distant galaxies was generally shifted toward the red. Furthermore, he observed that the greater the distance, the greater the shift. This phenomenon, called the red shift, demonstrated that the universe is expanding. Hubble's constant, 18 km/s per million light years, is the proportionality constant for the uniform expansion rate. Thus two galaxies 1 million light-years apart travel away from each other at the relative speed of 18 km/s. Two galaxies 10 million light-years apart move away from each other at the relative speed of 180 km/s. A galaxy 1.67 billion light-years from Earth would recede at the speed

$$1.67 \times 10^9 \times 18/10^6 = 30{,}000 \text{ km/s}$$

which is one-tenth the velocity of light (i.e., $\beta = 0.1$). Referring to our sodium example, this galaxy would thus exhibit the sodium D line as received on Earth as an orange-red ($\nu_1 = 4.52 \times 10^{14}$). It follows that whenever we observe the sodium D line from a galaxy at orange-red frequency ν_1, we can conclude that the galaxy is 1.67 billion light-years from Earth. In this way the Doppler shift together with Hubble's relation allows us to find the distances to the far-away galaxies. Without the information carried by the Doppler effect, we could not have astronomy as we know it today. However, without the Doppler effect, we would not have the Doppler radar that the police use to tell how fast a car is going. But, then, without electromagnetics we would still be living in the horse-and-buggy days, so it would not matter anyhow.

NEWTON'S FAILURE AND MAXWELL'S SUCCESS

The notion of physical color originates in the simple but elegant experiments of Sir Isaac Newton in which he passed a beam of sunlight through a prism. The sunlight emerged as a band of colors, an artificial rainbow that he called the *spectrum*. Newton proposed the following explanation for the spectrum. A beam of white light is made up of a superposition of a whole variety of colored light beams. The prism merely spreads out these superposed colored beams into a spectrum of colors. The spreading is due to the fact that each beam of a different color is refracted through a slightly different angle when it enters and leaves the prism. This process of spreading is known as *dispersion*.

Next, Newton allowed a beam of white light to pass through two prisms (Figure 4.11). The second prism reverses the deflections caused by the first, and the beam emerges as white light again. The reason is that differently colored light rays that were spread out by the first prism are put back together by the second into the original white light. That is, the second prism acts as the inverse of the first prism.

Finally, Newton asked whether the component colored beams could be broken down further, by means of other experiments. Newton tried many experiments

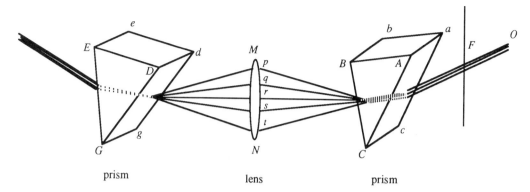

Figure 4.11. A beam of white light passing through two prisms.

and could not do this. He therefore concluded that these beams of pure color are fundamental. Newton's words follow:

> When any one sort of Rays hath been well parted from those of other kinds, it hath afterwards obstinately retained its colour, notwithstanding my utmost attempts to change it. I have refracted it with Prisms, and reflected it with Bodies, which in Day-light were of other colours; I have intercepted it with the coloured film of Air interceding two compressed plates of glass, transmitted it through coloured Mediums, and through Mediums irradiated with other sorts of Rays, and diversly terminated it; and yet never produced any new colour out of it.

The theory of relativity has changed this final conclusion of Newton. The pure color of one of these component beams of light is merely a perception. When viewed in another inertial frame, the color will be different. The reason is the Doppler shift, discovered and explained by Christian Doppler (1803–1853), but actually inherent in the work of Olaf Roemer (1644–1710).

It is an interesting facet of history that the Doppler effect was indeed a part of Roemer's momentous discovery in 1676. Roemer found that the speed of light is finite. His reasoning was as follows. By observing the eclipses of the four most prominent moons of Jupiter, he measured the periods of their orbits around Jupiter. (These moons were first observed in 1610 by Galileo.) However, Roemer obtained different results when he measured the periods six months later. Let us see what happened. Jupiter and Earth both revolve around the sun. Since one Jupiter year is so much greater than one Earth year, let us neglect the motion of Jupiter and consider it stationary. Also let us neglect the elliptic orbit of the Earth and, for simplicity, consider that the Earth rushes in a straight line away from Jupiter for six months, and then rushes directly back toward Jupiter the remaining six months of the year. The distance covered by the Earth in this hypothetical outward path is thus the diameter of the Earth's orbit, and the same distance is covered in the reverse path. In Roemer's time, the diameter of the Earth's orbit was not

Plate 4.2. Olaf Roemer (1644–1710)

known as accurately as it is today, so for the sake of this discussion we will use the modern figure of 300,000,000 km.

The moon Io orbiting Jupiter acts like a clock on Jupiter. Roemer measured the period of this moon by timing how long it takes for the moon to make one full revolution of Jupiter, that is, the time from one eclipse to the next for this moon. He found this period to be approximately 42.5 hours when the Earth was closest to Jupiter. As the Earth traveled in its orbit from the planet Jupiter, Roemer predicted that succeeding eclipses would occur (as seen on Earth) also every 42.5 hours. However, Roemer found that the eclipses occurred later and later, until in six months, when the Earth was most distant from Jupiter, the eclipse was too late by a sizable time lag. Using modern figures, the amount of this time lag is about 1000 seconds. The only logical conclusion that Roemer could make was that the time lag is in fact the additional time that it takes for light from Jupiter's moon to travel the extra distance across the diameter of the Earth's orbit. Thus the velocity of light must be finite, and equal to this diameter divided by the time lag; that is,

$$c = \frac{300{,}000{,}000}{1000} = 300{,}000 \text{ km/s}$$

This conclusion was stupendous. The division which gives the value of c was actually done by Huygens in 1677. It seems that Roemer was satisfied just to establish the fact that the velocity was not infinite (in which case the time lag would

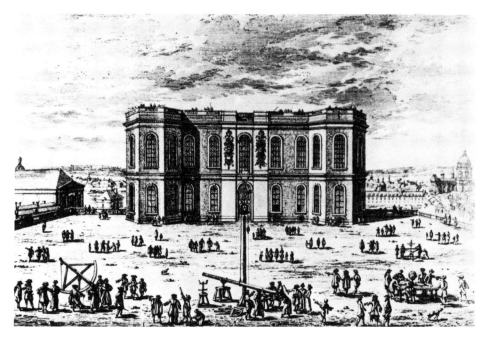

Plate 4.3. The Paris Observatory when Roemer discovered the movement of light in 1676.

have been zero). These results came to the attention of Newton by 1680, and now comes one of the greatest of the lost opportunities which has ever occurred in science. Newton had all the ingredients required to establish relativity theory, but instead he turned his mind to the mundane task of reforming English currency in his position as Master of the Mint. The missed opportunity was this: Roemer had discovered the *relativistic Doppler effect*, but Newton did not recognize it as such.

Let us explain. Designate the period of Jupiter's moon Io by t_0, which for this discussion we take to be 42.5 days, or 3,672,000 seconds. The Earth actually travels an elliptic orbit around the sun, but for simplicity we have assumed that the Earth goes back and forth in a straight path. In other words, we have supposed that the Earth-Jupiter system is like a yo-yo (a toy consisting of a flattened spool wound with string that is spun down from and reeled up to the hand by unwinding and winding the string). The Earth corresponds to the spool and Jupiter to the hand. The Earth in this straight path travels a distance equal to its orbital diameter of 300,000,000 km in six months (182.5) days. Six months is equal to 15,768,000 seconds. Thus the yo-yo speed of the Earth is

$$V = \frac{300{,}000{,}000 \text{ km}}{15{,}768{,}000 \text{ s}} = 19 \text{ km/s}$$

Plate 4.4. Christiaan Huygens (1629–1695), a portrait by C. Netscher in 1671.

which in natural units is

$$\beta = \frac{V}{c} = 0.0000634 \text{ (a pure number)}$$

(Note that the yo-yo speed is equal to $2/\pi$ times Earth's orbital speed of 30 km/s.) Newton was the first one to formulate accurately and use the Galilean principle of relativity. If he had merely applied this principle to Roemer's analysis of Jupiter's moons, Newton easily would have come up with the relativistic Doppler factor k. Using the modern values of Roemer's figures, the value of the relativistic Doppler factor is

$$k = \sqrt{\frac{1+\beta}{1-\beta}} \approx 1 + \beta = 1.0000634$$

Thus on the outward path of the Earth, the period of Io as seen on Earth is increased (red-shifted) to the value

$$kt_0 = 1.0000634 \, (3{,}672{,}000) = 3{,}672{,}233 \text{ seconds}$$

Thus the relativistic Doppler effect adds 233 seconds to the period of Io; that is, the difference $kt_0 - t_0$ is $3{,}672{,}233 - 3{,}672{,}000$, or 233 seconds. In six months

Newton's Failure and Maxwell's Success 131

Plate 4.5. Sir Isaac Newton (1642–1727)

there are 182.5 days. The period of Io is 42.5 days. Thus in six months Io makes 182.5/42.5, or approximately 4.294 revolutions around Jupiter. The Doppler effect increases each period by 233 seconds as seen on Earth. Thus in six months, the cumulative increase is 4.294(233) = 1000 seconds, which is the time lag observed by Roemer. (As said, we are using modern figures for these values instead of the inaccurate ones of Roemer.)

It is amazing that Newton did not in fact use this relativistic reasoning to explain the time lag detected by Roemer. If Newton had done this simple analysis, he would have originated the theory of relativity in its space-time context. The mathematical theory of relativity in a geometric context had already been discovered at the same university, Cambridge University, where Newton had spent so many of his mathematically active years. This great discovery had been made by Edward Wright, mathematical tutor to the Prince of Wales, in the year 1599, about 80 years before Newton became aware of Roemer's work.

Roemer had the theory of relativity in his hands. He (with the help of Huygens) postulated the finite constant velocity of light, and he had correctly analyzed the relativistic Doppler effect. On the basis of these two things, the special theory of relativity can be readily established. But Newton failed to do this. However, a worthy successor to Newton at Cambridge University, James Clerk Maxwell, did not miss this opportunity. Maxwell in 1879, the year of his death, realized that the pressing space-time questions at the forefront of physics could find their answers in the work of Roemer. To study the constancy of the velocity of light, Maxwell returned to Roemer's original astronomical experiment on the eclipses of the moons of Jupiter. However, Maxwell soon found out that even the then present-day observations of the times of the eclipses of these moons, and their positions in space, did not contain enough accuracy for his purposes. As a result, Maxwell conceived an Earth-based experiment to take the place of Roemer's astronomical experiment. This Earth-bound experiment was carried out, after Maxwell's death, in its definitive form by Michelson and Morley in 1887. This experiment gave the decisive evidence that, as we will see in the next chapter, led to the special theory of relativity.

The work at Cambridge University had come full circle, from Wright's discovery in 1599, to Newton's failure, and finally to Maxwell's success in 1879. Maxwell's equations inherently contain special relativity. Maxwell, stricken at a young age by cancer, did not live long enough either to see the Michelson-Morley experiment or to hear radio messages, both in a sense theoretical inventions of his.

LIGHT SPEED AND LIGHT COLOR

Roemer's seventeenth-century experiment involves the two key ideas of relativity theory, namely, the *speed of light* and the *color of light*. Pick one photon of light. It travels in vacuum at the constant and ultimate speed c, the same in all inertial frames. However, its color is governed by the relativistic Doppler effect, so its color changes in every different inertial frame. The color of a photon is determined by its wavelength λ. Its wavelength λ is equal to velocity c times its period τ, that is, $\lambda = c\tau$. In natural units $c = 1$, so wavelength equals period. Consider now two inertial frames, with relative velocity β. If the color of the photon in the source frame is given by the period t_0, then its color in the receiver frame is given by the shifted period

$$\tau = kt_0$$

where k is the relativistic Doppler factor

$$k = \sqrt{\frac{1 + \beta}{1 - \beta}}$$

If the receiver frame is receding from the source frame (i.e., if $\beta > 0$), then $k > 1$, so the period τ in the receiver frame is greater than the period t_0 in the source frame. This means that the photon's color in the receiver frame is redder than in the source frame (i.e., a red shift in the receiver frame). On the other hand, if the receiver frame is approaching the source frame (i.e., if $\beta < 0$), then $k < 1$, so the period τ in the receiver frame is less than the period t_0 in the source frame. This means that the photon's color in the receiver frame is bluer than in the source frame (i.e., a blue shift in the receiver frame).

Let us recall Newton's statement (given in the preceding section) in which he listed the various experiments he performed on light rays of pure color, and yet he was unable to change the color. Now we know that there is a way to change

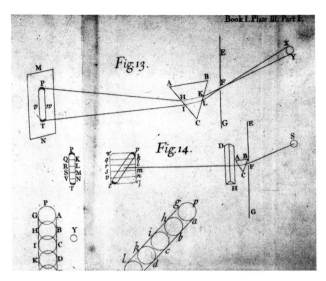

Plate 4.6. Newton's prism, an illustration from his *Opticks*.

the color, namely, by movement. The Doppler effect represents a shift in pure color produced by motion. A pure red beam of light appears blue to a person moving rapidly toward the source of light. A pattern of colors (the spectrum) emitted by a source of electromagnetic radiation is systematically shifted by motion. If the source and receiver are moving apart, the pattern at the receiver is shifted to the red; if moving together, to the blue.

Because pure colors are transformed into each other by the Doppler shift, it means that they are all equivalent to one another as viewed by suitably moving observers. In this way, the basic space-time symmetry as manifested by the Doppler effect means that the infinity of different colors in the electromagnetic spectrum are all representations of one underlying reality.

Let us look at the Doppler factor

$$k = \sqrt{\frac{1+\beta}{1-\beta}}$$

It is entirely democratic (i.e., relativistic) because its value does not depend upon whether the source is moving toward the receiver, or whether the receiver is moving toward the source. All that matters is the relative speed of the source and receiver. If they are moving together, then β is less than zero, in which case the received light has a smaller period than the emitted light (i.e., $k < 1$, so a blue shift). If they are moving apart, then β is greater than zero, in which case the received light has a greater period than the emitted light (i.e., $k > 1$, so a red shift). (See Figure 4.12.)

Suppose now that the Earth is moving toward the sun at nearly the speed of

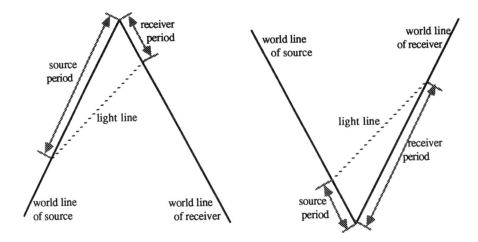

Figure 4.12. Left: Source and receiver moving together ($\beta < 0$) produces blue shift. Right: Source and receiver moving apart ($\beta > 0$) produces red shift.

light. Then β is almost equal to -1, so k is almost zero. This means that the light from the sun as received on Earth would be almost completely blue-shifted. The received waves of light would be compressed almost together (i.e., have almost zero wavelength) so their frequency ν would be nearly infinite. We recall that Planck's equation

$$E = h\nu$$

gives the energy E of a photon in terms of its frequency ν, where h is Planck's constant. Since ν is close to infinity, the energy content of this received light would also be nearly infinite.

On the other hand, suppose that the Earth is moving away from the sun at nearly the speed of light. Then β is nearly 1, so k is almost infinite. This means that the rays received on Earth from the sun would be almost completely red-shifted. The received waves of light would be nearly flat (i.e., have almost infinite wavelength), so their frequency and energy content would be nearly zero.

We see that the Doppler effect is not symmetrical with respect to energy. The energy gained in the blue shift is greater than the energy lost in the red shift. For example, let the sun emit one unit of energy. If the Earth moves toward the sun at speed β, then the blue shift in frequency (and energy) is given by

$$\sqrt{\frac{1 + \beta}{1 - \beta}}$$

which is greater than 1. On the other hand, if the Earth moves away from the sun at speed β, then the red shift in frequency (and energy) is given by

$$\sqrt{\frac{1-\beta}{1+\beta}}$$

which is less than 1. The average change in energy is thus

$$\frac{1}{2}\left(\sqrt{\frac{1+\beta}{1-\beta}} + \sqrt{\frac{1-\beta}{1+\beta}}\right)$$

which comes out to be

$$\frac{1}{2}\left(\frac{1+\beta}{\alpha} + \frac{1-\beta}{\alpha}\right) = \frac{1}{\alpha} = \gamma$$

where $\alpha = \sqrt{1-\beta^2}$. The average of the blue-shifted and red-shifted light is the familiar gamma of the Greek trio α, β, γ.

A stationary observer sees the sun emit one unit of energy. However, a moving observer, who travels toward the sun, passes it, and then travels away from the sun, sees the sun emit an average of γ units of energy. Thus there is an average gain of energy, given by $\gamma - 1$, emitted by the sun, as judged by a person in motion with respect to the sun.

Let us now summarize the argument and see what conclusions we can draw. As seen from a stationary Earth, the sun gives out light in equal quantities in all directions. Suppose now that the Earth moves toward the sun, passes it, and then moves away from it. The light as seen on the moving Earth will be altered by the Doppler effect. As the Earth approaches the sun, it will appear bluer. As the Earth goes away from the sun, it will appear redder. But there is a discrepancy in energy. The average of the energy of the blue-shifted and red-shifted light, as we have just shown, is proportional to the gamma factor. The gamma factor is greater than unity. In this context, unity (or one) corresponds to the amount of energy that would be seen if the Earth were stationary near the sun and seeing it in its true colors, that is, unaffected by the Doppler shifts. From the perspective of the moving Earth (i.e., traveling to and then away from the sun), the sun gives off more energy than the amount given off as seen from the perspective of a stationary Earth.

If the sun sheds energy at a greater rate for a moving Earth, then where does this extra energy come from? The only possible source is the energy of motion. This is the motion of the sun relative to the Earth, for we can consider the Earth as stationary and the sun rushing by the Earth at high speed. The relative motion of the sun, as seen by the stationary Earth, increases the output of light energy of the sun by the amount $\gamma - 1$. That extra energy must be supplied from the sun's energy of motion.

There are only two ways that a moving object can reduce its energy of motion, namely, (1) by reducing its speed β or (2) by reducing its mass. But we have assumed that speed β is constant. Therefore, the sun must be losing mass to supply this extra energy. In short, the sun has to reduce its mass to give off the extra energy indicated by the quantity γ − 1. But we cannot distinguish between energy and extra energy. If radiating the extra energy requires a loss of the sun's mass, so then radiating the energy in the case of a stationary Earth involves a loss of mass.

Now comes Einstein's famous result. The sun must lose a certain mass for each amount of energy it radiates. In fact, the sun is losing mass at the rate of 4 million tons per second. Thus radiant energy from the sun has a certain mass associated with it. Einstein realized that the same argument applies to all forms of energy: kinetic energy (i.e., energy of motion), heat energy, chemical energy, and so on. Thus mass and energy must be regarded as the same thing. Einstein's formula

$$E = mc^2$$

expresses the equivalent of energy E and mass m. The factor c^2 is required when unnatural units (SI) are used. In natural units ($c = 1$), we have simply $E = m$.

SPACE-TIME SYMMETRY

The first historical treatment of special relativity in its space-time context is implicit in the work done in Paris in 1676 by Roemer. Roemer, with the help of Christian Huygens, postulated the constancy of the speed of light and introduced the concept of the relativistic Doppler effect. In this section, we show that Roemer's results give the required space-time symmetry as required by the Poincaré group.

This section will also serve as an historical introduction to the work of Edward Wright (1560–1615). His mathematics will be presented in Chapter 6, so here we will only outline the results. The story starts with the map projection developed by the ancient Greeks and known as the stereographic projection. In this projection a point on a globe (of unit radius) at latitude angle φ is mapped into the point k given by the equation

$$k = \sec \phi + \tan \phi \qquad (4.21)$$

We will refer to k as the *stereographic projection* of latitude φ. Next comes the Mercator projection developed in the sixteenth century for ocean navigation. The critical breakthrough was made by Wright in 1599 who derived the equation

$$u = \log(\sec \phi + \tan \phi) \qquad (4.22)$$

SPACE-TIME SYMMETRY

This equation, *Wright's equation*, says that in the Mercator projection a point on a globe (of unit radius) at latitude ϕ is mapped into the point u. We will refer to u as the *Mercator projection* of latitude ϕ. These two equations show that the Mercator projection u is related to the stereographic projection k by the logarithmic relation

$$u = \log k \qquad (4.23)$$

The ancient Greek stereographic projection has the important property that points at opposite latitudes, ϕ and −ϕ, on the unit globe are projected into points symmetric with respect to the unit circle. In other words, if k is the stereographic projection of latitude ϕ, then k^{-1} is the stereographic projection of −ϕ. This result can be derived as follows. Let k be given by equation (4.21). Then

$$k^{-1} = \frac{1}{k} = \frac{1}{\sec \phi + \tan \phi}$$

By use of the trigonometric identities, this equation becomes

$$k^{-1} = \sec \phi - \tan \phi = \sec(-\phi) + \tan(-\phi) \qquad (4.24)$$

which is the desired result.

We now want to relate the relativistic results of Roemer to those of Wright. Roemer deals with distance x, time t, and velocity $v = x/t$. If body C is moving from body A at velocity v, then a period t_0 emitted from body A is received on body C with shifted period τ given by

$$\tau = kt_0$$

where k is the *relativistic Doppler factor* given by

$$k = \sqrt{\frac{1+v}{1-v}} \qquad (4.25)$$

Let us now make the linkage between the space-time results of Roemer with the cartographic results of Wright. The linkage is given by the following equations:

$$x = \tan \phi, \qquad t = \sec \phi, \qquad v = \frac{x}{t} = \sin \phi \qquad (4.26)$$

where ϕ is the *latitude angle*, and also as we will see in Part II of this book, ϕ is the *Gudermannian angle*.

Plate 4.7. Christian Johaan Doppler (1803–1853)

With this linkage, the stereographic projection k is identical to the relativistic Doppler factor k, which explains the reason why we use the same symbol k for both. Let us now verify this identity. If we let $v = \sin \phi$, then the relativistic Doppler factor becomes

$$k = \sqrt{\frac{1+v}{1-v}} = \frac{1+v}{\sqrt{1-v^2}} = \frac{1+\sin\phi}{\cos\phi} \qquad (4.27)$$

which is the same as equation (4.21) for the stereographic projection.

Let us next turn to the work of Kepler in 1609, as given in Chapter 3. Kepler's equation

$$\sin\phi' = \frac{\sin\phi - \sin\phi_0}{1 - \sin\phi \sin\phi_0}$$

relates three angles: the Copernicus angle ϕ, the eccentricity angle ϕ_0, and the Kepler angle ϕ'. By the use of some algebra, we can manipulate Kepler's equation to obtain it in the (reflexive) form

$$\sin\phi = \frac{\sin\phi_0 + \sin\phi'}{1 + \sin\phi_0 \sin\phi'} \qquad (4.28)$$

Let us now link this equation with space-time. We make the same linkage between

Space-Time Symmetry

angle and space-time as before, but now we have three angles. We therefore have

$$x = \tan \phi, \quad t = \sec \phi, \quad v = \sin \phi$$
$$x_0 = \tan \phi_0, \quad t_0 = \sec \phi_0, \quad v_0 = \sin \phi_0 \quad (4.29)$$
$$x' = \tan \phi', \quad t' = \sec \phi', \quad v' = \sin \phi'$$

The unprimed quantities refer to one inertial frame, and the primed quantities refer to another inertial frame. The primed frame is moving with velocity v_0 from the unprimed frame. Usually we use the symbol β for v_0. This is the β of the Greek trio α, β, γ.

With the linkage (4.29), the (reflexive) *Kepler's equation* (4.28) is the (reflexive) *physical velocity addition formula*

$$v = \frac{v_0 + v'}{1 + v_0 v'} \quad (4.30)$$

This is the *law of the composition of velocities*, namely, if body B moves from body A with velocity v_0, and if body C moves from body B with velocity v', then body C moves from body A with velocity v given by equation (4.30).

Similarly, there is a *law for the composition of Doppler factors*. It can be obtained by this reasoning. A signal sent from A to C is equivalent to a signal sent from A to B and then sent from B to C. If the Doppler factor from A to C is k, from A to B is k_0, and from B to C is k', then it follows that

$$k = k_0 \, k' \quad (4.31)$$

This is *Roemer's multiplicative equation* for the composition of Doppler factors.

We now want to show that Roemer's equation (4.31) is equivalent to Kepler's equation (4.30). We will do this by showing that they are both derived from *Wright's additive equation* for the Mercator projection

$$u = u_0 + u' \quad (4.32)$$

As we will soon demonstrate, this equation of Wright is the most basic equation that contains the symmetry of the Poincaré group. It gives the essence of the space-time makeup of nature.

Exponentiate Wright's additive equation (4.32). The result is

$$e^u = e^{u_0} \, e^{u'}$$

which we recognize as Roemer's multiplicative formula (4.31). Conversely, if we take the logarithm of Roemer's formula (4.31), we obtain Wright's equation (4.32).

The relationship is simply the logarithmic relationship between the Mercator and stereographic projections, which we have already noted.

In terms of angles, Roemer's equation (4.31) becomes

$$(\sec \phi + \tan \phi) = (\sec \phi_0 + \tan \phi_0)(\sec \phi' + \tan \phi') \qquad (4.33)$$

Similarly, the inverse of Roemer's equation

$$k^{-1} = k_0^{-1} k'^{-1}$$

becomes

$$(\sec \phi - \tan \phi) = (\sec \phi_0 - \tan \phi_0)(\sec \phi' - \tan \phi') \qquad (4.34)$$

If we take one-half of the difference of equations (4.33) and (4.34), we obtain an equation for $\tan \phi$. If we take one-half of the sum of equations (4.33) and (4.34) we obtain an equation for $\sec \phi$. The final results (after some algebra) are

$$\begin{aligned} \tan \phi &= \sec \phi_0 \tan \phi' + \tan \phi_0 \sec \phi' \\ \sec \phi &= \sec \phi_0 \sec \phi' + \tan \phi_0 \tan \phi' \end{aligned} \qquad (4.35)$$

To put these two equations in a form with which we are familiar, let us introduce the Greek trio α, β, γ, that is,

$$\alpha = \cos \phi_0, \qquad \beta = v_0 = \sin \phi_0, \qquad \gamma = \sec \phi_0, \qquad \gamma\beta = \tan \phi_0$$

Using the linkage (4.29), equations (4.35) thus become

$$\begin{aligned} x &= \gamma(x' + \beta t') \\ t &= \gamma(t' + \beta x') \end{aligned} \qquad (4.36)$$

These are the (reflexive) *Lorentz equations*, which provide the space-time shear of the Poincaré group. Finally, Kepler's equation (4.30) follows from equations (4.36) by dividing the first equation by the second equation. (See Appendix 1.)

Let us summarize. The most basic equation that expresses the shear symmetry of space-time is Wright's additive equation (4.32). From this, the following equations can be derived: the Roemer multiplicative formula (4.31) for the composition of Doppler factors, the Lorentz transformation (4.36), and the Kepler formula (4.30) for the composition of physical velocities. All these are but different ways of looking at the same space-time structure.

Let us use an analogy. Suppose that you have to guess the shape of the foundation of a building. You are told that the building will sit on the same foundation if the building is rotated 90° or by any multiple of 90° around its central point. This rotation symmetry constitutes valuable information, because it immediately narrows down the possibilities for the shape to a square (i.e., 4-gon), or to an octagon (i.e., 8-gon), and so on. The simplest guess is the square. Likewise, the knowledge that nature imposes a space-time symmetry narrows down the mathematical possibilities for the shape of physical laws. Ockham's razor says to choose the simplest possible mathematical form that meets all the physical requirements. The simplest possible mathematical model of space-time symmetry is the Poincaré group. The choice of the Poincaré group corresponds to the choice of the square in the analogy.

All the laws of physics must possess this symmetry. The necessity of space-time symmetry in nature serves as a powerful constraint on the forms of these laws. Maxwell's equations are so intricately interrelated by this symmetry that, given one of the equations, the others can be deduced. The symmetry that unites space and time into space-time is the same symmetry that unites electric and magnetic fields into the electromagnetic field. We cannot have an equation standing alone describing, say, the variation of the electric field in space. That equation can be only one piece of the unified Maxwell equations describing the electromagnetic field in space-time. If you are told that the shape of a building is the simplest shape with a 90° rotation symmetry, you can deduce that the building is square, and you can design it from a blueprint of one wall. In the case of Maxwell's equations, the situation is mathematically more complex, but the guiding principle is the same.

In our historical treatment of special relativity, we have traced back its development even to the stereographic projection of the ancient Greeks. In fact, the stereographic projection is mathematically equivalent to the relativistic Doppler factor upon which the symmetry of space-time can be based. Over the centuries, scientists and mathematicians perceived that nature follows certain patterns, and physical laws were formulated to describe them. The paths followed were generally inductive in content. A large collection of experimental facts were required to be summarized for the formulation of each of these laws which eventually were expressed by equations. Chief among these were Newton's laws of motion and Maxwell's equations. These equations revealed a symmetry in nature's design. The most simple mathematical model of this symmetry is the Poincaré group, which includes the Lorentz transformation. This symmetry, once revealed, could then be used to follow paths that are generally deductive in content to find new equations. The best example of an equation produced by such deductive reasoning is Einstein's equation $E = mc^2$ for the equivalence of mass and energy. This equation was actually not experimentally verified until many years later. The verification in the form of nuclear power made the general public aware of Einstein's theory in a decisive way and made his equation the most famous mathematical equation of all time.

Objects are said to be symmetric when they can be rearranged or manipulated without changing their overall appearance. Look at a solid brick wall. Suppose the bricks in the wall were rearranged. There is no way to tell that this happened, because the new wall looks the same as the old. The bricks are symmetric. The new arrangement and the old arrangement are completely interchangeable.

When we say a physical law is symmetric we mean that the pertinent physical entities can be manipulated without changing the import of the law. Basic tenets are that physical nature is built from indistinguishable parts and that physical laws are local. This locality simply means that the influence of bodies far enough away is small. Both the indistinguishable and local properties come from empirical findings and not theory. Because of these properties, we can create or isolate identical systems such as the inertial frames we speak about in relativity theory. It is not necessary to manipulate the entire physical universe because we have available identical isolated systems which we can deal with and compare.

For example, in relativity theory, we speak of physical experiments in a given inertial frame, such as an isolated space vehicle. Suppose that the astronaut does extensive physical experiments within the space ship. The results of every experiment are reported to Earth, such as the lifetimes of radioactive particles and the spectra of atoms and molecules. In this case, what is meant in the statement that physical laws are symmetric? In essence, it means that no matter how many facts are reported about conditions within the space ship, there are important other facts about the ship that cannot be deduced by any of these reported facts.

In this book we are concerned with the symmetries imposed on the space-time continuum by the Poincaré group of transformations. One important symmetry of this group is the symmetry of physical laws under *translation in space*. This symmetry means that the position of the space vehicle cannot be established from the reported description of conditions within the vehicle. In other words, space vehicles that differ merely in location in space are interchangeable. These vehicles cannot be distinguished on the basis of even the most complete possible internal description of each vehicle.

Another symmetry of the Poincaré group is the symmetry under *rotation in space*. The space vehicle may be pointed in any direction in space. This symmetry means that it is not possible to deduce, from the most complete internal description, how the ship is oriented in space. The symmetry under *shear in space-time* of the Poincaré group means that even from a complete description of conditions within the ship, it is not possible to determine its velocity.

Let us now return for a moment to the brick wall analogy. Suppose that one brick is taken out of the brick wall. It is examined in detail, so every fact about it is known. It is then put back into the wall. From the reported facts we cannot tell where it is in the wall. This is what is meant by the symmetry of isolated bricks in the wall. To know what brick it was, we must know about facts outside of the brick itself, that is, facts that can be used to tell the location of this specific brick in the wall. In a similar manner, an astronaut must report facts about events outside of the space ship itself, such as its coordinates with respect to the Earth, planets, and stars, for its position in space to be known on Earth. The meaning of symmetry is that these outside facts cannot be obtained from information gathered within the isolated environment of the space ship itself. Certain knowledge of things outside the ship is required to determine its location, orientation, and velocity with respect to Earth.

Although the idea of symmetry is basic to physics, its formulation as a fundamental principle did not come easily. The ancient Greeks did not even recognize the rotation symmetry, that is, the symmetry of different directions in space. Because of gravity, the vertical direction on Earth seems favored. The realization that the three dimensions of space are equivalent came only after considerable abstraction and imagination. Because of friction, a moving body on Earth slows down unless a motive force is applied. The concept of inertia, that is, the fact that a body travels at a constant velocity in a straight line in the absence of external forces, took the geniuses of Galileo and Newton to formulate. It took most of civilized time for people to realize that such things as the Earth's gravitational field and friction hide the basic symmetries of nature. In conclusion, it is the recognition of the essential role which symmetry plays that makes relativity theory important not only for physicists but also for anyone interested in the intrinsic beauty of nature.

The beauty of space-time symmetry is seen in the relativistic Doppler factor or, equivalently, the stereographic projection of the ancient Greeks. If we take the Roemer multiplication equation

$$k = k_0 k'$$

for the composition of relativistic Doppler factors as fundamental in expressing the symmetry of the Poincaré group, then the following derived equations also express the same symmetry.

Wright: $\log k = \log k_0 + \log k'$ (addition equation)

Lorentz: $\dfrac{k - k^{-1}}{2} = \dfrac{k_0 k' - k_0^{-1} k'^{-1}}{2}$ (offset equation)

$\dfrac{k + k^{-1}}{2} = \dfrac{k_0 k' + k_0^{-1} k'^{-1}}{2}$ (midpoint equation)

Kepler: $\dfrac{k - k^{-1}}{k + k^{-1}} = \dfrac{k_0 k' - k_0^{-1} k'^{-1}}{k_0 k' + k_0^{-1} k'^{-1}}$ (ratio equation)

We close this chapter with the words of William Blake[8]

Tyger! Tyger! burning bright
In the forests of the night,
What immortal hand or eye
Could frame thy fearful symmetry?

---

[8] Blake, William. "The Tyger." From *The Poetical Works of William Blake*, Chatto & Windus, London, 1906.

Appendix 1: Derivation of Physical Velocity Addition Formula from Lorentz Transform

We know that the speed of light must have the same value in every inertial frame, no matter what the relative velocities are between the frames. Let us consider two inertial frames K and K' such that K' moves with a velocity β in the x direction relative to K. Let us now consider an object in uniform rectilinear motion, and let its velocity be v as measured in frame K, and v' as measured in frame K'. We suppose that frames K and K' have the same origin O and that the world line of the moving object passes through the origin O. Another point on this world line represents an event with coordinates x, t in the K frame and coordinates x', t' in the K' frame. In terms of these coordinates, the velocities of the moving object in the K frame and the K' frame are simply the average velocities (Figure 4.13)

$$v = \frac{x}{t}, \qquad v' = \frac{x'}{t'}$$

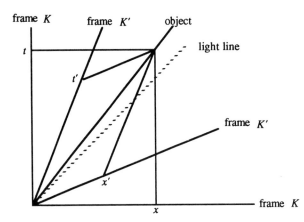

Figure 4.13. Minkowski diagram of two inertial frames with relative velocity β.

However, we know that the coordinates are related by the Lorentz transformation

$$x' = \gamma(x - \beta t)$$
$$t' = \gamma(t - \beta x)$$

where $\gamma = 1/\sqrt{1 - \beta^2}$. If we divide these two equations, the γ cancels out, and we are left with

$$\frac{x'}{t'} = \frac{x - \beta t}{t - \beta x}$$

We now divide both the numerator and denominator of the right-hand fraction by t, and thus obtain

$$\frac{x'}{t'} = \frac{(x/t) - \beta}{1 - \beta(x/t)}$$

Recognizing x'/t' as v', and x/t as v, we see that this equation is

$$v' = \frac{v - \beta}{1 - v\beta}$$

This formula is the *physical velocity addition formula*, or *Kepler's formula*.

The speed of light in the K frame is $c = 1$. What is the speed of light in the K' frame? We let $v = c = 1$ in the physical velocity addition formula:

$$v' = \frac{1 - \beta}{1 - \beta} = 1$$

Thus, light has the same speed in both reference frames. Consequently, the constancy of the velocity of light is maintained by the physical velocity addition formula.

As another illustration, consider an object in the K frame moving with velocity $v = 1/2$. Let K' move relative to K with velocity $\beta = -1/2$. The speed of the object relative to K' is then

$$v' = \frac{(0.5) - (-0.5)}{1 - (0.5)(-0.5)} = 0.8$$

We thus see that v' is less than the velocity of light ($c = 1$).

Appendix 2: Derivation of the Relativistic Doppler Factor from Lorentz Transform

Consider a light wave with wavelenght x_0 and period t_0 in some inertial frame. Wavelength and period are related by the equation $x_0 = ct_0$, or $x_0 = t_0$, which is the light line. Consider the same light wave in a second frame moving at relative velocity β with respect to the first frame. In the second frame let λ denote the wavelength and τ the period of the given light wave. The event x_0, t_0 in the first frame and the event λ, τ in the second frame are related by the Lorentz transform

$$\lambda = \gamma(x_0 + \beta t_0)$$
$$\tau = \gamma(t_0 + \beta x_0)$$

Since $x_0 = t_0$ these equations become

$$\lambda = \gamma(1 + \beta)x_0$$
$$\tau = \gamma(1 + \beta)t_0$$

But the factor $\gamma(1 + \beta)$ is the relativistic Doppler factor k; that is,

$$k = \gamma(1 + \beta) = \sqrt{\frac{1 + \beta}{1 - \beta}}$$

Thus the Lorentz transform gives the equations

$$\lambda = kx_0$$
$$\tau = kt_0$$

for the Doppler shift.

CHAPTER 5

MICHELSON-MORLEY EXPERIMENT

To me every hour of the light and dark is a miracle
Every cubic inch of space is a miracle.

Walt Whitman (1819–1892)
Miracles

THE AETHER

In theory at least, the three forms of government, monarchy, democracy, and aristocracy, cannot coexist in the same country at the same time. Monarchy is government by the monarch, a sole and absolute dictator. Democracy is govern-

Plate 5.1. Albert A. Michelson (1852–1931)

ment by the people under principles of social equality and respect for the individual. Aristocracy is government by the nobility or by a privileged minority or upper class.

Yet, in practice, we know that these forms of government do coexist in political systems. So, too, with the special theory of relativity. Light is the monarch. There is no inertial frame in which light is at rest. The monarchy is expressed by the mathematical fact that light is perfective. In physics, this is called the light principle, namely, light travels in every inertial frame at the ultimate signal velocity (equal to unity in natural units), no more and no less.

The various inertial frames represent a democracy in special relativity. Each inertial frame sees any other inertial frame from the same point of view. This democracy is expressed by the mathematical fact that inertial frames are perspective. In physics, this is called the relativity principle, namely, all the laws of physics are the same in every inertial frame.

To understand these two principles better, let us go back into history. Christian Huygens in the seventeenth century said that light was wave motion. The water waves that he observed on the canals of Holland are propagated as disturbances in the water. With this model, he postulated a "subtle fluid" or "aether" spreading throughout space in which light is propagated. The aether concept went back to Aristotle's perfect aether, that quintessence that composed the heavens. On the other hand, Newton had argued that light is a stream of little particles, and under the weight of Newton's authority, the wave theory of light fell into oblivion. However, in the early nineteenth century, the wave theory together with the "aether" was resurrected by Thomas Young and Augustin Fresnel. Since all other kinds of waves, including sound waves, can be transmitted only in a physical medium, it seemed natural to suppose that light waves would also require some transmitting medium.

Unlike sound, however, light can travel through a vacuum; it can travel faster, in fact, than through air or water. Thus the light-bearing aether had to be radically different from any known physical substance. It was assumed that the aether is at rest in Newton's absolute space; so as far as the propagation of light is concerned the inertial frames were not all equivalent. Instead, the aether frame was specially privileged and thus represented the monarch. At the time of Maxwell's work in the 1860s, the theory of light might be described in terms of monarchy: one privileged frame stood apart from all the other inertial frames.

In Newtonian mechanics, all states of uniform straight-line motion are mechanically equivalent to a rest frame in absolute space. Thus in Newtonian mechanics there is no privileged frame. This is the content of Newton's principle of relativity, which says that all the laws of mechanics are the same in every inertial frame. Thus Newtonian mechanics may be described in terms of a democracy. The situation where light represented a monarchy and mechanics a democracy, both standing side by side in theoretical physics in the 1860s, represented a fatal crack in the structure that had to be repaired.

The first step in resolving this difficulty would be a search for the aether frame through measurements on the velocity of light. All through the middle years of the nineteenth century, there were many attempts both experimental and theoretical to detect the luminiferous aether and to determine its properties. These early experiments had failed to come up with any evidence for the existence of the aether. However, with the introduction in 1864 of James Clerk Maxwell's celebrated equations, new hopes were raised that the Earth's motion through a stationary aether could be detected by a suitable optical experiment. Maxwell had shown that light could be described as an electromagnetic wave. But light could travel through space, and the regions between the planets and stars seemed empty of any medium to carry the waves. Yet we see the sun and stars. The accepted explanation in Maxwell's time was that outer space was filled with an extremely fine, imponderable substance, the aether, which is the carrier or medium of these electromagnetic waves. As we have seen in Chapter 1, Maxwell had successfully computed the speed of light from the electromagnetic constants as $c = 1/\sqrt{\mu_0 \varepsilon_0}$. But what was this speed to be measured against? Speeds are always relative to measuring posts and instruments which make up the reference frame.

Nineteenth-century physicists expected that, if they measured the speed of light taking the Earth as the frame of reference, they would obtain varying results. The reason for their conclusion was that the Earth is constantly changing its velocity with respect to the fixed stars. Imagine we are measuring the speed of light coming from a fixed star. At the moment we make our measurement, the Earth occupies some position in its orbit around the sun and is moving at its orbital speed of 30 km/s. But six months later, the Earth would be halfway around the sun and moving in the opposite direction. With the fixed stars as the aether reference frame, the

Plate 5.2. David Peck Todd (1855–1939)

relative velocity of that frame with the Earth frame would depend upon the Earth's position in orbit and the speed of the Earth about the sun, which is 30 km/s. This speed in natural units (where $c = 1$) is $\beta = 30/300{,}000 = 1/10{,}000$.

The term "fixed star" is derived from the common experience that the relative position of the stars remains unchanged, in marked contrast to the "wandering stars," the planets. Greek astronomers were by no means convinced, however, that the apparent invariability of the positions of the fixed stars was in fact mathematically accurate. As we know from Macrobius (around 400 A.D.), there was a school of astronomers who thought that only the vastness of the universe and the length of time prevent observing the motion of individual stars. But only modern astronomy could furnish the proof of the correctness of the ancient hypothesis. There are some 50 stars known whose proper motion (i.e., motion in a direction perpendicular to our line of sight) exceeds 1 second of arc per year. Sirus has a proper motion of about 1.33 seconds per year. Thus the displacement of Sirus during the past two millennia amounts to about 40 minutes of arc (or two-thirds of a degree).

In 1879, James Clerk Maxwell wrote an acknowledgment of some astronomical tables he had received from the American astronomer David Peck Todd. These tables contained many observations of the planet Jupiter. Maxwell was interested in measuring the velocity of light in various reference frames and described in his letter an ingenious method based upon observations of the eclipses of Jupiter's moons. We recall that many years earlier Roemer was the first to detect the velocity of light by studying the time lag in observing these eclipses. Unfortunately, the astronomical data available to Maxwell were not accurate enough for use in Maxwell's method. Maxwell's method was based upon one-way times from sources to distant receivers. In his letter to Todd, Maxwell remarked that this distinguished it from methods based upon two-way times, that is, methods for which the source and receiver are at the same point, and the signal is returned by reflection. In other words, a two-way experiment is one that uses a beam of light that is returned to its starting point. The two-way optical experiment would have to be capable of measuring the extremely small but finite quantity represented by the square of the ratio of the Earth's orbital speed V to the speed of light; in symbols the quantity is $(V/c)^2 = (30/300,000)^2 = (1/10,000)^2$, which represents 1 part in 100 million. These two-way experiments could be performed on Earth, but with current instrumentation Maxwell remarked that the results would be undetectably small because no known optical device could approach this sensitivity. Maxwell died that same year, and in respect and admiration for him, Todd had the letter published in *Nature*, Vol. 21, 1880.

Maxwell's letter was read by A. A. Michelson. The part of the letter that particularly attracted Michelson's attention was the statement in the final paragraph that "in all terrestrial methods of determining the velocity of light, the light comes back along the same path again, so that the velocity of the Earth with respect to the aether would alter the time of the double passage by a quantity depending on the square of the ratio of the Earth's velocity to that of light, and this is quite too small to be observed." Michelson immediately began thinking about better instrumentation to achieve Maxwell's two-way experiment. In 1881, Michelson had some results. In 1887, he performed a refined version of the experiment working in collaboration with E. W. Morley. The Michelson-Morley experiment, as far as the textbooks are concerned, is regarded as the main experimental pillar of special relativity, and almost invariably it makes up the lead-off discussion. The purpose of this chapter is to carry on this tradition, but with a somewhat different path than usual.

THE MICHELSON-MORLEY EXPERIMENT

Michelson played a leading role in developing methods of interferometry. He constructed instruments of extraordinary sensitivity that displayed the interference of light waves with one another. If a train of light waves is split into two and subsequently recombined, Michelson's instruments could reveal minute differences in the travel times of these two split wavetrains.

The basic idea of Michelson's experiment can be described in terms of two swimmers of equal ability in a river. Each takes a different course. The boy swims down the river and back, the girl swims across the river and back. Both courses are the same length, so if there is no current, both swimmers will take the same time. However, suppose instead that the current is flowing down the river. Now the time it takes to swim each course will be increased. First, consider the course downstream and back. On the way down, the speed of the boy will be increased by the speed of the current, whereas on the way back upstream, it will be decreased by the same amount. Since the boy will travel at the lower speed for a longer time than at the higher speed, the net effect will be that his time will be increased. Next consider the course across the river and back. The girl taking this course will also take a longer time than before, since she must now swim into the current on each leg by a certain amount in order not to be swept down the river. However, her average speed turns out to be higher than that of the boy, so the girl, although delayed, still returns before the boy. In consequence, when there is a current, the two swimmers of equal swimming speed will return at different times despite the fact that both swim the same distance.

Now let us make the analogy. The river banks represent the aether frame, the swimmers represent light, and the river water represents the Earth frame. The current corresponds to the aether wind. If the Earth is at rest in the aether (no current in the river), then light would take the same time to travel two perpendicular courses of equal length. However, if the Earth is traveling through the aether (current in the river), then light would take a longer time to travel the course parallel to the wind than the equilength course across the wind.

Michelson's apparatus was designed to detect the effects of the aether wind which he thought would result from the Earth's motion with respect to the fixed stars. It consisted of a pair of arms at right angles. Light was split by a half-silvered mirror and sent down each of the arms. After reflection by mirrors set at the end of each arm, the light signals were recombined by the half-silvered mirror and redirected into a telescope through which interference fringes could be seen.

Although $(V/c)^2$ is a small quantity to measure, still in 1887 it was within the grasp of Michelson's interference techniques. Therefore, if the fixed stars were the reference frame of the aether, Michelson knew that his instruments were accurate enough to detect it. With this assurance, he set about to find the unknown aether frame.

Now that we have described the idea of the Michelson-Morley experiment, let us translate the verbal description into mathematics. Imagine a river in which there is a current flowing with velocity β in the direction indicated by the arrow (Figure 5.1).

Now, which of two equally fast swimmers would win when (1) the boy goes from A to B and back to A and (2) the girl goes from A to C and back to A? We denote the distances by $\lambda = AB$ and $\lambda' = AC$. The line AB is parallel to the current, and line AC is perpendicular to it. Let each person's speed of swimming in still water be c. We assume c is expressed in units which make c come out to be equal to one ($c = 1$). Then by the addition theorem for apparent velocities, when the boy swims with the current from A to B, his apparent velocity would be $c + \beta$, whereas when he swims against the current from B to A, his apparent velocity would be $c - \beta$. (We assume that β is less than $c = 1$, for otherwise he could not swim against the current.) Therefore, the time required for him to swim from A to B would be $\lambda/(c + \beta)$, where λ represents the distance AB. The time required for him to swim from B to A would be $\lambda/(c - \beta)$. Consequently, the

The Michelson-Morley Experiment

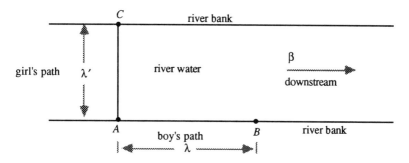

Figure 5.1. The river bank represents the aether frame, the river water represents the Earth frame, the swimmers represent light, and the current corresponds to the aether wind.

time required for his round trip from A to B and back to A would be

$$t_B = \frac{\lambda}{c+\beta} + \frac{\lambda}{c-\beta} = \lambda \frac{c-\beta+c+\beta}{(c+\beta)(c-\beta)}$$

$$= \frac{2\lambda c}{c^2-\beta^2} = \frac{2\lambda}{1-\beta^2} \quad \text{(since } c=1\text{)}$$

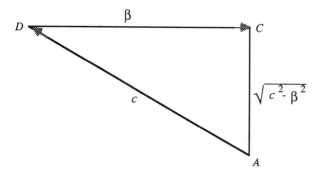

Figure 5.2. The effect of the current is to reduce the velocity of the girl swimmer in a passage across the river (AC) to the value $\sqrt{c^2-\beta^2}$.

Next, let us see how much time it takes for the girl to make the round trip from A to C and back to A (Figure 5.2). If she heads directly toward C, the current would carry her downstream, and thus she would land at some point to the right of C. To arrive at C, she should head upstream just enough to counteract the effect of the current. We can think of this action of the girl as follows. The water is kept still until she swims at her own speed c from A to D, and then the current is suddenly allowed to operate, carrying her at speed β from D to C (without her making any further effort). The effect would be the same as if she swims directly from A to C with a velocity equal to $\sqrt{c^2-\beta^2}$ as shown by the right triangle in Figure 5.2. Consequently, the time required for her to go from A to C would be $\lambda'/\sqrt{c^2-\beta^2}$, where $\lambda' = AC$. Likewise, in going back from C to A, we see, by the same line of reasoning, the time again would be $\lambda'/\sqrt{c^2-\beta^2}$. Thus, the time for her round trip from A to C and back to A is

$$t_G = \frac{2\lambda'}{\sqrt{c^2 - \beta^2}} = \frac{2\lambda'}{\sqrt{1 - \beta^2}} \quad \text{(since } c = 1\text{)}$$

To compare his time t_B and her time t_G more readily, let us make use of γ, which we define as

$$\gamma = \frac{1}{\sqrt{1 - \beta^2}}$$

Since $\beta < 1$, it follows that $\gamma > 1$. We have

$$t_B = 2\lambda\gamma^2, \quad t_G = 2\lambda'\gamma$$

If the condition $\lambda = \lambda'$ is imposed (i.e., both courses of equal length), then $t_B = (2\lambda\gamma)\gamma = t_G\gamma$. Thus, $t_B > t_G$, which says that it takes longer for the boy to swim downstream and back than the girl to swim the same distance across the stream and back.

Let us now discuss the Michelson-Morley experiment. In the experiment, a light beam (photon) is sent from A to B and also from A to C. At B, there is a mirror which reflects the light back to A so that the photon makes the round trip from A to B and back. At C, there is another mirror which also reflects the light back to A so that another photon makes the round trip from A to C and back. Because the entire apparatus is at rest on the Earth, it shares the inertial frame of the Earth. Let us now consider the inertial frame of the aether. Let the relative velocity between these two frames be given by β, where this velocity is in the direction of the line AB. Thus, line AC will be perpendicular to the direction of the relative velocity.

In modern terminology, Michelson wanted to find the velocity β between the Earth frame and the aether frame. If we use the analogy of the swimmer in which the swimmer becomes a photon moving with the velocity $c = 1$ of light, then time t_B gives the boy's round trip time from A to B and back, and time t_G gives the girl's round trip time from A to C and back. Thus, if this analogy were true, then

$$t_B = 2\lambda\gamma^2, \quad t_G = 2\lambda'\gamma$$

If the condition $\lambda = \lambda'$ could be guaranteed, then

$$t_B = 2\lambda\gamma^2, \quad t_G = 2\lambda\gamma$$

Michelson's intent was first to find t_B and t_G experimentally by his interferometry techniques. Then dividing t_B by t_G he would obtain the numerical value of γ, because from the foregoing expressions for t_B and t_G we have

$$\frac{t_B}{t_G} = \gamma$$

Because γ is given by the formula

$$\gamma = \frac{1}{\sqrt{1 - \beta^2}}$$

the value of β could then be calculated as $\beta = \sqrt{1 - (1/\gamma)^2}$. Thus Michelson would be able to compute the velocity β of the Earth relative to the aether. Such was the plan of his experiment. For small β, the factor γ has the approximation $\gamma = 1 + \beta^2/2$. Thus Michelson was searching for an effect that, with respect to 1, is of the order of $\beta^2/2$, that is, $(V/c)^2/2$, or $(30/300,000)^2/2$. This is the quantity that Maxwell thought was too small to be detected. Now we ask ourselves what actually did happen?

Michelson, when he performed the experiment, found that the empirical values of t_B and t_G were the same numbers, instead of t_B being greater than t_G as given by the swimmer analogy. This was a most disturbing result, one that Michelson could never accept, so he kept repeating the experiment over the years with more accurate apparatus. However, he kept finding that $t_B = t_G$, which to him was a negative (a bad) result, called a null result.

THE NULL RESULT AND THE AETHER

The null result of the Michelson-Morley experiment greatly troubled the world of physics. However, in 1889 G. F. FitzGerald made the crucial breakthrough (*Science*, 1889; *Philosophical Transactions of the Royal Society*, Vol. 184, pp. 727–804, 1893). With remarkable insight, FitzGerald said that the negative result was due to the real or apparent effect of the motion through the aether on the size of the apparatus. That is, the apparatus in moving through the aether frame contracted in the direction of motion precisely enough to nullify the expected result ($t_B = \gamma t_G$) of the Michelson-Morley experiment to give instead the observed result ($t_B = t_G$). This FitzGerald contraction factor is denoted by the symbol α (Greek letter alpha).

FitzGerald's insight provided the ultimate reason why the Michelson-Morley experiment failed in the inertial frame of the aether. Relative to the aether, the Michelson apparatus on the Earth is in motion. Here we assume that the apparatus is at rest in the Earth frame and that the experiment is staged in the aether frame. Because the apparatus is at rest on the Earth, each of the two perpendicular distances represents a proper distance. We assume that these two distances are equal and denote them each by λ'. In the frame of the aether, FitzGerald said that the distance λ' in the direction of motion contracts to λ, which is less than λ'. Because the FitzGerald contraction factor is denoted by α, the FitzGerald result can be expressed as

$$\lambda = \alpha \lambda'$$

However, in the frame of the aether, the distance λ' perpendicular to the direction of motion does not shrink. Thus

$$t_B = 2\lambda\gamma^2 = 2\alpha\lambda'\gamma^2$$

$$t_G = 2\lambda'\gamma$$

The Michelson-Morley null result says that $t_B = t_G$, which is

$$2\alpha\lambda'\gamma^2 = 2\lambda'\gamma$$

so the FitzGerald contraction factor is

$$\alpha = \frac{1}{\gamma} \quad \text{or} \quad \alpha = \sqrt{1 - \beta^2}$$

Thus the negative result that Michelson found is explained by the FitzGerald contraction factor α. Because of the null result, there is no way that one can find the velocity β of the Earth with respect to the aether by means of the Michelson-Morley experiment.

The Larmor time dilation factor appears as a consequence of the FitzGerald contraction. In the Earth frame, the common travel time for each leg is

$$t' = 2\lambda'c = 2\lambda'$$

whereas in the aether frame the common travel time is

$$t = t_B = t_G = 2\lambda'\gamma$$

Therefore

$$t = \gamma t'$$

Because $\gamma > 1$, this equation is the time dilation equation.

THE RECOGNITION PRINCIPLE

The Michelson-Morley experiment in 1887 played the crucial role in the overthrow of the classical aether theory and in the genesis of the theory of relativity. The experiment was designed to determine the absolute motion of the Earth through the privileged inertial frame, the aether frame. However, the null result of the experiment could not be explained in terms of classical physics. The impasse was finally broken in 1889 when FitzGerald introduced an entirely new concept, the idea that the length of a moving rod contracts in the direction of motion. This was the conceptual beginning of special relativity, for the FitzGerald contraction model is the starting point in the explanation of every observable special relativistic effect known. In particular, it is the explanation of why the muon in its rest frame sees the moving Earth's atmosphere contracted. It also explains the dilation of time, that is, why the moving muon appears to live longer than a muon at rest.

In the mathematics of the Michelson-Morley experiment, the other frame does not have to be the aether, but in fact it can be any other frame, such as that of the fixed stars or even a muon, in constant relative motion with respect to the Earth. In this regard, the aether frame, if it exists, is unobservable. In this way, the aether is deposed from its monarchy of a privileged frame. With no privileged frame, it follows that the relativity principle applies not only to mechanics but to all laws of physics. Thus, the monarchy of light in classical physics is replaced by a democracy in which all inertial frames are of equal status. However, the deposed monarch, light, now without its privileged frame of the aether, finds refuge in no frame at all, the vacuum. Belonging to no frame, light has the same velocity in all inertial frames. This is the light principle, which reestablishes the monarchy of light but in a different guise.

By completely omitting the concept of the aether, the essential difference between the propagation of electromagnetic waves (light waves, radio waves, etc.), on the one hand, and material waves (sound waves, seismic waves, water waves, etc.), on the other hand, is clearly brought out. Whereas matter is not required for the propagation of electromagnetic waves, matter (air, rocks, water, etc.) is required for the propagation of material waves. Electromagnetic waves can propagate in vacuum, that is, in the absence of matter possessing rest mass. Thus, in current physical reasoning there is no reason for the unobservable material medium known as the aether at all, because it actually confuses the essential difference between electromagnetic waves and material waves.

Now we can see why electromagnetic waves play the essential role in relativity theory, whereas no material wave could play such a role. We say that a wave propagates isotropically if it travels with the same velocity in all directions. A material wave can propagate isotropically only in the rest frame of its medium. Let the rest frame of the medium be K, and let frame K' have relativity velocity β with respect to K. In frame K, the medium is at rest, so the material wave propagates with the same speed to the right and to the left. However, in frame K', these speeds will be different. In contrast, the vacuum has no rest frame, so electromagnetic waves propagate isotropically (the same speed c in all directions) in every inertial frame. There is no privileged frame for an electromagnetic wave (such as light) as there is for a material wave (such as sound). This is the light principle.

The light principle is the most difficult aspect of special relativity. In classical physics light was a monarch with a kingdom, the privileged frame of the aether. In special relativity, light is a monarch without a kingdom. Light has a unique characteristic velocity in spite of the fact that there is no underlying materialistic medium in which it travels. Light travels in vacuum, which means that it travels in all inertial frames with the same ultimate velocity. In this sense, special relativity has reinstated light as a monarch, but in the contradictory sense of both having no kingdom and all kingdoms.

Newton's first law of motion is called the law of inertia. It states that a body in uniform motion in a straight line will maintain that motion unless an external resultant force acts upon it. We are faced with the problem of understanding how it is that a force-free massive body manages to maintain its velocity for all future time, cut off from all contact with its surroundings. This fundamental problem is closely analogous to the problem of the propagation of light. In this case, the problem is to understand how light can travel at a constant velocity in vacuum, cut off from any underlying medium to support the wave motion.

In the first section of this chapter, we introduced the systems of monarchy, democracy, and aristocracy. We know that all the requirements for the special theory of relativity are contained in the relativity principle and the light principle. Why do we need to drag the aristocracy into the picture? Why do we not just say that the special theory of relativity is a case of monarchy and democracy coexisting in physics, as we do know that they coexist in politics such as in the constitutional monarchies of Europe. The correct answer is yes, the special theory of relativity is the coexistence of monarchy and democracy.

Unfortunately, the correct answer is not going to allow us to understand the special theory of relativity, any more than we can account for the true workings of a government by completely omitting the consideration of a privileged upper class, the aristocracy. So let us admit it: in special relativity, all inertial frames are equal, but some are more equal than others. This aristocracy is expressed by the mathematical fact that certain inertial frames are rendered distinctive due to some contrived imbalance. This new principle, not usually explicitly mentioned, can be called the recognition principle in physics, namely, symmetrical physical relationships must be broken into at some point to make certain inertial frames distinguishable from the others in some sense.

In summary, the special theory of relativity can be summed up as follows:

Government	Mathematics	Physics
Monarchy	Perfective aspect	Light principle
Democracy	Perspective aspect	Relativity principle
Aristocracy	Distinctive aspect	Recognition principle

There is one inertial frame that is distinctive. It is the one defined by our planet Earth. We know that the Earth rotates about its axis once a day and revolves around the sun once a year, so it is always subject to acceleration. Thus, it is not an inertial system in the pure sense, but it is the best that we have. One way or another we have to be content with the Earth as an inertial frame, or better as a sequence of inertial frames in its journey around the sun.

On Earth, we are at rest in our home inertial frame. Out there, in space, everything is moving. Muons are crashing through the Earth's atmosphere at nearly the speed of light. But way out there are the fixed stars. Of course, they are not really fixed, but from the Earth they seem to be at rest in their own inertial frame. Thus, the aristocracy consists at least of these members, the inertial frame or frames defined by our planet, and the inertial frame defined by the fixed stars.

But still, the Earth frame is more aristocratic because we can physically measure distance and duration on Earth, but we have no way of ever being present on a fixed star. Nor can we ever be present on a muon on its trip through our atmosphere. We can only infer things about these other inertial frames. This represents the mathematical imbalance with which we must contend. Although the relativity principle says that the laws of physics are the same in all inertial frames (a democracy), the recognition principle says that our physical measurements are confined to only certain of the aristocratic inertial frames. Thus, the theory of special relativity uses any trick possible to elevate a democratic frame to aristocratic status. These are the tricks that must be learned to see what is going on in the special theory of relativity.

Appendix: Light Clocks and the FitzGerald Length Contraction

In this appendix, we want to reformulate the Michelson-Morley experiment in terms of light clocks. First, we want to consider a stick that is always perpendicular to the direction of motion. The stick moves with constant velocity β in this direction (like the mast of a ship moving horizontally on the sea). Using the principle of relativity, it can be shown that the stick neither stretches nor shrinks. The only possible outcome is that the moving stick has the same length as a stationary stick. We summarize this result as

> **Result 1.** A stick moving with a constant velocity in a direction perpendicular to itself has the same length as a stick at rest.

Next, we want to construct a special clock to find how fast-moving clocks run compared to stationary clocks. The clock is made from a stick with a mirror at each end, and a beam of light is made to flash back and forth between the two mirrors. Each time the bottom mirror is hit, the clock ticks. We measure a time duration by counting how many times the clock ticks.

Figure 5.3 shows such a light clock at rest. If the length of the stick is λ', then it takes time λ'/c for light to go from the bottom mirror to the top mirror, and it takes another λ'/c for light to go from the top mirror to the bottom one. Because we are using natural units, we have $c = 1$, so we can simply write each of these time durations as λ', the length of the stick. Thus a complete round trip takes a time $t' = 2\lambda'$. This is the time interval between ticks of the clock when the clock is at rest.

Now let this clock move with velocity β. The stick is perpendicular to the direction of movement, so by Rule 1 the length of the stick does not change. The question is: How long does it now take between the ticks? We let t be the length between ticks of the moving clock. The light spends half the tick time t going from the bottom to top mirror, and the other half going back from the top to bottom. Because the clock moves to the side during time t, the light has a longer

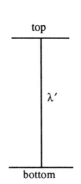

Figure 5.3. Light clock at rest.

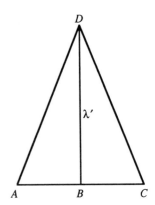

Figure 5.4. Light clock moving to the right.

distance to travel than it did when the clock was at rest. Therefore, the moving clock takes a longer time between ticks; that is, t is greater than t'.

Let us now calculate t. See Figure 5.4. In time $t/2$ the bottom of the stick moves from A to B, and in $t/2$ again from B to C. Since the stick moves at velocity β, the distance AB and also the distance BC are each equal to $\beta t/2$. In going from the bottom mirror to the top mirror, the light travels from A to D. But AD is the hypotenuse of a right triangle, one side of which is λ' and the other $\beta t/2$. Thus, by the Pythagorean theorem, we have

$$AD = \sqrt{\lambda'^2 + \left(\frac{\beta t}{2}\right)^2}$$

The distance light travels in going from the top mirror back to the bottom mirror is DC, which is equal to AD. Thus, the total distance d that light travels in making one round trip is $d = 2AD$. The time for light to travel the distance d is the time t between ticks, so $t = d/c$. Since $c = 1$, we have $t = d$, or

$$t = 2\sqrt{\lambda'^2 + \left(\frac{\beta t}{2}\right)^2}$$

We solve this equation to find the unknown quantity t. (Square the equation to remove the radical sign and then collect the terms in t^2. Solve for t^2 and then take the square root.) The result is

$$t = \frac{2\lambda'}{\sqrt{1 - \beta^2}}$$

Since we know that the time t' between ticks of the rest clock is $2\lambda'$, we have

$$t = \frac{t'}{\sqrt{1-\beta^2}}$$

Since $\sqrt{1-\beta^2}$ is less than 1, it follows that t is greater than t', as we previously asserted. Thus, the motion has dilated the time from t' to t. We have

Result 2: If a light clock takes time t' between ticks when it is at rest, then when its stick moves perpendicular to the motion at velocity β it takes the longer (or dilated) time $t = t'/\sqrt{1-\beta^2}$ between ticks.

Let us now relate the slowness of a clock to the time between ticks. We say that the slowness of a clock is proportional to the time between ticks. Because $t > t'$, we see that the clock runs slower when it is moving than when it stands still. The rate of a clock is inversely proportional to the time between ticks. Because $1/t$ is less than $1/t'$, we can say that an observer at rest finds that the rate of a clock moving past him is less than the rate of a clock at rest with him.

Up to this point we have used a special kind of clock, namely, a light clock whose stick is perpendicular to the direction of motion. However, experiments show that this result is true for any clock. In particular, it is true for a light clock whose stick is parallel to the direction of motion. Therefore, we have

Result 3: If a light clock takes time t' between ticks when it is at rest, then when the clock moves with velocity β, it takes the longer (or dilated) time

$$t = \frac{t'}{\sqrt{1-\beta^2}}$$

between ticks. This result does not depend upon whether the clock moves perpendicular or parallel to its length. Furthermore, this result is true for any clock whatsoever.

As we know, the premier experiment of special relativity was due in concept to Maxwell and carried out by a young professor at the Case School of Applied Sciences, Albert A. Michelson, and a chemist at Western Reserve College, Edward W. Morley. Both institutions are in Cleveland, Ohio, and are now merged into Case Western Reserve University. The results were announced in 1887. Michelson had the ability to conceive and build instruments of high precision on a grandiose scale that transcended the traditional idea of laboratory device. Physics research has benefited from the forward look of Michelson. Today for high-energy research it is necessary to build instruments on such a stupendous scale that it would make Michelson happy indeed.

The interesting aspect of the Michelson-Morley instrument is that it can be described as follows. It consists of *two perpendicular light clocks*. The two sticks form a right angle where they are joined together. As constructed, the two sticks had different lengths, but for pedagogical purposes we suppose they are equal, each being of length λ'. The Michelson-Morley experiment verified that both clocks kept the same time, regardless of their orientation in respect to the Earth's motion around the sun, or the sun's motion around the center of the galaxy, and so on. Their experiment results were the first empirical verification of Result 3.

Let us now do some analysis. We know that when the instrument is at rest, the two clocks tick at the same rate. This is also true when they are in motion. Now let one of the sticks move parallel to the motion, and necessarily the other stick moves perpendicular to the motion. Both clocks tick at the same rate. When the moving instruments pass a stationary observer, the perpendicular stick (by Result 1) has rest length λ'. We now want to derive again the FitzGerald contraction. Therefore, let us assume that, as yet, we do not know what the observer will measure as the length of the moving stick parallel to the direction of motion. We therefore assign the unknown value λ to designate this unknown moving length.

Our problem is to deduce the value of λ from the knowledge that the parallel stick clock keeps the correct time.

In Figure 5.5, we see the moving parallel stick clock at three different instants, namely, 0, t_1, and $t_1 + t_2$, measured by the observer at rest. At instant 0, the pulse of light leaves the left-hand mirror. At instant t_1 the pulse arrives at the right-hand mirror. Finally, at instant $t_1 + t_2$ the light arrives back at the left-hand mirror. Thus the total time elapsed between the first and third pictures is just the time t for a single tick, that is, $t = t_1 + t_2$.

First, let us find t_1 in terms of the unknown length λ and velocity β. In time t_1, light travels the distance d_1 from the left-hand mirror (in the first picture of the figure) to the right-hand mirror (in the second picture of the figure). This distance is equal to

$$d_1 = ct_1$$

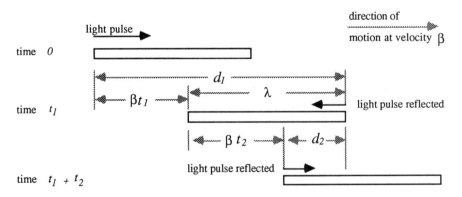

Figure 5.5. Three positions of a light clock moving in the same direction as the direction of its stick.

In time t_1, the left-hand mirror has moved a distance βt_1. Therefore, as seen in the figure,

$$d_1 = \lambda + \beta t_1$$

The two equations give

$$ct_1 = \lambda + \beta t_1$$

which gives

$$t_1 = \frac{\lambda}{c - \beta}$$

Next let us find t_2, the time required to go from the second to the third picture in the figure. In this time duration, the reflected light pulse travels to the left from the right-hand mirror (in the second picture in the figure) to the left-hand mirror (in the third picture in the figure). This distance d_2 is equal to the speed of light times the elapsed time t_2; that is,

$$d_2 = ct_2$$

The distance d_2 is made up of the difference of two parts. The larger part is the unknown length λ of the stick. The smaller part is the amount βt_2 that the left-hand mirror moves to the right during the time duration t_2. Their difference is

$$d_2 = \lambda - \beta t_2$$

The foregoing two equations give

$$ct_2 = \lambda - \beta t_2$$

which yields

$$t_2 = \frac{\lambda}{c + \beta}$$

We can now use the fact that the time of one tick is equal to the sum $t_1 + t_2$; that is

$$t = t_1 + t_2$$
$$= \frac{\lambda}{c - \beta} + \frac{\lambda}{c + \beta} = \frac{2c\lambda}{c^2 - \beta^2}$$

We have carried the c, but now let us use the fact that $c = 1$ (in natural units). Thus

$$t = \frac{2\lambda}{1 - \beta^2}$$

This expression gives the clock tick time t in terms of the unknown length of the stick moving in the direction of motion. Both measurements λ and t are made in the rest frame.

In Result 3, we found that if the stick has rest length λ', then the clock tick time for the moving clock is

$$t = \frac{t'}{\sqrt{1 - \beta^2}} = \frac{2\lambda'}{\sqrt{1 - \beta^2}}$$

These two equations for t therefore give

$$\frac{2\lambda}{1 - \beta^2} = \frac{2\lambda'}{\sqrt{1 - \beta^2}}$$

If we solve for the unknown length λ, we obtain

$$\lambda = \lambda'\sqrt{1 - \beta^2}$$

Here λ' is the length of the stick at rest, and λ is the length of the stick while moving as seen by an observer at rest. The factor $\sqrt{1 - \beta^2}$ is less than 1. The equation tells us that a stick moving parallel to its length shrinks by that factor. This shrinking of moving things as seen by an observer at rest is called the *FitzGerald contraction* and sometimes also the *Lorentz contraction*. We thus have

Result 4: A stick of length λ' (at rest) moving with a constant velocity β in a direction parallel to itself has the contracted length

$$\lambda = \lambda'\sqrt{1 - \beta^2}$$

when seen by an observer at rest.

The contraction factor is usually denoted by the Greek letter alpha:

$$\alpha = \sqrt{1 - \beta^2}$$

CHAPTER 6

WRIGHT'S EQUATION AND THE LORENTZ TRANSFORM

I was thinking the day most splendid,
   till I saw what the not-day exhibited;
I was thinking this globe enough,
   till there sprang out so noiseless around me myriads of other globes.
Now while the great thoughts of space and eternity fill me,
   I will measure myself by them.

Walt Whitman (1819–1892)
Night on the Prairies

TRIGONOMETRY

The ancient Greek civilization developed a remarkably accurate geometric model of the universe. Greek philosophers began to make geometric representations of celestial objects early in the sixth century B.C. Over the ensuing years, they

Plate 6.1. Hendrik A. Lorentz (1853–1928)

determined that the Earth was a sphere and that the sun and moon were spherical bodies each at a fixed distance from the Earth. Eratosthenes (276–198 B.C.) and Aristarchus of Samos (ca.310–ca.230 B.C.) both working in Alexandria made major contributions to our understanding of the world. Eratosthenes was the first to estimate the circumference of the Earth, and Aristarchus asserted that the sun was the center of the universe and that the Earth revolved around the sun. Aristarchus in a treatise *The Sizes and Distances of the Sun and Moon* gave values that were accurate in principle, and his results were accepted by Archimedes.

Hipparchus (ca.180–ca.125 B.C.) of Nicaea was the most eminent of Greek astronomers. It is probable that he spent some years at Alexandria, but finally settled at Rhodes where he made most of his observations. His astronomical investigations naturally led to trigonometry, and Hipparchus must be credited with the invention of that subject. The pinnacle of Hipparchus' mathematics is his theorem on similar triangles. The theorem states that if two triangles are similar, then the ratio of the lengths of any two sides of one triangle equals the corresponding ratio of the other. On the basis of this theorem, the trigonometric functions, sine, cosine, tangent, secant, are defined. Unfortunately, the greater part of the works of Hipparchus is lost, so we must infer his contributions from the works of Menelaus (first century) and Ptolemy (ca.75–unknown). Ptolemy drew principally on the work of Hipparchus, and the system of the universe that he obtained from Hipparchus is now referred to as the Ptolemaic system. In particular, the elegant theorem (in *Euclid*, VI D) generally known as Ptolemy's theorem is due to

TRIGONOMETRY

Plate 6.2. Ptolemy guided by the muse of astronomy. Woodcut done in 1508.

Hipparchus. It contains implicitly the addition formulas for $\sin(\phi \pm \phi')$ and $\cos(\phi \pm \phi')$.

Instead of the usual algebraic definitions, let us give a geometric method of defining the trigonometric functions. Let AOP in Figure 6.1 be the angle ϕ which is measured from line AO in a counterclockwise direction. Let P be a point on the circle with radius OA or OP. Furthermore, let the radius be of unit length. From the point P, let the perpendicular PS be drawn to the radius OA. Then PS is the *sine* of the angle AOP, and OS is its *cosine*. Also at P, let a perpendicular be drawn to OP until it meets the extension of OA at point T. Then PT is the *tangent* of the angle AOP, and OT is its *secant*. In summary (with $OP = 1$ and angle $AOP = \phi$),

$$\sin \phi = PS$$
$$\cos \phi = OS$$
$$\tan \phi = PT$$
$$\sec \phi = OT$$

Two applications can be made of the Pythagorean theorem, namely,

$$PS^2 + OS^2 = 1 \quad \text{or} \quad \sin^2 \phi + \cos^2 \phi = 1$$

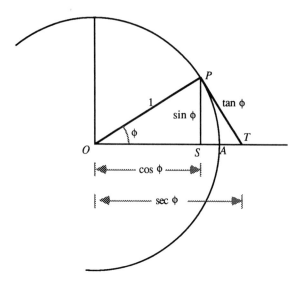

Figure 6.1. Geometric definitions of the sine, cosine, tangent, and secant of the central angle ϕ.

and

$$OP^2 + PT^2 = OT^2 \quad \text{or} \quad 1 + \tan^2 \phi = \sec^2 \phi$$

These represent two valuable trigonometric identities.

Because right triangles OSP and OPT have a common angle ϕ, they are similar to each other. Thus

$$\frac{OT}{OP} = \frac{OP}{OS} \quad \text{or} \quad \sec \phi = \frac{1}{\cos \phi}$$

which says that sec ϕ is the reciprocal of cos ϕ.

ROUND GLOBE

Hipparchus was the first to indicate that the position of any point on the Earth's surface can be accurately and fully defined by reference to a parallel of latitude and a meridian of longitude. Each parallel is a circle running around the globe and maintaining uniform distance from the poles. The central parallel is the equator, which is the great circle equidistant from the two poles. Each parallel lies wholly in its own plane, which is at right angles to the north-south axis of the globe.

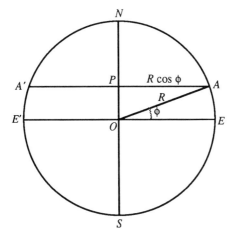

Figure 6.2. Opposite meridians.

The meridians are semicircles, running north-south from one pole to the other. Two opposite meridians make a complete great circle. In Figure 6.2, the two semicircles *NAES* and *NA'E'S* are opposite meridians. Together they make a great circle, which is the circumference of the globe. There is no obvious central meridian, but an arbitrary choice of the prime meridian has been made, the Meridian of Greenwich. Each meridian lies wholly in its own plane, and the planes of the different meridians all intersect the axis of the globe. The longitude of any particular meridian is defined as the angle between the plane that contains the prime meridian and the plane that contains the meridian in question. The meridians are all of equal length. They are widest apart at the equator, and they converge toward the poles. Further, each meridian intersects each parallel at right angles.

The latitude of any particular parallel is defined as the angle subtended at the center of the globe by the arc of a meridian, the arc being bounded by the equator and the particular parallel in question. Referring to Figure 6.2 again, we see that *NS* represents the axis of the globe, and *E* and *E'* are opposite points on the equator. The radius of the equator is given by *OE*, which is equal to the radius *R* of the globe. The meridian arc *AE* runs north-south and subtends an angle ϕ at *O*, the center of the globe. The point *A* therefore has latitude ϕ. Since the plane of the parallel through *A* is parallel to the plane of the equator, it follows that *A* and *A'* are opposite points on the parallel of latitude ϕ. The radius of this parallel is given by *PA*, which is $R \cos \phi$, and the length of the parallel is therefore $2\pi R \cos \phi$. Thus, the lengths of the parallels of latitude vary directly as the cosine of the latitude.

The spacing of the parallels along the meridians can be obtained from Figure 6.2. The arc distance *AE* is equal to $R \phi$, where *R* is the radius of the globe and φ is the angle of latitude expressed in radians. The spacing of parallels on a globe of unit radius is given in the table:

Latitude	Arc Distance from Parallel to Equator		
0° (equator)	0.0000		
15°	0.2618	or	π/12
30°	0.5236	or	π/6
45°	0.7854	or	π/4
60°	1.0472	or	π/3
75°	1.3090	or	5π/12
90° (North Pole)	1.5708	or	π/2

We see that the spacing of the parallels along the meridians is uniform. There is, for example, exactly the same arc distance (0.2618) between the equator and parallel 15°, as there is between parallel 75° and the North Pole.

FLAT MAP

Relativity theory has to do with symmetry. Maxwell obtained his four electromagnetic equations on the basis of considerations of symmetry. Poincaré obtained the fundamental space-time symmetry predicated on the demand for the equivalence of all physical laws in each and every inertial frame, together with the concomitant demand that light travels at the ultimate signal speed. We must now try to justify special relativity. We can only do so by metaphor. Our metaphor will use the globe and the flat map.

The true shape of the Earth is approximately a sphere, and the only true map is therefore a reduced Earth, or globe, which has been modeled on the Earth. When the Earth is represented, either in part or as a whole, on a flat surface, certain difficulties inevitably arise. A very small part of the Earth's surface is approximately flat, and for most practical purposes no serious error is introduced into mapmaking of limited areas if it is regarded as actually so. The larger the area surveyed, however, the more difficult is the problem of representing it, with any pretense to accuracy, on a flat map. It is this problem which is the essence of map projections. The difficulty is that a spherical shell cannot be slit open and laid out on a flat surface without badly distorting it. For example, if an orange is slit open and the peel is flattened, the peel will stretch and crack. Distances, directions, or areas must be distorted to produce a flat map.

When drawing a network, or grid, of parallels and meridians on a flat surface, it is impossible to preserve all the properties that hold on the globe. Distortion, in some form or other, is inevitable. Broadly speaking, the most important factors to be considered in the preparation of a map are (1) the position of the region, (2) the direction which one point bears to another, (3) the distance between two points, (4) the shape of the region, and (5) the area of the region. Since any flat map must be a compromise, the main problem is to select the properties essential to each particular case and find a careful balance that preserves the properties as faithfully as possible. A map projection is simply a device for representing the parallels and meridians of the globe on a flat map.

Hipparchus in the second century B.C. developed the method of mapmaking known as the *stereographic polar projection*. Such a map is a conformal representation of the spherical surface. A *conformal mapping* is one that preserves angles. That is, if two curves meet at an angle θ on the sphere, the images of these curves on the map will meet at the same angle θ. For example, the circles of latitude cross the meridians at right angles on the sphere. The projections of these curves also meet at right angles on a conformal map. For a map to be conformal, a small figure (e.g., a triangle) at a place on the sphere must be almost similar to the corresponding figure on the map and must be more similar the smaller the figures are. In other words, a necessary and sufficient condition that a mapping be conformal is that it be a similarity in the infinitesimal. A conformal map of a small region is therefore very nearly accurate in the angles as well as in the ratio of distances, although the map may give a very distorted picture of the region in the large.

The mathematical development of the stereographic projection is given in Appendix 1 to this chapter.

TOSCANELLI'S MAP AND THE DISCOVERIES OF COLUMBUS

In the second century A.D., Ptolemy wrote his *Geography*, which summarized all ancient Greek learning on the subject. This book included the various methods of projection known to the Greeks for making flat maps of the round Earth. During the Dark Ages, Ptolemy's book was lost to European civilization, and cartography became an unknown science.

Cartography represents the first use of the mathematical technique of perspective in world history. As we know, relativity theory is also based on perspective but with the concomitant condition of perfectivity. As we would expect, cartography and relativity theory are intimately related. But to understand this connection, we must go back to the fifteenth century and to Toscanelli of Florence, Italy, and the Portuguese king. Toscanelli's family had been traders in spices for several generations and were afraid that the Turks would capture Constantinople and cut off the route to the East. The king of Portugal was also keenly interested in the spice trade. In trying to find a sea route to the Spice Islands, off the Malay peninsula, the Portuguese started exploring the west coast of Africa. In 1419,

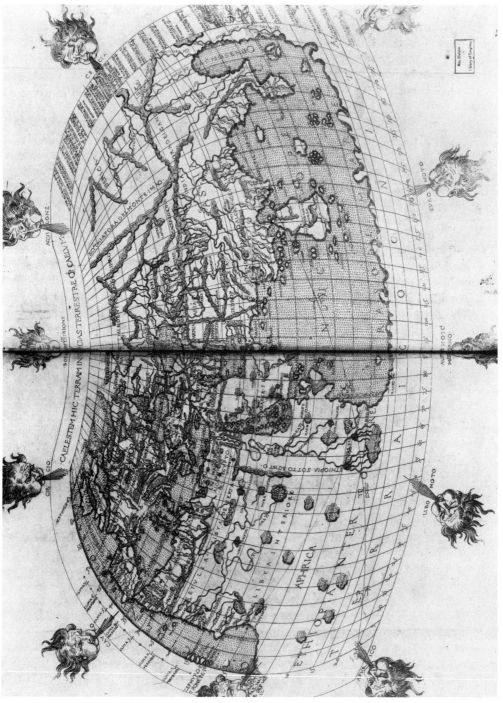

Plate 6.3. Map of the world from Ptolemy's *Geographia* published in 1482.

Prince Henry of Portugal, known as "the Navigator," set up a navigation school at Sagres. In 1425, Henry's brother, Don Pedro, visited Florence to pick up maps, and there he made contact with Toscanelli.

The revelation which Don Pedro experienced in Florence was the knowledge of the world as put forth in Ptolemy's *Geography*. The Florentines had obtained this priceless work in Constantinople in 1400 on a search for culture and classical texts. This discovery of Ptolemy's book had created intense excitement in Florence. It opened up the lost knowledge about the Earth known to the Greeks, and the maps in the book were unlike anything seen before by the fifteenth-century Europeans.

At the time of Don Pedro's visit, the European sailors only had portolan charts, developed after the introduction of the compass in the late thirteenth century. The earliest portolan charts are of the Mediterranean. The portolans gave careful outlines of the coast, with distances between landmarks carefully determined. A network of fine lines radiating from a series of compass roses permitted the sailor to work out his approximate course from one compass rose to the next. However, portolan charts were not intended to represent large areas of the Earth's surface and were indeed inadequate in this respect. In short, they were sort of a schematic representation of the possible routes, something like the directions you might take over the telephone to find the way to someone's house. They did not give an accurate pictorial representation of the surface of the Earth.

The maps in Ptolemy's book were extraordinary because they covered the entire known world. They were drawn in a consistent and standardized way based upon the projections invented by the Greeks. One of these projections was the Hipparchus stereographic projection, which we treat in Appendix 1 to this chapter. For the first time, the Europeans saw maps with grid lines of latitude and longitude. This gridding anticipated the analytic geometry of Descartes in the next century. It meant that all points on the map, even those in unknown locations, could be given coordinates from which distances and directions could be computed. This is essentially what the theory of relativity is all about, namely, the giving of coordinates to events from which time durations and spatial distances can be computed.

Toscanelli was a physician and also he had studied mathematics. He used this mathematics in his study of cartography. At the request of Don Pedro and other Portuguese, Toscanelli gathered all information about the Far East from people arriving in Florence. The Portuguese desire to develop long-distance navigation became a matter of urgency by the middle of the 1400s. Toscanelli worked in close contact with Nicholas, a German from Kues, who became a Cardinal. In the 1440s, Nicholas wrote in his *Reconciliation of Opposites* the first clear statement of relativity, a view that would influence Galileo more than half a century later.

> If the universe is infinite then the Earth is not necessarily, or even possibly at its center. And if that is so the Earth may well be circling the Sun. It is only the viewpoint of the observer as he stands on the Earth that makes him think it is the

Plate 6.4. A portolan chart of the fourteenth century showing the Atlantic at the top and the Mediterranean vertically downward.

center of the universe. The same could be true of anybody standing on the Moon or any one of the stars or planets there might be in the universe. And if everything were relative to everything else, the only way to know where you were, on Earth, or on a planet, would be to find a way to measure the elsewhere.

Toscanelli's work in cartography culminated in his letter of June 24, 1474, to the Portuguese:

> I am pleased to hear that the King is interested in a shorter route than the African one now being attempted. I enclose a chart showing all the islands from Ireland to India. If you go west from Lisbon you will get to the fine and noble city of Quinsay [Cathay, China] and to Chipango [Japan], full of gold, pearls, and precious stones.

In his chart, Toscanelli used 50 miles for the value of a degree of longitude at the equator. He assumed that Quinsay was at the same latitude as Lisbon but about a third of the way around the world. Thus he concluded that the voyage from Lisbon to China by a western route would be 4600 miles. However, his estimate of the Earth's circumference was too small. Also, because he used the exaggerated size of Eurasia reported by Marco Polo, his estimate of the eastward land distance from Lisbon to China was proportionally too great. As a result, the westward route from Lisbon to China was too short on Toscanelli's map.

An Italian sea captain, Christopher Columbus, took Toscanelli's chart to Portugal with a proposal for a westward journey. (Some historians believe that Columbus was actually a Spanish sea captain who had obtained Toscanelli's map while working in the shops of mapmakers in Lisbon.) He was turned down because the scholars knew that the distance Columbus proposed was too short. He then took the map to Spain and was again turned down. When Columbus was about to take it to the French, the Spanish king and queen, against the wishes of their advisors, agreed to support his voyage. Columbus set sail for China and Japan (the Indies). He never arrived, but to his death Columbus believed that he had found the Indies, and not a new world. He never gave up his faith in Toscanelli's map, a legacy of the ancient Greeks.

Columbus opened up the age of discovery. His predecessors, the early fifteenth-century seamen, who had ventured blindly but hopefully into the South Atlantic, provoked the learned men safe at home into a frenzy of dismay over the inadequacy of their navigational methods. As we know, the spherical surface of the Earth cannot be represented properly on a plane. However, the makers of the portolans proceeded as if one could, and as a result they contained dangerous errors. Many sea captains used these portolans and terrible disasters resulted. It was of foremost importance to find how to portray accurately the Earth's round surface on a flat map. This was made possible by the Florentine discovery of Ptolemy's *Geography*, which revealed the ancient Greek methods and maps. Those mapmakers who read Ptolemy became aware that various solutions were possible, and they started using the methods of projection given in that ancient book. Toscanelli's chart used by Columbus is the most famous example. As they mastered the

Plate 6.5. King Ferdinand V of Spain and the discoveries of Columbus.

methods described by Ptolemy, the more enterprising cartographers started to invent new and better projections. This development took place in the sixteenth century, and the most renowned cartographer of all in this enterprise was Mercator. The result was the introduction of significant improvements in how to find the way into unknown seas. The great achievements in cartography as exemplified by Mercator mark the foremost scientific advances in the first part of the sixteenth century.

MERCATOR, DEE, AND WRIGHT

The most original mathematical creation of the sixteenth century was inspired by the needs of cartography, that is, mapmaking. The search for world trade routes involved extensive geographical explorations, and good maps were required to keep pace with the discoveries. The new explorations revealed the inaccuracies and inadequacies of existing maps and thereby created a need for better ones. These shortcomings are illustrated by the plane chart, so called because it treated the spherical Earth as if it could be mapped exactly on a flat plane. At the beginning of the sixteenth century, the plane chart was still often used at sea, even though many land maps were based on some form of projection. Mathematically, the problem of making a map is that of projecting figures from a sphere onto a flat sheet. The principles involved are those of perspective, the subject matter of projective geometry. In the sixteenth century, mapmakers started to use these ideas to develop new methods.

The first great advance over the work of the ancient Greeks as found in Ptolemy's book was due to the Portuguese mathematician Pedro Nuñez (1502–1578). Nuñez discovered that on a sphere a rhumb line, or loxodrome (that is, a line of constant compass heading), is not a straight line, as it is on a plane, but a spiral terminating at the pole. However, Nuñez was unable to solve the problem of finding a projection that would make the rhumb lines straight. This is the problem solved in the most famous map of all, the one developed by the Flemish cartographer, Gerard Mercator (1512–1594). It is still known as the Mercator projection.

The maps in use before the time of Gerard Mercator were not designed for long-distance navigation. On a long voyage of exploration, the course to be pursued by the ship was marked off by a straight line joining the ports of arrival and departure. This method of navigation on the old maps resulted in considerable

Plate 6.6. Gerhardus Mercator (1512–1594). Portrait of 1574.

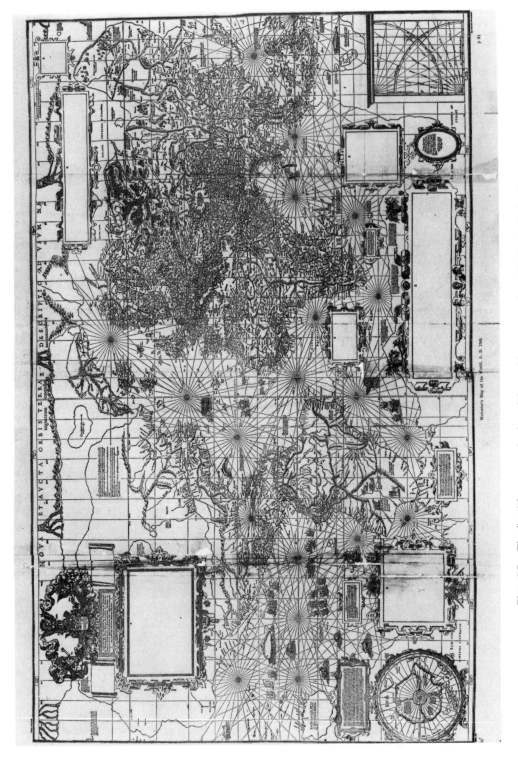

Plate 6.7. The first Mercator projection of the world, constructed empirically by Gerhardus Mercator in 1569.

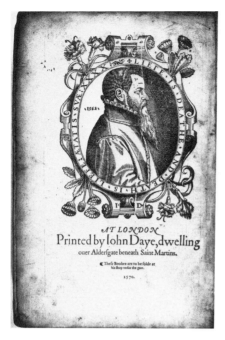

Plate 6.8. John Dee (1527–1608)

errors, and many ships were sunk or lost because of the wrong courses they took. Mercator realized that to make this straight-line method of tracing the course of the ship on the map at all accurate, the distance assigned on the flat map to a degree of latitude ought progressively to increase as the latitude increased. Using this principle, Mercator empirically constructed his world map in 1569. This map "for the use of mariners" represented the long sought-after projection that makes

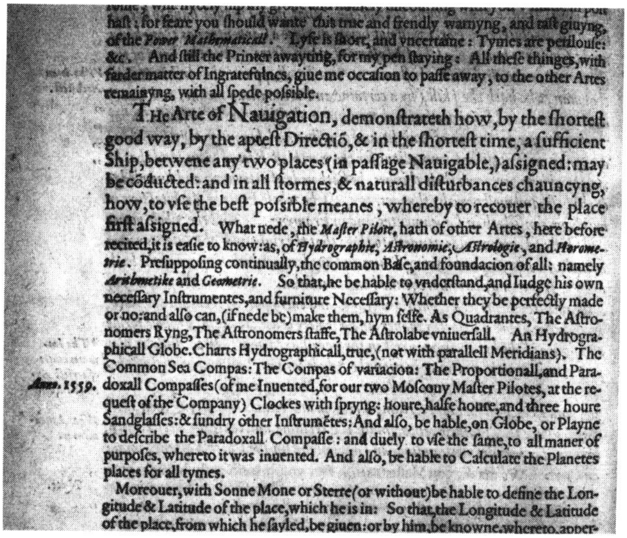

haſt; for feare you ſhould wante this true and frendly warnyng, and taſt giuyng, of the *Power Mathematicall*. Lyfe is ſhort, and vncertaine: Tymes are perilouſe: &c. And ſtill the Printer awaytyng, for my pen ſtaying: All theſe thinges, with farder matter of Ingratefulnes, giue me occaſion to paſſe away, to the other Artes remainyng, with all ſpede poſsible.

THe Arte of Nauigation, demonſtrateth how, by the ſhorteſt good way, by the apteſt Directiō, & in the ſhorteſt time, a ſufficient Ship, betwene any two places (in paſſage Nauigable,) aſsigned: may be cōducted: and in all ſtormes, & naturall diſturbances chauncyng, how, to vſe the beſt poſsible meanes, whereby to recouer the place firſt aſsigned. What nede, the *Maſter Pilote*, hath of other Artes, here before recited, it is eaſie to know: as, of *Hydrographie*, *Aſtronomie*, *Aſtrologie*, and *Horometrie*. Preſuppoſing continually, the common Baſe, and foundacion of all: namely *Ariſthmetike* and *Geometrie*. So that, he be hable to vnderſtand, and Iudge his own neceſſary Inſtrumentes, and furniture Neceſſary: Whether they be perfectly made or no: and alſo can, (if nede be) make them, hym ſelfe. As Quadrantes, The Aſtronomers Ryng, The Aſtronomers ſtaffe, The Aſtrolabe vniuerſall. An Hydrographicall Globe. Charts Hydrographicall, true, (not with parallell Meridians). The Common Sea Compas: The Compas of variacion: The Proportionall, and Para-
Anno. 1559. doxall Compaſſes (of me Inuented, for our two Moſcouy Maſter Pilotes, at the requeſt of the Company) Clockes with ſpryng: houre, halfe houre, and three houre Sandglaſſes: & ſundry other Inſtrumētes: And alſo, be hable, on Globe, or Playne to deſcribe the Paradoxall Compaſſe: and duely to vſe the ſame, to all maner of purpoſes, whereto it was inuented. And alſo, be hable to Calculate the Planetes places for all tymes.

Moreouer, with Sonne Mone or Sterre (or without) be hable to define the Longitude & Latitude of the place, which he is in: So that, the Longitude & Latitude of the place, from which he ſayled, be giuen: or by him, be knowne, whereto, apper-

Plate 6.9. Discussion of the art of navigation by John Dee published in 1570.

Plate 6.10. King James I of England. Patron of Edward Wright.

rhumb lines straight. Apparently Mercator obtained his solution by guesswork. Furthermore, and this is why we remember him today, he never explained how he derived his figures. Others could admire his product, but they could not duplicate his work. In his wisdom, Mercator never made another such map, although he published many other maps and globes during his career.

Next comes the inspiration of John Dee (1527–1608), England's leading mathematician of the time. Mathematics to him was not merely an abstract art for the specialist. Instead, to him as to the Greeks, the term meant all the sciences of magnitude and number with their practical applications. Dee kept in close touch with the work being done in navigation and cartography. He had traveled to the Low Countries in 1547 to speak with Gemma Frisius and Mercator. Dee brought back some of Mercator's globes. Dee also established a correspondence with Nuñez. Dee's work influenced Wright.

Edward Wright (1560–1615) was educated at Cambridge University, was a good sailor, and had a special talent for the construction of instruments. He became mathematical tutor to Henry, Prince of Wales, the son of King James I. Wright's intention was to determine all the errors commonly associated with the usual methods of dead reckoning. In particular, he treated the errors inherent in the commonly used plane chart. Most importantly, he solved the problem of the rhumb line and in so doing determined the mathematical principle on which the Mercator map should be drawn. His discovery of the law of the scale of the map is one of the greatest achievements in the history of mathematics and physics. All this and more are included in his monumental book, published in 1599, *Certaine Errors in Navigation Arising either of the ordinarie erroneous making of the Sea Chart, Compasse, Crosse staff, and Tables of declination of the Sunne, and fixed starres, detected and corrected*. Because the mathematical theory of Wright is identical with that of special relativity, we present it in Appendix 2 to this chapter. It can be said that it took Einstein (the best physicist of the twentieth century) and

Plate 6.11. Henry, Prince of Wales, 1612. Student of Edward Wright.

Poincaré (the best mathematician of the twentieth century) to duplicate the work of Edward Wright in the sixteenth century.

MERCATOR PROJECTION

The principle of the Mercator projection cannot be presented in terms of simple geometric projection as can be done in the case of Hipparchus' stereographic projection. However, the principle can be described approximately by a related projection. This other method, known as the perspective cylindrical projection, employs a cylinder that surrounds the globe, and the cylinder is tangent to the globe along some great circle. Ordinarily, this circle is chosen to be the equator. The rays that produce the projection emanate from the center of the globe, cut the globe's surface, and terminate on the cylinder. Thus, point P on the globe's surface is projected onto P' on the cylinder. The cylinder is next slit along a vertical line (meridian) and laid flat. On the flat map, the meridians of longitude appear as vertical lines and the parallels of latitude as horizontal lines. No finite points on the map correspond to the North and South Poles, which are at infinity.

On both the cylindrical projection and the Mercator projection, the grid of longitude and latitude is composed of straight lines at right angles. Such a grid is the most easy one for people to use. The essential difference between the two projections is in the spacing of the parallels of latitude. As we know, the problem of the spacing was empirically solved by Mercator. The theoretical solution was obtained by Edward Wright in 1599, a feat that requires integral calculus, which was officially invented by Sir Isaac Newton 70 years later. But, of course, inte-

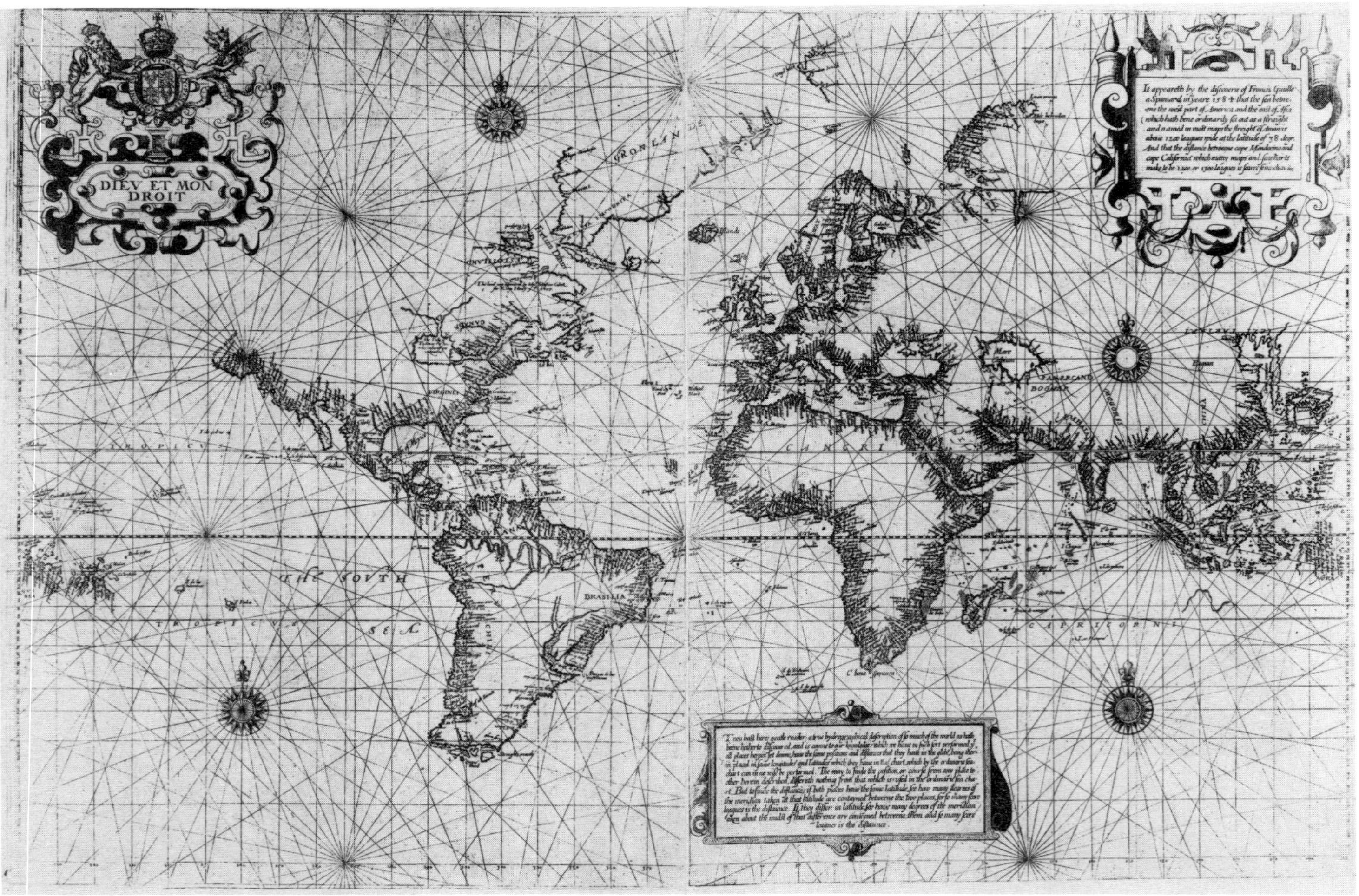

Plate 6.12. The second Mercator projection of the world, constructed mathematically by Edward Wright. All subsequent Mercator projections are based on this one.

gration was known to Archimedes in ancient times and was in the process of being rediscovered by mathematicians in the sixteenth century. However, in every sense, the mathematical feat of Wright is one of the great achievements in the history of science. In technical language, the mathematical theories of the Mercator projection and of special relativity are isomorphic. The advantage to us is that we can visualize a Mercator projection, as everyone is familiar with Mercator's world map from childhood, whereas special relativity is new and unfamiliar.

The shortcoming of the cylindrical projection is that it is not conformal; that is, it does not preserve on the map the angles which appear on the globe. The stereographic projection of Hipparchus is conformal; it does preserve angles. However, on the Hipparchus projection, the medians and parallels do not form a right-angled grid of straight lines. The problem Wright solved is

> To draw a conformal geographic map whose grid is composed of right-angled compartments of straight lines.

The importance of this map is twofold. First, as in the case of Hipparchus' stereographic projection, it preserves angles. Second, in navigation it is convenient to follow a course with constant compass bearings. This means a course that crosses successive meridians on the sphere at the same angle. This kind of course is known as a rhumb line. Such a course appears as a straight line on the Mercator map. As a result, it is especially easy to lay out a ship's course, and follow it, on the Mercator map. Navigators who use a magnetic compass need a map on which compass bearings appear as straight lines.

Rhumbs are routes of constant compass readings, and therefore are extremely useful to the navigator (Figure 6.3). If the craft is headed along one of these lines, the course is maintained without the necessity of continually figuring out new headings and making turns. The navigator merely checks the compass to be sure that each meridian is crossed at the angle of the rhumb line. To preserve constant bearings on maps, a special projection is required. Mercator's achievement was the construction of a map which showed a constant compass direction between any two points as a straight line, a considerable feat on a round Earth. Mercator solved the problem by pulling the meridians and parallels apart in a special way so that rhumb lines became straight lines. This valuable trait makes navigation by compass and straightedge quite easy. There are no complicated computations to make, or in other words, the Mercator projection has already made them. Rhumb lines that run east-west along the equator, or rhumb lines that run north-south along a meridian, are great circle routes, and so are the shortest-distance routes. However, other rhumb lines are not great circle routes. As a result for these rhumbs, the compass course is not the shortest-distance path. By replacing great circle azimuths with constant rhumb azimuths, the job of the navigator is easier but the course is lengthened.

A long-distance navigator is presented with a dilemma. Should great circle azimuths be used and save distance, or should a constant rhumb azimuth be used

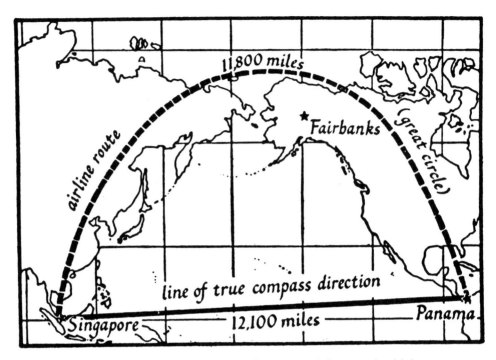

Figure 6.3. A great circle route (curved) and a rhumb-line route (straight) on a Mercator projection.

and simplify navigation? The solution to the navigator's dilemma is to use both types of azimuth together. The great circle (shortest-distance) path is transferred to a Mercator projection as a curve. This curve, which is always concave toward the equator (Figure 6.4), is then approximated with a series of straight-line segments. The compass is then used to sail or fly in a straight line on the Mercator projection along each leg of the route, making only a few turns. When you fly in a commercial plane, you can often sense when the pilot is making one of these corrections at the end of each leg.

Mercator's projection is unquestionably one of the most renowned and familiar of all map projections, the one most widely used for world maps, and the one used for navigation of airplanes and ships. The underlying principle is briefly as follows. All parallels of latitude are projected on the map equal in length to the generating globe, namely, $2\pi R$, where R is the radius of the globe. The scale along the equator is true, but away from the equator the scale along the parallels is exaggerated. The true length of the parallel at latitude ϕ is $2\pi R \cos \phi$; the length on the Mercator projection of this same parallel is $2\pi R$. The magnification factor of the scale along the parallels is thus

$$\frac{2\pi R}{2\pi R \cos \phi} = \sec \phi$$

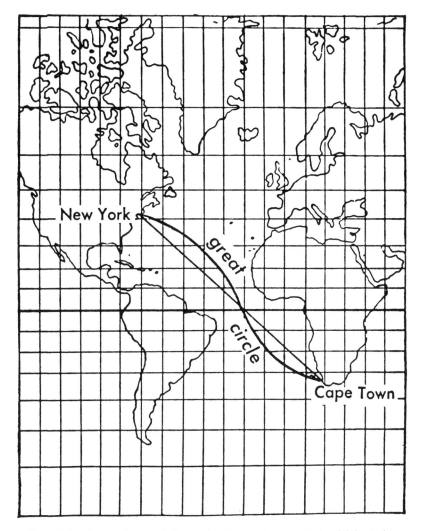

Figure 6.4. A great circle route (curved) and a rhumb-line route (straight) crossing the equator on a Mercator projection.

That is, every parallel is projected sec ϕ times its true length. The trick used in the Mercator projection is to adjust the distances of the parallels from the equator. The adjustment makes the scale along the meridians at any point equal to the scale along the parallels at the same point. In other words, the inevitable east-west stretching is accompanied by an equal north-south stretching at every point over the entire projection. The actual amount of stretching (namely, sec ϕ) varies from one latitude ϕ to another.

Thus at every point, the representation of shape is true, but a different scale is required for each parallel of latitude. It is in respect of the correct representation of shape at any point that Mercator's projection is said to be conformal. Conformal, however, is a property that requires careful interpretation when large areas are under consideration. Thus a small square on the equator of the globe will be projected as a square. An equally small square at latitude 60° will also be projected as a square, but as a square on a very different scale. For latitude 60°, we have sec 60° = 2, so the parallel is projected at twice its true length, and so also the meridian. The sides of the small square at latitude 60° are accordingly stretched to twice their true length, so its area is stretched to four times its true area. Because sec 75.5° = 4, we see that at latitude 75.5° the linear scale is magnified 4 times and the area scale by 16 times on the Mercator projection. Because sec 90° = ∞, the poles, of course, cannot be projected, for the magnification factor is infinite.

Provided the areas under consideration are small, shape is accurately projected over the entire Mercator projection. In the case of large areas, however, the limitation of conformality is at once apparent, because of the changing scale over a continuous surface. South America is about ten times the size of Greenland, but on the Mercator projection Greenland appears somewhat larger than South America. Further, because Greenland lies wholly to the north of latitude 60°, that is, in a region where the amount of stretching is becoming rapidly more and

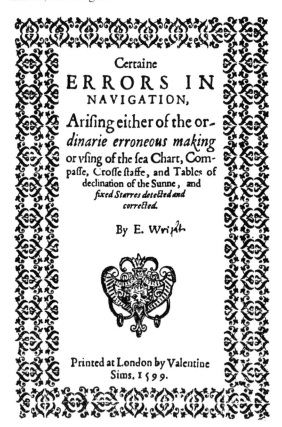

Plate 6.13. Title page of *Certaine Errors in Navigation* by Edward Wright, 1599.

more exaggerated, the shape of Greenland as a whole is very distorted on the Mercator projection. In the more northerly parts of Greenland, the exaggeration of the linear scale is four to five times as great as in the more southerly parts. South America, because it lies mainly in the tropics, where the amount of stretching is so much less, possesses a very reasonable shape on the projection.

Wright's calculation of the spacing of the parallels on the Mercator projection is treated in Appendix 2. The result is that on the Mercator projection the distance u of the parallel of *latitude* ϕ from the equator is given by *Wright's equation*

$$u = \log(\sec\phi + \tan\phi)$$

where log denotes the natural logarithm, where sec denotes the secant, and tan the tangent. This relationship is shown graphically in Figure 6.5, in which case the distance u of the parallels from the equator is plotted against the angle ϕ of latitude. The distance u is called the *exaggerated latitude*. Alternatively, u may be called the *Mercator projection* of a point on a unit globe at latitude ϕ. The quantity

$$k = \sec\phi + \tan\phi$$

is called the *stereographic projection* of a point on a unit globe at latitude ϕ. (See Appendix 1.) Wright's equation $u = \log k$ relates these two projections.

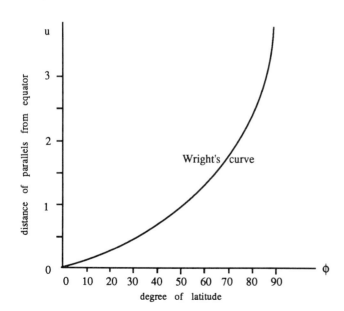

Figure 6.5. The curve of Wright's equation, with latitude ϕ on the horizontal axis, and with exaggerated latitude u on the vertical axis.

The complete projection to latitude 80° north and south is shown in Figure 6.6. The poles cannot be projected, for infinite magnification of scale is entailed,

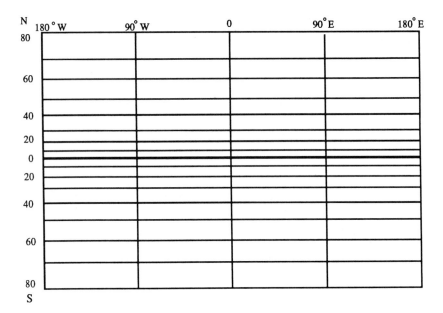

Figure 6.6. Scale of Mercator's projection.

which can only be effected at infinite distance. Moreover, in the vicinity of the poles the exaggeration in the scales assumes excessive proportions. In latitude 87° north and south, for example, linear dimensions are projected at about 20 times their true value, with the result that areas are magnified about 400 times.

TRANSVERSE MERCATOR PROJECTION

The Mercator method of map projection is so commonly used that often people hardly realize the distortion it introduces in areas. Canada appears twice as large as the United States, but it is only about one and one-sixth as large. These discrepancies can easily be seen by first looking at Mercator map of the world and then a globe of the world. The Mercator map is quite faithful near the equator, but gives a greatly exaggerated impression of areas near the polar regions. This faithfulness near the equator can be exploited, as we now will describe.

The equator of course is a great circle of a sphere. The remedy is to choose a different great circle, one that crosses the geographic region in which we are interested. We then use that great circle in place of the equator, and the resulting flat Mercator map will be relatively accurate for the geographic region in question. As a matter of terminology, we will call any Mercator projection based upon a great circle other than the equator a *transverse* or *oblique Mercator projection* (Figures 6.7 and 6.8). No flat map can give a true picture of the round world. However, a series of oblique Mercator projections, each developed on the basis

TRANSVERSE MERCATOR PROJECTION

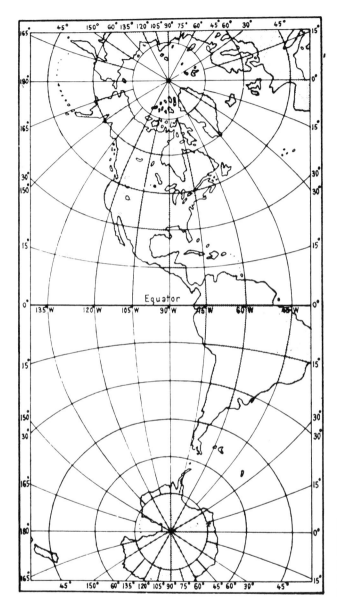

Figure 6.7. The transverse Mercator projection based on the great circle given by the 90° W meridian. This gives a conformal representation with the least deformation along this chosen (90° W) meridian.

of its own great circle, is today the best flat representation that can be made of the globe.

The convenience of the Mercator projection in this way can be enjoyed at the global level by the use of not just one map but by a system of overlapping maps. Each map portrays the globe accurately in one zone. Enough zones must be used to ensure a reasonable accuracy for surveying and navigation. Probably

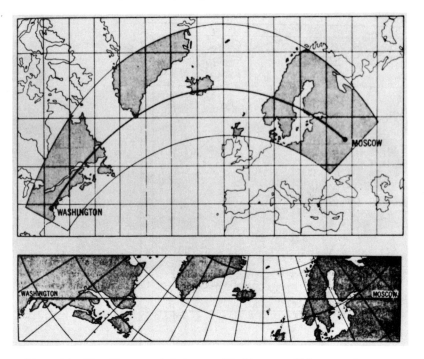

Figure 6.8. Top: The great circle between Washington and Moscow as it appears on a conventional Mercator projection. Bottom: The oblique Mercator projection based on the great circle through Washington and Moscow.

the best known plane coordinate system of international scope is the Universal Transverse Mercator (UTM) grid, which extends around the world. Sixty north-south zones are used. Each one covers 6° of longitude (for $6 \times 60 = 360$), with an overlap of 30 minutes with the zones on each side. As a result, there are 60 Mercator projections required. None of these is based on the Earth's equator. For each one, a meridian on the globe takes the place of the equator, and thus the projections are vertical instead of horizontal. Each zone is individually numbered from west to east, starting at the International Date Line (180° west or east).

One source point on the round world will map into a separate image point on each Mercator projection in a series of transverse Mercator projections. However, we would consider the best image point to be the one on the Mercator projection whose great circle was closest to the source point on the sphere. The other Mercator projections are not wrong; they are just different. Each is best for its own great circle.

In summary, the representation of the round world by a series of transverse Mercator projections can be described as a one-to-many transformation. That is, each point on the round world maps into many image points, one image point for each transverse Mercator projection in the series. Consider now the inverse problem in which we are given a flat Mercator map and want to represent it by a series of round globes. This can also be described as a one-to-many transformation. That is, each point on the flat map maps into many image points, one image point for each round globe in the series.

THE KING IS RIGHT

In this section, we want to present in simple language the mathematical concepts given in the next three sections. One should have ready a globe and the conventional Mercator map of the world based on the equator. As we know, the Mercator map spreads out all lines of latitude in such a way that the farther a country is from the equator, the more exaggerated it appears in size. The reason is that the latitude ϕ on the globe corresponds to the exaggerated latitude u on the flat map.

Now one has two choices. One choice is that Columbus is right, so the true world is round, and the Mercator map is an exaggerated image of the round world. The other choice is that the king of Spain is right, so true world is flat, and the globe is a contracted image of the flat world. In either case, we see that the image must be a distortion of the truth. However, there can be only one truth, but many possible images. Let us explain.

Suppose Columbus is right. The usual Mercator projection is drawn through the equator so geographic areas near the equator are relatively accurate on the map. As we get farther from the equator, the exaggerations on the map become increasingly pronounced. For example, look at South America and Greenland on the globe. South America is about ten times the area of Greenland. These are their true sizes. Now look at these two land masses on the map. It is seen that the size of Greenland is greatly exaggerated, so it is about the same size as South America. Because the round world is true, we have no other choice for the globe. But we can draw other maps. For example, take the transverse Mercator projection for a great circle crossing Greenland. On this map, the situation is reversed, for now Greenland is about its true size but the size of South America is absurdly swollen.

Suppose a South American was restricted to seeing only the conventional Mercator map. Also suppose a Greenlander was restricted to seeing only the transverse Mercator map through Greenland. Neither is allowed to see the globe, which represents the truth. Both maps are drawn to the same scale, but no scale of any kind is marked on the maps. We now get the two together, and let them compare distances. The South American sees Greenland the same size as South America on his map, say, both middle size. The Greenlander sees Greenland as tiny (much smaller than the Greenland of the conventional Mercator map), and he sees South America hugely swollen (much greater than the South America of the conventional Mercator Map). (See Figure 6.9.) Each sees his own land as approximately the true size, but sees the other's land as greatly exaggerated. Who is right? Actually, neither is right because both are using flat maps, which are only approximations to the true globe. The map of each is a good approximation

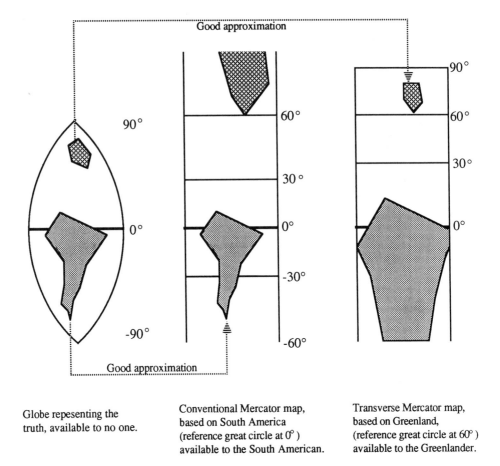

Figure 6.9. The round globe as the truth, and two flat Mercator representations of the truth.

for his land, but not for the other. There is no contradiction here once we understand that the truth is in the globe and that there are an infinite number of maps that approximate the globe, each best for its own transverse great circle. This is like the theory of relativity, but not quite. The reason is that Columbus is wrong and the king of Spain is at long last vindicated. In perseverance there is victory.

The king of Spain is right. The true world of space-time (in special relativity) is flat. However, each of us in our inertial frame of reference sees space and time in terms of a round globe. Each possible globe approximates the space-time continuum, and each approximation is best for a given frame and becomes worse the faster another frame recedes from the given frame. There are many round approximations but only one flat truth.

The King is Right

The flat world goes on for an infinite distance. There is no edge as imagined by some of the sailors of Columbus. No one can fall off. In the middle of this flat world, there is of course Spain. To the west lies the Indies, and for the sake of argument let us say that the Indies is the same size as Spain. Of course, this is on the flat Mercator projection, which now represents the truth. Let us consider two approximating globes. The first belonging to the king of Spain is a globe centered on Spain. This globe shows Spain as about the correct size, but shows the Indies as greatly shrunk. The other Globe belonging to the Indians is a globe centered on the Indies. The globe shows the Indies as about the correct size, but shows Spain as greatly shrunk. See Figure 6.10.

The king and an Indian each sees his own land as approximately the true size on his own globe, but sees the other's land as terribly shrunk or contracted. Who is right? Actually, neither, because both are using globes that are only approximations to the true flat Mercator space-time continuum. The globe of each is a good approximation for his own time and spatial coordinates, but not for the other's. There is no contradiction here once we understand that the truth is the flat Mercator map, not the many possible approximating globes. This is the theory of relativity. In mathematical language, this example is isomorphic to special relativity. This isomorphism is established in the next three sections, which are mathematical in

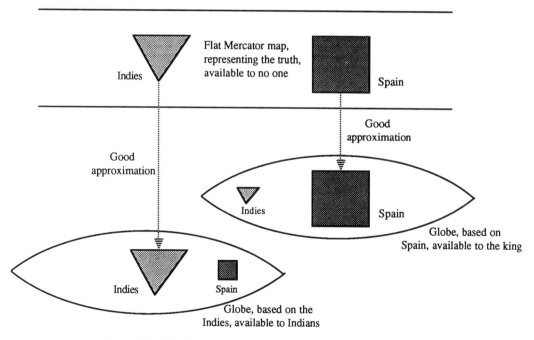

Figure 6.10. The flat Mercator map as the truth, and two round global representations of the truth. (This is relativity theory.)

character. However, all the principles are elucidated in this section, so one can proceed on to the next chapter without loss of continuity.

Let us now summarize. The theory of relativity, or at least the mathematical structure of special relativity, was first developed in the fifteenth and sixteenth centuries. The visionary who gave the impetus was Toscanelli, the explorer who obtained the critical observations was Columbus, the discoverer who found the practical solution was Mercator, the inspirer who had the insight was Dee, and the explainer who produced the dénouement was Wright. In the same epoch of humankind the mathematical structure of special relativity was again uncovered. The visionary was Copernicus, the explorer was Tycho, the discoverer was Kepler, the inspirer was Galileo, and the explainer was Newton. In the nineteenth and twentieth centuries, special relativity was formulated for the third time, again with identical mathematics. The visionary was Maxwell, the explorer was Michelson, the discoverer was Lorentz, the inspirer was Poincaré, and the explainer was Einstein. Who can say which of the three achievements was the greatest? Because of their priority, it seems that Dee and Wright were the greatest mathematicians, and who can gainsay Columbus as the greatest explorer?

ROUND WORLD

The king of Spain believed the world was flat, but in 1492, Columbus proved that the world is round. However, ship captains still required flat maps for navigation, but the old charts led to navigation errors and the resulting terrible shipwrecks. In 1569, Mercator empirically developed a flat map that made possible accurate navigation. Finally, in 1599 Edward Wright produced the mathematical theory of navigation (the use of a flat map to represent a round world). *Wright's equation* is

$$u = \log(\sec \phi + \tan \phi)$$

where ϕ denotes the true latitude on the round world and u the corresponding exaggerated latitude on the flat map.

To keep our discussion as simple as possible, we will consider the round world as a circle and the flat map as a straight line. We assume that the circle has a radius equal to 1. See Figure 6.11. If we take E as the equator and N as the North Pole, then the angle ϕ is the latitude of P. The angle ϕ (in radians) is equal to the circular arc length EP, that is, the distance along the surface of the world to point P. If we straighten out the half circle $SEPN$, we obtain the vertical straight line $SEPN$ to the right of the circle. The scale on this vertical straight line goes from $\phi = -\pi/2$ at the South Pole S, to $\phi = 0$ at the equator E, and then to $\phi = \pi/2$ at the North Pole N.

Let the horizontal line in Figure 6.11 represent the Mercator map, which in

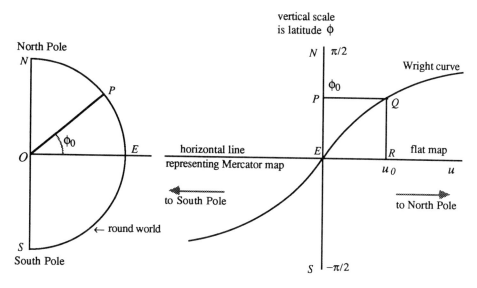

Figure 6.11. Construction of the flat Mercator map from the round world.

the case of a circular world reduces to a straight line. The coordinate of the Mercator map is the exaggerated latitude u. The relationship between ϕ and u is given by Wright's equation and is shown graphically by the curve in the figure. The whole curve cannot be shown, as u is negative infinity ($u = -\infty$) at the South Pole S ($\phi = -\pi/2$) and u is positive infinity ($u = \infty$) at the North Pole N ($\phi = \pi/2$). Close to the equator u is approximately equal to ϕ. In all cases the magnitude of u is greater than the magnitude of the corresponding true latitude ϕ, and it is for this reason that u is called the exaggerated latitude. As the true latitude approaches that of the poles, the exaggeration of u gets bigger and bigger. Thus the Mercator map is very accurate near the equator, and the accuracy deteriorates at an accelerated rate as the poles are approached.

Let us find the exaggerated latitude u_0 corresponding to the true latitude ϕ_0 of point P on the round world. We draw a horizontal line from the point P to where it intersects the Wright curve (say, point Q). Then draw a vertical line from point Q to the horizontal exaggerated latitude axis u (say, point R). Then point R on the flat Mercator map corresponds to point P on the round world. Denote the value of the exaggerated latitude at point R by u_0. Then u_0 is the exaggerated latitude corresponding to the true latitude ϕ_0.

Let us now construct an actual Mercator projection by this method as originally put forth by Edward Wright. Ecuador is on the equator $\phi = 0$, Disneyworld (in Florida) is at true latitude $\phi = 0.50$ radians, and London (the one in Canada, not England) is at true latitude $\phi = 0.75$ radians. The Mercator projection will be based on the equator. Using the method outlined, we find the Ecuador Mercator projection:

	Round World True Latitude ϕ	Flat Map (Based on Ecuador) Exaggerated Latitude u
Ecuador	0.000	0.000
	0.250	0.253
Disneyworld	0.500	0.522
London	0.750	0.832
	1.000	1.226
	1.250	1.821
	1.500	3.341
	1.570	7.829
North Pole	1.571	∞

The left-hand column depicts the round world and the right-hand column the flat Mercator map. We see that the flat map (u) agrees well with the round world (ϕ) from Ecuador to Disneyworld, but beyond Disneyworld to the north, the agreement becomes worse. This discrepancy might not be acceptable to people who want to travel from Disneyworld to London. What can we do about that?

We turn to the principle of the transverse Mercator projection. The Mercator projection just constructed is based on the equator ($\phi = 0$), so it is very accurate at near Ecuador but less accurate at Disneyworld. However, if we are particularly interested in navigation close to Disneyworld on the round world, then we would instead use a Mercator projection based on Disneyworld, not Ecuador. Thus, we would construct a Mercator projection based on the great circle (a new "equator") through Disneyworld. Using primes to denote this new Disneyworld system, the latitude at Disneyworld is now $\phi' = 0$, and the corresponding exaggerated latitude is $u' = 0$. A latitude ϕ' is converted into the corresponding exaggerated latitude u' by use of the Wright curve, exactly as earlier, except now the variables have primes. The result is the Disneyworld Mercator projection:

	Round World True Latitude ϕ'	Flat Map (Based on Disneyworld) Exaggerated Latitude u'
Ecuador	−0.500	−0.522
	−0.250	−0.253
Disneyworld	0.000	0.000
London	0.250	0.253
	0.500	0.522
	0.750	0.832
	1.000	1.226
North Pole	1.071	1.365

On this Disneyworld flat map, we see the flat map (u') agrees well with the round world (ϕ') from Disneyworld to London, so this is the map that people traveling that route would want to use.

We are now ready to take the critical step made by Edward Wright in 1599. He placed the two maps side by side:

	Round World		Flat Map	
	ϕ	ϕ'	u	u'
Ecuador	0.000	−0.500	0.000	−0.522
Disneyworld	$\phi_0 = 0.500$	0.000	$u_0 = 0.522$	0.000
London	$\phi = 0.750$	$\phi' = 0.250$	$u = 0.832$	$u' = 0.253$

He wants to explain what is going on to the people traveling back and forth between Disneyworld and London.

In Ecuador coordinates (unprimed), the true latitude of Disneyworld is $\phi_0 = 0.500$, and the true latitude of London is $\phi = 0.750$. However, in Disneyworld coordinates (primed), the true latitude of London is but $\phi' = 0.250$. How is this possible? Wright simply explains that the relationship is

$$\phi = \phi_0 + \phi' \quad \text{or} \quad 0.750 = 0.500 + 0.250 \quad \text{(RIGHT)}$$

Wright says right. The real world is true. The equation results form the shift ϕ_0 due to making Disneyworld the origin of coordinates in the primed system. The good people immediately infer that their maps must bear the same relationship, namely,

$$u = u_0 + u' \quad \text{or} \quad 0.832 = 0.522 + 0.253 \quad \text{(WRONG)}$$

But Wright says wrong. Actually, $0.522 + 0.253 = 0.775$, not 0.832 as the people desperately want. The reason is that the maps are distorted. Physical relations in the real round world cannot have the same form on the distorted flat maps. However, all the foregoing is based on the proposition that Columbus is right, that the real world is round.

FLAT WORLD

Columbus may have been right in the fifteenth century, and he did discover America. But in the twentieth century, we know that we live in a space-time world, and the space-time world of special relativity is as flat as a pancake. (The space-

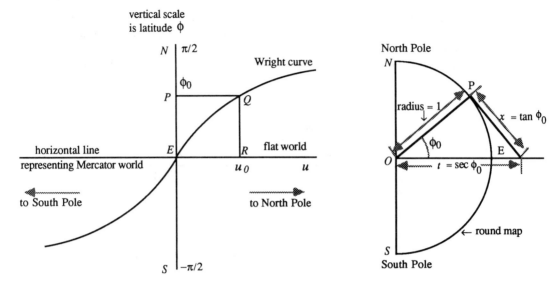

Figure 6.12. Construction of the round map from the flat Mercator world.

time world of general relativity, however, is curved.) The king of Spain was living in space-time, and he was right. The world is flat.

For simplicity, we will take the flat world as a horizontal straight line, running from negative infinity to infinity. The king of Spain did not believe that Columbus would sail off the edge of the world, but instead he knew Columbus would keep sailing west. With Spain at the origin ($u = 0$), Columbus would first reach the Indies ($u_0 = 1.5$), and then Columbus would discover El Dorado ($u = 2.5$) on the flat world. For navigation, Columbus needed a map, and so Wright constructed it as follows.

In Figure 6.12, let the real flat world be a straight line with abscissa the exaggerated latitude u. Let the vertical axis have ordinate ϕ, the true latitude of the round map. From point R on the real flat world, draw a vertical line to where it intersects the Wright curve (say, point Q). Then draw a horizontal line from point Q to the vertical axis (say, point P). Then point P gives the true latitude ϕ on the round map.

The round map is used as follows. On the flat world, Spain is at the origin ($u = 0$). Suppose Columbus wants to sail from Spain to the Indies. The king tells Columbus that the Indies is at $u_0 = 1.5$ on the flat world. Columbus then consults Wright, who determines that $u_0 = 1.5$ corresponds to the latitude $\phi_0 = 1.13$ on the round map. Wright tells Columbus that taking into account winds, tides, and sea monsters, the sailing distance is

$$x_0 = \tan \phi_0 \quad \text{or} \quad x_0 = \tan 1.13 = 2.13$$

FLAT WORLD

and the sailing time is

$$t_0 = \sec \phi_0 \quad \text{or} \quad t_0 = \sec 1.13 = 2.35$$

Suppose Columbus does not go anywhere, but stays in Spain. The coordinate $u = 0$ for Spain on the flat world corresponds to the latitude $\phi = 0$ on the round map. The sailing distance is

$$x = \tan 0 = 0$$

as expected, but the sailing time is

$$t = \sec 0 = 1$$

This sailing time of unity represents the time required to fit out the ship, even when the ship never sails. Fitting out the ships was a major problem for Columbus, and in fact the queen of Spain had to sell her jewels.

Thus, Wright constructs the map:

	Flat World	Round Map (Based on Spain)			
	Exaggerated Latitude u	True Latitude ϕ	Distance x ($x = \tan \phi$)	Time t ($t = \sec \phi$)	Velocity v ($v = \sin \phi$)
Spain	0.00	0.00	0.00	1.00	0.00
	0.50	0.48	0.52	1.13	0.462
	1.00	0.87	1.18	1.54	0.762
Indies	1.50	1.13	2.13	2.35	0.905
	2.00	1.30	3.63	3.76	0.964
El Dorado	2.50	1.41	6.05	6.13	0.987
	3.00	1.47	10.02	10.07	0.995
	3.50	1.51	16.54	16.57	0.998
	4.00	1.53	27.29	27.31	0.999
	. . .				
Ultima Thule	∞	1.57	∞	∞	1.000

Using this navigational information prepared by use of Wright's equation, Columbus drew the following conclusions. The sailing distance from Spain to the Indies is $x_0 = 2.13$ with a sailing time of $t_0 = 2.35$, thereby giving a velocity of $x_0/t_0 = 2.13/2.35 = 0.905$. The sailing distance from Spain to El Dorado is $x = 6.05$ with a sailing time of $t = 6.13$, thereby giving a velocity of $x/t = 6.05/6.13 = 0.987$. From Wright's table, Columbus saw that he would have to sail all the way to Ultima Thule (the pole on the map) to achieve the ultimate sailing velocity of unity.

When Columbus reached the Indies, the first thing he asked was how much farther was it to El Dorado. The Indians had their own navigational information, also prepared by Edward Wright. The coordinates used by the Indians are indicated by primes.

	Flat World	Round Map (Based on the Indies)			
	Exaggerated Latitude u'	True Latitude ϕ'	Distance x'	Time t'	Velocity v'
Spain	−1.50	−1.13	−2.13	2.35	−0.905
	−1.00	−0.87	−1.18	1.54	−0.762
	−0.50	−0.48	−0.52	1.13	−0.462
Indies	0.00	0.00	0.00	1.00	0.000
	0.50	0.48	0.52	1.13	0.462
El Dorado	1.00	0.87	1.18	1.54	0.762
	1.50	1.13	2.13	2.35	0.905
	2.00	1.30	3.63	3.76	0.964
	2.50	1.41	6.05	6.13	0.987
	. . .				
Ultima Thule	∞	1.57	∞	∞	1.000

(Actually, in compiling the navigational information for the Indians, Wright did not have to make any new computations, but merely shifted the origin of coordinates. He also used the symmetry existing between coordinates with negative signs and those with positive signs.)

Columbus traded information with the Indians to obtain the table:

	Flat World		Round Map	
	u	u'	ϕ	ϕ'
Spain	0.00	−1.50	0.00	−1.13
Indies	$u_0 = 1.50$	0.00	$\phi_0 = 1.13$	0.00
El Dorado	$u = 2.50$	$u' = 1.00$	$\phi = 1.41$	$\phi' = 0.87$

The Indians want to know why Columbus has El Dorado at $u = 2.50$, whereas they have it at $u' = 1.00$. Columbus says that the relationship is

$$u = u_0 + u' \quad \text{or} \quad 2.50 = 1.50 + 1.00 \quad \text{(RIGHT)}$$

Wright says right, and says it is because of the shift of coordinates by $u_0 = 1.50$, the distance on a flat world from Spain to the Indies. The Indians immediately infer that the two maps must be also so related. They infer that

$$\phi = \phi_0 + \phi' \quad \text{or} \quad 1.41 = 1.13 + 0.87 \quad \text{(WRONG)}$$

Wright says wrong. Actually, $1.13 + 0.87 = 2.00$, not 1.41, as the Indians desperately want. Physical relations in the real flat world of space-time cannot have the same form on the distorted round maps. Of course, all of the foregoing is based on the proposition that the king of Spain (and Einstein) is right—that the real world of space-time in special relativity is flat.

LORENTZ TRANSFORM

Columbus now looks at the sailing distances, the sailing times, and the velocities. He makes the table

	Flat World		Round Map			
	u	u'	x	x'	t	t'
Spain	0.00	-1.50	0.00	-2.13	1.00	2.35
Indies	$u_0 = 1.50$	0.00	$x_0 = 2.13$	0.00	$t_0 = 2.35$	1.00
El Dorado	$u = 2.50$	$u' = 1.00$	$x = 6.05$	$x' = 1.18$	$t = 6.13$	$t' = 1.54$
Ultima Thule	∞	∞	∞	∞	∞	∞

If we let $x = \tan \phi$ and $t = \sec \phi$, then Wright's equation

$$u = \log(\sec \phi + \tan \phi)$$

becomes

$$u = \log(t + x)$$

Using trigonometric identities, it can be shown that

$$\frac{1}{\sec \phi + \tan \phi} = \sec \phi - \tan \phi$$

so

$$\frac{1}{t + x} = t - x$$

We now want to consider a displacement u_0. We have

$$u = u_0 + u'$$

which is

$$\log(t + x) = \log(t_0 + x_0) + \log(t' + x')$$
$$= \log[(t_0 + x_0)(t' + x')]$$

Taking antilogs, we get

$$(t + x) = (t_0 + x_0)(t' + x') \qquad (6.1)$$

Taking reciprocals, we obtain

$$\frac{1}{t + x} = \frac{1}{t_0 + x_0} \frac{1}{t' + x'}$$

which is

$$t - x = (t_0 - x_0)(t' - x') \qquad (6.2)$$

If we expand the right hand sides of equations (6.1) and (6.2), we obtain

$$t + x = t_0 t' + x_0 x' + t_0 x' + x_0 t'$$
$$t - x = t_0 t' + x_0 x' - t_0 x' - x_0 t'$$

If we subtract the second equation from the first, we obtain

$$2x = 2 t_0 x' + 2 x_0 t'$$

If we add the second equation to the first, we obtain

$$2t = 2 t_0 t' + 2 x_0 x'$$

Dividing each of these two equations by 2, we obtain

$$x = t_0 x' + x_0 t'$$
$$t = t_0 t' + x_0 x'$$

Factor out t_0 from the right-hand sides. We obtain

$$x = t_0 \left(x' + \frac{x_0}{t_0} t' \right)$$

$$t = t_0 \left(t' + \frac{x_0}{t_0} x' \right)$$

Let $\gamma = t_0$ and $\beta = x_0/t_0$. Then these equations are the (reflexive) Lorentz transform

$$x = \gamma(x' + \beta t')$$
$$t = \gamma(t' + \beta x')$$

We have therefore shown that Wright's equation is mathematically equivalent to the Lorentz transformation. The Lorentz transformation is a mapping from one round globe to another, both of which are approximations to a true flat world of space-time. Wright's equation shows the underlying structure of space-time; in this sense it is more descriptive than the Lorentz transformation.

Appendix 1: Hipparchus' Stereographic Projection

Of all maps, globes give us the most realistic picture of the Earth as a whole. Basic spatial attributes such as distance, direction, shape, and area are preserved. However, it would be ideal if the Earth's surface could be mapped onto a flat medium, such as a sheet of paper. Unfortunately, the Earth is not a so-called developable surface. As a result, it cannot be flattened without distorting such geometrical properties as direction, distance, area, and shape. Thus, all flat maps distort reality. This is the map projection problem. There is no true solution; only approximations are possible.

Stereography is the art or technique of depicting solid bodies on a plane surface. The *stereographic polar projection* is very important in cartography. In all probability, the source of this method is Hipparchus, one of the most amazing persons of antiquity. One selects as the projection plane or image plane (map plane) a plane through the center of the globe. For illustrative purposes, let us pick the equatorial plane as the map plane. The north pole N is chosen as the projection center. A projection ray is a straight line originating at the north pole, cutting the surface of the globe, and terminating on the map plane. Let P be the point where the ray cuts the globe, and let P' be the point where the ray cuts the map plane. Then P' is called the stereographic image of P. See Figure 6.13.

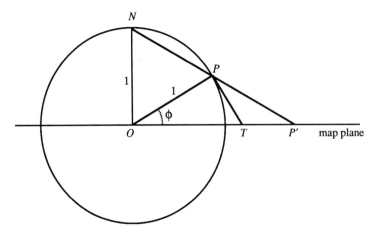

Figure 6.13. Construction of the stereographic polar projection.

In the figure, the angle ϕ is the latitude of point P. We let the radius of the globe be equal to 1 (i.e., $ON = OP = 1$). Also we draw the tangent PT. Then

$$PT = \tan \phi$$
$$OT = \sec \phi$$

From geometry, it can be shown that $PT = TP'$. Let k be the length of line segment OP'. Thus we have

$$OP' = OT + TP' \quad \text{or} \quad k = \sec \phi + \tan \phi$$

Because the quantity k is the distance from the origin to the projected point P' on the map plane, the quantity k may be called *stereographic projection* of the point P with latitude ϕ on the unit globe.

The map is shown in Figure 6.14. The circle $k = 1$ represents the equator, and outside this circle are mapped all points on the globe in the northern hemisphere. Circles on the map represent parallels of latitude. The figure shows the circles corresponding to latitude ϕ_1 and the adjacent latitude ϕ_2. Radii on the map represent meridians. The figure shows the radii corresponding to longitude λ_1 and the adjacent longitude λ_2. On the map, the distance from the origin to the circle of latitude ϕ_1 is k_1; the distance to the circle of latitude ϕ_2 is k_2.

The intersection of the two radii with the two circles gives a small compartment $ABCD$. See Figure 6.15. The sides AD and BC are each equal to dk. The sides AB and CD are each equal (approximately) to $kd\lambda$. That is, we consider the small compartment as a rectangle.

We now want to show that this small rectangle has approximately the same

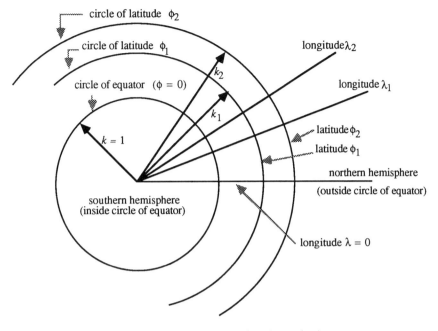

Figure 6.14. A stereographic polar projection.

shape as the corresponding compartment on the globe's surface at the given latitude ϕ and longitude λ. That is, we want to show that in the mapping from the globe to the plane, shape is preserved for small areas. This shape preserving quality is required for a conformal map. The ratio of the sides DA to AB is

$$\frac{dk}{kd\lambda}$$

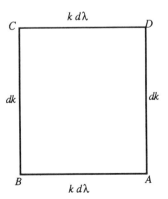

Figure 6.15. Small compartment on stereographic polar projection. The calculus symbols dk and $d\lambda$ represent, respectively, small increments (or differentials) in the projection parameter k and in the longitude parameter λ.

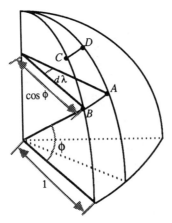

Figure 6.16. Small compartment $ABCD$ on globe.

Because $k = \sec \phi + \tan \phi$, we have $dk = (\sec \phi \tan \phi + \sec^2 \phi)d\phi$. Thus

$$\frac{dk}{kd\lambda} = \frac{\sec \phi(\tan \phi + \sec \phi)d\phi}{(\sec \phi + \tan \phi)d\lambda} = \frac{\sec \phi \, d\phi}{d\lambda}$$

In Figure 6.16, we see the corresponding area $ABCD$ on the globe, and in Figure 6.17 we see an enlarged version. Sides AB and CD are each equal to $\cos \phi \, d\lambda$ and sides BC and AD are each equal to $d\phi$. The ratio of sides AD to AB is

$$\frac{d\phi}{\cos \phi d\lambda} \quad \text{or} \quad \frac{\sec \phi \, d\phi}{d\lambda}$$

Thus the ratio of sides is the same, namely, $\sec \phi \, d\phi/d\lambda$, for a small compartment on the globe as for the corresponding compartment of the map. This means that the stereographic projection is conformal.

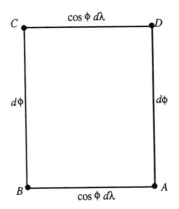

Figure 6.17. Enlarged view of the small compartment in Figure 6.16. The calculus symbols $d\phi$ and $d\lambda$ represent, respectively, small increments (or differentials) in the latitude parameter ϕ and in the longitude parameter λ.

Appendix 2: Wright's Equation

Let us now turn to Wright's problem of drawing a conformal map whose grid is composed of right-angled compartments of straight lines. On the Mercator map (Figure 6.18) the equator is segment AB, the length of which agrees with the length (2π) of the globe equator. (The radius R of the globe is taken to be unity.) If we divide AB into 360 equal parts and erect at the dividing points perpendiculars to AB, we thereby obtain the map meridians. If we let λ denote longitude, then the meridian spacing $d\lambda$ would be 1° or $2\pi/360$ radians.

On the map, the lines of equal latitude (the parallels) are horizontal straight lines parallel to the equator. The latitude parallel on the map that corresponds to the globe parallel of latitude ϕ has the distance u from the map equator. This vertical distance u on the map is called the *exaggerated latitude*. The core of Mercator's problem consists of representing the exaggerated latitude u as a function of the geographical latitude ϕ.

To solve this problem, we will compare the Mercator map with the (also conformal) Hipparchus (stereographic polar projection) map. On the left in Figure 6.19, we reproduce Figure 6.15, which shows a small compartment on the Hipparchus map. On the right is the corresponding compartment on the Mercator map. The ratio of adjacent sides (AD/AB) for each of the two compartments must be equal; that is,

$$\frac{dk}{kd\lambda} = \frac{du}{d\lambda}$$

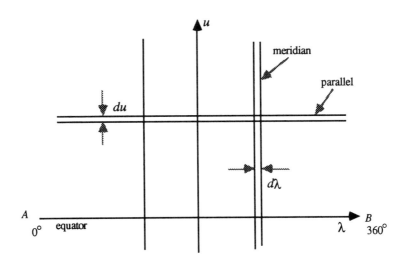

Figure 6.18. Mercator map. The calculus symbols du and $d\lambda$ represent, respectively, small increments (or differentials) in the exaggerated latitude parameter u and the longitude parameter λ.

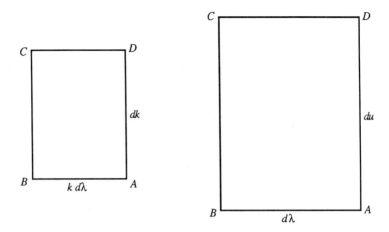

Figure 6.19. Left: Small compartment on stereographic polar projection (Hipparchus map). Right: Corresponding small compartment on Mercator projection.

Canceling the $d\lambda$ from each side, we obtain

$$\frac{dk}{k} = du$$

This is a differential equation that must be integrated. Edward Wright performed this integration (70 years before Newton and Liebnitz) and obtained the solution

$$u = \log k$$

This equation states that the exaggerated latitude u on the Mercator map is equal to the logarithm of the radial distance on the Hipparchus map. Because the stereographic projection k is given by

$$k = \sec \phi + \tan \phi$$

we have the final result that the Mercator projection u is given by

$$u = \log (\sec \phi + \tan \phi)$$

This equation gives the exaggerated latitude u as a function of the geographic latitude ϕ. This celebrated result is known universally as *Wright's equation*. With this equation, Edward Wright put the art of navigation on a scientific basis, as well as relativity theory, because as we know from Chapter 4, the stereographic projection k is the *relativistic Doppler factor*.

CHAPTER 7

EINSTEIN, MASS, AND ENERGY

When I heard the learn'd astronomer,
When the proofs, the figures, were ranged in columns before me,
When I was shown the charts and diagrams, to add, divide, and
    measure them,
When I sitting heard the astronomer where he lectured with much
    applause in the lecture-room,
How soon unaccountable I became tired and sick,
Till rising and gliding out I wander'd off by myself,
In the mystical moist night air, and from time to time,
Look'd up in perfect silence at the stars.

Walt Whitman (1819–1892)
When I Heard the Learn'd Astronomer

MASS

The philosopher Heraclitus taught that nothing is eternal; everything is in a constant state of flux. Experience shows that over a period of time, the objects around us do change in such respects as form and position. However, despite the transfor-

Plate 7.1. Albert Einstein (1879–1955)

mations of objects and the impermanence of most earthly structures, there has been from earliest times a great interest in stability and permanence. Certainly this is why the Egyptians built the pyramids 5000 years ago. There is a strong tradition in Western philosophy, at least from Plato's time, that regards as real that which is permanent, eternal, and unchanging. The physical sciences have been successful in finding various forms of constancy in nature. Many examples come from astronomy in the motion of heavenly bodies. Also there are examples of what may be called the principle of perseverance, namely, that certain entities continue to exist despite external changes. There also evolved principles that were arithmetical or geometric in character. They specify that the shape, size, or number describing a physical entity remains constant throughout time. For example, it was believed that bodies were made up of ultimate particles, called atoms, which were each supposed to be indestructible. Each atom satisfied the principle of preservation. It was argued that the arithmetic and geometric principles of constancy then showed that the total mass of a body must remain constant, because its mass was made up of a fixed number of indestructible atoms.

Physical quantities such as force, momentum, energy, velocity, and torque can be expressed in terms of only three physical dimensions: the dimension of spatial distance (with unit the meter in SI units), the dimension of time duration (with unit the second in SI units), and the dimension mass (with unit the kilogram in SI units). Mass shows its effect in two important ways. First is by its weight, which depends on the force of gravity. However, gravity varies from place to place. For example, the force of gravity is less on the moon than on Earth, so the same object weighs less on the moon than on Earth. Second, mass shows its effect by inertia, which is a measure of how difficult it is to start or stop the motion. Inertia is an intrinsic quality of mass and does not depend upon an outside effect such as gravity.

Galileo and later Newton made the idea of mass or inertia fundamental in their mechanics. Others before them had asked questions about what made objects move and had given various reasons. Galileo turned the question around and asked why a moving body stops, if and when it does. He saw that the same property of inertia which makes it hard to move a massive body makes it hard to stop the same body when it is moving. The more massive the body, the more damage it will do when it hits something, speed being held constant.

The concept of mass as formulated by Galileo and Newton sufficed for all purposes for nearly 250 years. It was the most firmly grounded of all physical concepts. The main property of matter was that it had mass. However, in 1881 J. J. Thomson showed theoretically that if a metal sphere were electrified, it would behave as though it had a small extra mass. This extra mass was too small for there to be any hope of detecting it experimentally, but still Thomson's result was the first hint that the concept of mass was not quite so simple as believed. Then came the discovery in the early twentieth century that the masses of electrons in the form of beta rays depend upon their speed. That is, the mass of an electron increases as its speed increases. The variation in mass only becomes appreciable when the velocity approaches that of light, but then it becomes large. This effect is quite contrary to classical ideas.

CONSERVATION OF MOMENTUM

Most basic of all the laws of nature are the conservation laws. First, let us look at the law of the conservation of momentum. Suppose that two children, one with twice the mass of the other, wearing ice skates, stand on an ice pond and push against each other. They move apart in opposite directions across the ice. The one with twice the mass rebounds with half the velocity of the other. In general, when two bodies of mass m_1 and m_2 are pushed apart by some force, their respective velocities v_1 and v_2 are related by the equation

$$m_1 v_1 = m_2 v_2$$

That is, when two bodies are pushed apart, the mass times the velocity of one body equals the mass times the velocity of the other body.

The product of mass m and velocity v is an important property of an object and is called the *momentum*. The usual symbol for the momentum is the letter p, which stands for mv. Thus the balance of momentum between two bodies can be written simply as

$$p_1 = p_2$$

Taking p_2 to the other side of the equation, we obtain

$$p_1 - p_2 = 0$$

This equation says that the total momentum of the two bodies is zero. Because the two bodies are moving in opposite directions, we have to put the minus sign in front of p_2 to add it to p_1. The reason is that momentum has direction, so it is a vector, just like velocity.

Our discussion has shown that if two bodies start out with zero momentum, then when they are pushed apart the total momentum of the two bodies remains equal to zero. The same idea can be applied to any number of particles. When a stationary bomb explodes, the sum of the momenta of all the pieces, taking all the directions into account, is equal to zero. If the bomb were moving before the explosion, then total momentum of the fragments would add up to the original momentum. Although the fragments are flying in various directions, their entirety carries the same momentum as the moving bomb before the explosion.

We can now formulate the law of the conservation of momentum. The law states that no changes that take place as a result of forces operating within a closed system of objects can alter the total momentum of the system. More concisely, the *law of the conservation of momentum* is: *the total momentum of an isolated system is always constant.*

NEWTONIAN DYNAMICS

Newton's first law, also known as the law of inertia, implies that a body cannot change its velocity by itself without interaction with surrounding bodies. Every change in the body's velocity both in magnitude and direction is caused by the action of external bodies upon it. Newton's second law of motion expresses the relation between force and the change in momentum of interacting bodies. We now want to indicate how Newton's second law is derived. In this section we use calculus, so a reader might wish just to skim the mathematics.

The momentum of a given body is a vector quantity whose value is equal to the product of the mass of the body times its velocity. The direction of the momentum coincides with that of the velocity. Denoting by p the momentum of a body having a mass m and velocity v, we can write

$$p = mv$$

There are no special names for the unit of momentum. In the SI system, the unit of momentum is kg m/s. For example, a body having mass 20 kg and moving with a velocity 4 m/s has momentum (20)(4) or 80 kg m/s. However, in natural units in which distance and time are measured in the same unit (so $c = 1$), velocity is a pure number, so the unit of momentum is the same as the unit of mass.

Consider two bodies being pushed apart by a spring. The bodies start at rest and after the spring acts on them for a certain time, each body has a certain momentum. The law of the conservation of momentum says that the momentum of the first body has the same magnitude as the momentum of the second body. The final momentum of each body depends upon the length of time that the force of the spring is applied. If the spring pushes for 1 second, there results a certain momentum. If it pushes for 2 seconds, the final momentum is twice as great. Thus we can say that the momentum acquired by the body equals the force exerted by the spring multiplied by the time that the force acts; that is,

$$mv = Ft$$

Here m is the mass and v is the velocity, and their product is the momentum p. Also F is the force and t is the time. For this equation to be true, we must assume that the force F is constant.

In the foregoing we have assumed that the force starts acting at initial time zero and that the initial momentum is also zero. If instead we assume that the force starts acting at initial time t_0, and that the initial momentum is $m_0 v_0$, then the equation (which holds for the case of constant force) becomes

$$mv - m_0 v_0 = F(t - t_0)$$

Thus

$$F = \frac{mv - m_0 v_0}{t - t_0}$$

The Greek letter Δ denotes an incremental change in a quantity. Thus we write

$$\Delta(mv) = mv - m_0 v_0$$

for the incremental change in momentum. Similarly, we write

$$\Delta t = t - t_0$$

for the incremental change in time. Thus our equation for force becomes

$$F = \frac{\Delta(mv)}{\Delta t}$$

If we make the time increment very small, and in fact, infinitesimal, then the force F becomes the instantaneous force. Then the differential notation d replaces the Δ, and thus the equation becomes

$$F = \frac{d(mv)}{dt}$$

This equation says that instantaneous force (which now can be variable) is equal to the derivative of momentum mv with respect to time t. This is *Newton's second law* of motion.

Newton stated the second law as follows. The time rate of change of momentum (i.e., the derivative) is equal to the force F producing the motion, and its direction coincides with the direction in which the force acts. Using the symbol p for momentum, the second law is

$$F = \frac{dp}{dt}$$

In many applications, the mass m of the body may be considered constant. In such a case, the mass m can be taken outside the differentiation sign, so we can write

$$F = m\frac{dv}{dt}$$

The derivative dv/dt is the acceleration of the body, denoted by a. Thus in this special case (which, however, is the most frequently encountered), the formulation of the second law can be written as force equals mass times acceleration; that is,

$$F = ma$$

CONSERVATION OF ENERGY

The belief that there is a fundamental quantity, later called energy, that is conserved throughout every physical change is deeply rooted in physical thinking. The idea was dominant in physics in the middle nineteenth century and is associated with the names of Mayer, Joule, Kelvin, Helmholtz, and Clausius. The concept of energy is one of the most important in physics because it brings together all branches of physics. Energy is not just another physical quantity. There are various forms of energy not at all superficially alike. Heat and mechanical work are but two forms of energy. From the work of Joule the idea was put forth that, whereas energy in any one of its forms might not be conserved, the total energy in all its forms is conserved. The principle of conservation of energy, together with some theory of the forms of energy, became a cornerstone of nineteenth-century physics.

The phenomenon of energy is difficult to describe in a simple way, because it has so many aspects and exists in so many forms. Let us begin with the form known as kinetic energy. Let us consider a situation in which a constant force F acts on a body as it moves through a given distance x. The body was initially at rest, so its initial velocity and initial momentum are each zero. To avoid complications at this point, we assume that the mass of the body does not change appreciably during the motion. This allows us to assume that a constant force acting upon the body increases its velocity at a steady rate.

Next, let us multiply the force F by the distance x over which it acts. The result defines the work W done by the force in moving the body this distance; that is,

$$W = Fx$$

As we know, distance x is equal to the velocity times time t. If the initial velocity is zero, and the final velocity is v, then the average velocity is $v/2$. Therefore the distance is

$$x = vt/2$$

Therefore the foregoing two equations give

$$W = \frac{Fvt}{2}$$

We now use Newton's second law, which says that force is the rate of change of momentum. Because the force is constant, the rate of change of momentum is the average change in time period t, that is, mv/t. That is, the constant force is

$$F = \frac{mv}{t}$$

so $t = mv/F$. Substituting this for t in the equation for W, we obtain

$$W = \frac{Fv}{2} \frac{mv}{F}$$

which gives

$$W = \frac{mv^2}{2}$$

The quantity $mv^2/2$ is known as the *kinetic energy* K of the moving body. We have thus shown that $W = K$; that is, the kinetic energy of the body is equal to the work done to set it in motion. As we see, the kinetic energy is measured only in terms of mass and velocity. It says nothing about the force that produced it. It is simply a property belonging to a moving body.

We thus have two derived quantities, each involving mass m and velocity v.

One is momentum $p = mv$, and the other is kinetic energy $mv^2/2$. Each figures prominently in the history of science in the attempts to find a proper measure of "quantity of motion." The concept of momentum was introduced by Descartes, who lived a generation before Newton. By analyzing the problem of the collision of two bodies, Descartes arrived at the law of the conservation of momentum. With the publication of Newton's *Principia*, the conservation of momentum became an immediate consequence of Newton's third law (action and reaction).

Although the formula for kinetic energy looks deceptively similar to that of momentum, there is a crucial difference in that the velocity is squared. This makes the kinetic energy a nondirectional measure of motion, whereas the momentum is directional. Whether the velocity is positive or negative, the square is always positive. Motions in opposite directions do not cancel out in the total kinetic energy, but add.

In simple Newtonian dynamics, the problem of kinetic energy is slightly more complicated than is that of momentum. Kinetic energy is not conserved by itself, but only when added to another quantity called potential energy. As a frictionless pendulum swings, its energy changes gradually from all potential at the top of the swing to all kinetic at the bottom. The law of conservation of energy says that the total energy remains the same.

In the pendulum example, we did not admit friction. In the development of a broader picture of energy, it is recognized that when kinetic energy disappears, through the action of friction, heat is produced. Thus heat is another form of energy. In fact energy exists in, and can be transferred to, many other forms, such as chemical and electrical. Thus the law of energy conservation must include all the various types of energy.

Let us consider a certain system of bodies. First, we require that the system be *closed*, in that it does not interact mechanically with external bodies. Also we require that the system be *adiabatically isolated*, in that it has no heat exchange with external bodies. From the condition that the system is closed, it follows that it does no work to overcome external forces. From the condition of adiabatic isolation, it follows that the system neither gives up nor receives energy by heat exchange. Then from the first law of thermodynamics, the total energy of the system does not change, so total energy E is a constant. Thus the most general expression for the *law of conservation of energy* is: *the total energy of a closed and adiabatically isolated system is a constant*.

The law of conservation of energy does not require that the separate energies of the individual bodies making up the system remain unchanged. Both mechanical interaction and heat exchange may take place among the bodies within the system. As a result, the energies of the component bodies making up the system will change. All that is required for conservation of energy is that the system must not interact mechanically or participate in heat exchange with surrounding bodies. Then the total energy of the system will remain constant, even though the energies of its components change.

THE MUON AND THE MOUNTAIN REVISITED

In classical Newtonian mechanics the mass of a body is the same in every inertial frame of reference. A 1-kilogram mass on an airplane is the same as on the ground. With ordinary velocities (that is, velocities much less than that of light), all experiments seem to confirm this statement. No dependence of mass on velocity can be detected.

On the basis of this proposition, let us analyze the motion of a body with a constant mass subjected to a constant force F. Under such conditions, the body accelerates. The use of the classical form of Newton's laws shows that the velocity attained by this body of constant mass is directly proportional to the time that the constant force acts. As a result, the velocity of the body of fixed mass increases without limit if the constant force is applied long enough. Thus we see that classical Newtonian mechanics allows the body to attain a speed greater than the speed of light.

This conclusion disagrees with the theory of relativity, one of whose fundamental principles states that no body can move with a velocity exceeding the velocity of light in vacuum. It is therefore necessary to revise Newtonian mechanics so that its conclusions agree with the theory of relativity. It turns out that this can be done by assuming that the mass of a body is not constant, but instead is different in every different inertial frame. We will arrive at this relativistic expression for mass by use of a fable.

We recall the fable of the muon and the mountain given in Chapter 2. *If the mountain will not come to the muon, then the muon must go to the mountain.* A muon is created high up in the atmosphere. The Earth represents the mountain in the fable. If we consider the mountain at rest, then the muon spends its lifetime going to the mountain at speed $\beta = 199/200$. The distance of this lifelong journey, as seen by an observer at rest on the mountain, is $x = 19.9$ light-μs. The lifetime of the muon as seen by an observer at rest on the muon is $\tau = 2$ μs. These three observations are summed up in the table

Relative speed of muon and mountain	$\beta = 199/200$
Distance covered by muon as seen by mountain at rest	$x = 19.9$ light-μs
Lifetime of muon as seen by muon at rest	$\tau = 2$ μs

The purpose of the theory of relativity is to make these three numbers compatible. Speed, of course, is distance over time, but β is not equal to x/τ; that is, $199/200 \neq 19.9/2$. To make things work, just create in your mind another distance λ and another time t, defined by means of the equations

$$\beta = \frac{\lambda}{\tau} = \frac{x}{t}$$

That is, define $\lambda = \beta\tau = (199/200)2 = 1.99$ light-μs as the distance covered by the mountain as seen by the muon at rest. Likewise define $t = x/\beta = 19.9/(199/200) = 20$ μs as the lifetime of the muon as seen by the mountain at rest. Now everything works. If Tom (x) does not love Mary (τ), just create Dick (λ) and Jane (t), so now Tom loves Jane and Dick loves Mary, and everything is fine. In relativity theory the added quantities λ and t cannot be observed, which is like saying Dick and Jane cannot be seen either. Relativity theory is based upon the unobservable; since these things cannot be observed, they cannot be contradicted. Dick and Jane do what we say; so do λ and t. In summary;

Distance covered by mountain as seen by muon at rest: $\lambda = 1.99$ light-μs

Lifetime of muon as seen by the mountain at rest: $t = 20$ μs

Three fundamental physical entities are distance, time, and mass. In this fable we have without qualm defined a new distance λ and a new time t. Why should mass go unscathed? First, we need the metaphor of the fat man and the thin man on a lifeboat. The fat man has a lot of mass which is converted to energy, so he lives a long time. The thin man unfortunately has little mass to carry him through, so sadly expires early. Now the muon has two lifetimes:

Lifetime of muon as seen by muon at rest: $\tau = 2$ μs

Lifetime of muon as seen by mountain at rest: $t = 20$ μs

According to the lifeboat analogy, the short lifetime of $\tau = 2$ μs belongs with a thin muon. Let the mass of this thin muon be designated by μ. The long lifetime of $t = 20$ μs belongs with a fat muon. Let the mass of this fat muon be designated by m. Let us summarize:

	Thin Muon	Fat Muon
Mass	μ	m
Lifetime	$\tau = 2$ μs	$t = 20$ μs
Distance	$\lambda = 1.99$ light-μs	$x = 19.9$ light-μs

The right triangle in Figure 7.1 illustrates the distance-time relationships. The hypotenuse is t, the two sides are x and τ, and the altitude from the hypotenuse is λ. The velocity β is $\sin \phi$, where ϕ is the angle between sides t and τ. Also ϕ is the angle between side x and altitude λ. The two fundamental relationships are

$$\beta = \sin \phi = \frac{\lambda}{\tau}, \quad \beta = \sin \phi = \frac{x}{t}$$

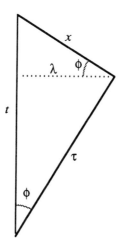

Figure 7.1. Right triangle with hypotenuse t, sides x and τ, and altitude λ.

The Pythagorean theorem gives

$$\tau = \sqrt{t^2 - x^2}$$

Since $x = \beta t$, it follows that

$$\tau = t\sqrt{1 - \beta^2}$$

Let us recall the definition of the Greek trio: α, β, γ. The given member is the velocity β. Then the alpha α (or *contraction*) factor and the gamma γ (or *dilation*) factor are defined as

$$\alpha = \sqrt{1 - \beta^2} = \cos \phi, \qquad \gamma = \alpha^{-1} = \sec \phi$$

The alpha factor is necessarily less than one, and the gamma factor greater than one, whenever β is not equal to zero.

The equation relating τ and t can be written $\tau = t\alpha$, or

$$t = \gamma \tau$$

This is the equation for the *dilation of time*. The time t is the dilated (or increased) value of time τ.

In Figure 7.1, the angle between x and λ is also ϕ. Thus

$$\lambda = x \cos \phi$$

Since $\alpha = \cos \phi$, this equation is

$$\lambda = \alpha x$$

Since α is less than one, this is the equation for the *contraction of length*. The length λ is the contracted (or decreased) value of length x. Thus the computed time t is dilated, and the computed length λ is contracted. The equation may also be written as

$$x = \gamma \lambda$$

An old parable says: *everything in proportion*. Thus the Greek letters μ, τ, and λ must be in proportion, respectively, to the Roman letters m, t, and x. We have just seen that

$$t = \gamma \tau$$

$$x = \gamma \lambda$$

Therefore, we conclude that

$$m = \gamma \mu$$

This is the equation for the *dilation of mass*. The mass of the fat muon is the dilated (or increased) value of the mass of the thin muon. The dilation factor γ is the familiar gamma factor which we have used throughout the book. Thus we conclude

Mass of muon as seen by muon at rest: μ

Mass of muon as seen by mountain at rest: $m = \gamma \mu$

The quantity μ is called the *rest mass* of the muon, and the quantity m is called the *relativistic mass* of the muon. Mass is no longer constant. It depends upon the speed β. If a muon has mass μ at rest, then the same muon moving at speed β has the *relativistic mass*

$$m = \frac{\mu}{\sqrt{1 - \beta^2}}$$

This is one of the most startling and decisive facts that came out of relativity theory. The rest of this chapter will be concerned with this result.

RELATIVISTIC MASS

A motion with a velocity almost as great as the speed of light was first encountered in physics in the investigation of the flow of electrons emitted by a radioactive substance. These experiments were carried out at the beginning of the twentieth century. The results showed that the inertial mass of an electron depends upon its velocity v, or more precisely on the ratio $\beta = v/c$ of the velocity to the speed of light. At first, it was assumed that this was only a property characteristic of charged particles like the electron. However, relativity theory shows that the dependence of mass on the velocity is a property of all material bodies. In fact, the variability of mass is a result that follows theoretically from the postulates of special relativity. We did not derive this result, but instead arrived at it in the preceding section by appealing to the fable of the muon and the mountain.

Let us first restate the results of the last section concerning mass and momentum. Instead of the symbol μ, we will now use the usual notation m_0 for rest mass. The *rest mass* m_0 is the mass of the body in the reference frame in which the body is at rest. It is also called *proper mass*. Let the quantity m be the mass of the same body in the reference frame in which the body is moving with velocity β. The velocity β is measured in natural units so its magnitude is less than one. The quantity m is called the *relativistic mass*. The relationship between these two masses is that the relativistic mass is equal to the dilated value of the rest mass; that is,

$$m = \gamma m_0$$

The dilation factor is the familiar gamma factor

$$\gamma = \frac{1}{\sqrt{1 - \beta^2}}$$

In Newtonian physics, the momentum p is defined as the product of mass times velocity. *Momentum* is defined in the same way in the theory of relativity,

$$p = m\beta$$

However, in contrast to Newtonian mechanics, the mass in this equation is the relativistic mass and not the rest mass. As the speed β approaches that of light ($c = 1$), the gamma factor approaches infinity, and hence both the relativistic mass $m = \gamma m_0$ and the relativistic momentum $p = \gamma m_0 \beta$ approach infinity. For this reason, the speed β can never actually reach the speed of light for any material body (i.e., for any body with rest mass m_0 greater than zero) because at such a speed, the mass and momentum would both be infinite.

The theory of relativity does not annul Newtonian mechanics, but incorporates it as a special case for relatively slow-moving bodies. It is entirely permissible to apply the classical laws of Newtonian mechanics to the case of a moving object with velocity much less than that of light. However, when dealing with high velocities, the theory of relativity must be used. Let us now discuss how one might decide which motions are slow and which ones have a velocity high enough to require relativistic treatment.

If m_0 is the rest mass, and m is the relativistic mass, then the error between the two is

$$m - m_0 = \gamma m_0 - m_0 = m_0(\gamma - 1)$$

The relative error in measuring the mass is then

$$\varepsilon = \frac{m - m_0}{m} = \frac{m_0(\gamma - 1)}{m_0 \gamma} = 1 - \frac{1}{\gamma}$$

or

$$\varepsilon = 1 - \sqrt{1 - \beta^2}$$

Let us assume that we are able to measure quantities with an accuracy of n significant digits. Then if the relative error ε is less than 10^{-n}, we shall not be able to detect it. Let us now calculate at which velocity it will still be impossible to detect any change in the mass of a body. We have

$$10^{-n} = 1 - \sqrt{1 - \beta^2}$$

or

$$\sqrt{1 - \beta^2} = 1 - 10^{-n}$$

If we square both sides of this equation, we have

$$1 - \beta^2 = 1 - (2 \cdot 10^{-n}) + 10^{-2n}$$

or

$$\beta^2 = (2 \cdot 10^{-n}) - 10^{-2n}$$

Because 10^{-2n} is much less than 10^{-n}, we can drop the 10^{-2n} term, so

$$\beta = \sqrt{2 \cdot 10^{-n}}$$

Since $\beta = v/c$ we finally have

$$v = c\sqrt{2 \cdot 10^{-n}}$$

If measurements are made with an accuracy of six significant digits, then $n = 6$, so

$$v = 300{,}000 \sqrt{2 \cdot 10^{-6}} = 423 \text{ km/s}$$

Thus, at velocities less than 423 km/s, the rest mass differs from the relativistic mass by less than 10^{-6}, that is, 1 part in 1 million. The velocity of a space vehicle is only about 10 km/s, or about one-fortieth of the calculated value 423 km/s. Also measurements are rarely made in engineering practice with accuracy greater than 1 in 1 million. It is quite clear then that the applications of Newtonian mechanics in the case of ordinary velocities yield more than satisfactory results as far as engineering accuracy is concerned.

In the nuclear world, however, velocities are often encountered that are close to the speed of light in vacuum. Correct results are obtained only by making use of the theory of relativity. In particular, it is possible to analyze the motion of fast-moving elementary particles to check experimentally the formula for relativistic mass. The results of the experiments agree excellently with the theory.

RELATIVISTIC ENERGY

The most widely known result of special relativity is that mass and energy are the same. Let us now indicate how Einstein obtained the result. We know that the mass of any body in motion is increased according to the formula $m = \gamma m_0$, or

$$m = \frac{m_0}{\sqrt{1 - \beta^2}}$$

Here m_0 is the mass of the body when it is at rest (that is, m_0 is the rest mass), β is the velocity (in natural units) of the body relative to the observer, and m is its relativistic mass as perceived by the observer. The relativistic mass is also called the *total mass*, or often simply the *mass*.

We see from the foregoing formula that as β gets larger, the relativistic mass m also gets larger. At ordinary speeds, the increase in mass is imperceptible, but when β approaches one (the speed of light), the relativistic mass increases greatly. In fact, if the velocity β of the body could equal the speed of light, the relativistic mass would become infinite. It is for this reason that no material body (that is, a body with rest mass m_0) can ever be accelerated to the speed of light. As we will see, any particle that travels at the speed of light must only travel at the speed of light, no faster and no slower. They are particles with no rest mass, such as photons and neutrinos.

Let us now see what happens when the speed β is small, that is, close to zero. We can then expand the γ factor in terms of the binomial theorem:

$$\gamma = \frac{1}{\sqrt{1 - \beta^2}} = 1 + \frac{1}{2}\beta^2 + \cdots$$

The terms represented by dots on the right are omitted because they are insignificantly small in comparison to the term with β^2. Thus the relativistic mass has the expansion

$$m = \gamma m_0 = m_0 + \frac{1}{2} m_0 \beta^2 + \cdots \qquad (7.1)$$

The second term on the right is the *Newtonian expression for the kinetic energy* of a moving body. If this part of the relativistic mass m of a moving body is energy, why not all of the relativistic mass? It follows then that each term in this equation can be interpreted as representing energy. The first term on the right, m_0, represents a new kind of energy intrinsic to the particle, which is called the *rest energy*, E_0.

Having introduced the concept of interpreting mass as energy, let us now no longer restrict ourselves to small values of the speed β. In general, let us define *kinetic energy* in terms of the increase in mass instead of using the Newtonian expression involving the square of the velocity β. Let the total mass m represent the *total energy* or *relativistic energy* E; that is,

$$E = m$$

As we have seen the rest mass m_0 represents the *rest energy* E_0; that is,

$$E_0 = m_0$$

The *kinetic energy* K is now defined as the increase in mass; that is,

$$K = m - m_0$$

The way in which the mass m varies with velocity, as given by equation (7.1), ensures that the Newtonian expression $K = m_0 \beta^2/2$ holds at low speeds. This means that this new relativistic definition of kinetic energy is compatible with the classic Newtonian expression for small β. The classic Newtonian definition in assuming a constant mass made it appear that kinetic energy is primarily due to velocity. The new relativistic definition shows that kinetic energy K is related to the increasing mass as velocity increases, and indeed is equal to the mass increase. In summary, *kinetic energy*, which is energy of motion, manifests ifself directly as the relativistic increase in mass due to motion. The relativistic definition is philosophically more satisfying than is the classic one.

The classical law of the conservation of energy states that energy can neither be created nor destroyed, but can only be changed from one form to another. The classical law of conservation of mass says that mass can neither be created nor destroyed; that is, the mass of the input of any process is equal to the mass of the output. As we have just seen, special relativity reveals that energy and mass are indeed the same thing. Thus neither classical law is correct in itself, because classical energy and classical mass are interconvertible.

Mass can become energy. For example, electrons and positrons can annihilate each other, converting their rest mass completely into electromagnetic energy in the form of gamma rays. Another example is the fusion of two nuclei of heavy hydrogen to give a helium nucleus. The fusion releases an enormous amount of energy (in SI units) due to the decrease in rest mass. Thus rest mass energy is a real quantity, representing stored energy locked up in the particle that can be released under certain circumstances.

Conversely, *energy can become mass.* However, it takes a lot of energy (in SI units) to create even a small amount of mass. It takes giant accelerators to produce beams of energetic particles. In the collisions of these particles, some of this energy is converted into the mass of newly created particles. For example, a proton must have speed half that of light to be energetic enough to produce a pion (that is, a pi meson) in a collision.

From all the experimental evidence, it is seen that energy E and mass m are in fact the same thing. The equation $E = m$ says that the relativistic energy E of a system is the same as its relativistic mass m. Hence if E is conserved, so is m. Hence there are not two independent conservation laws, but only one. In other words, the law of conservation of mass is the same law as the law of conservation of energy. Thus we speak of a unified law of conservation of mass-energy. The generally accepted practice is to forget about the law of conservation of mass and just speak of the *law of conservation of energy* in the relativistic sense, namely, *the relativistic energy of a closed and adiabatically isolated system is a constant.*

In closing this section, we want to derive an essential relationship in the theory of relativity. It is the relation between the total energy E of a body and its linear momentum p. We recall that the equation for the relativistic mass is

$$m = \frac{m_0}{\sqrt{1 - \beta^2}}$$

so

$$m_0 = m\sqrt{1 - \beta^2}$$

If we square this equation, we obtain

$$m_0^2 = m^2(1 - \beta^2)$$

or

$$m^2 = m_0^2 + m^2 \beta^2$$

Since (in natural units) the total energy of a body is $E = m$, the rest energy is

Plate 7.2. Albert Einstein in 1931.

$E_0 = m_0$, and the linear momentum is $p = m\beta$, we can write the foregoing equation as

$$E^2 = E_0^2 + p^2$$

This equation indicates that there are two ways of changing the total energy of a body. In the one way, the momentum of the body can be changed without changing the rest energy. This can be done by changing the velocity of the body with respect to the chosen reference frame. In the other way, the rest energy can be changed, which means that the rest mass will naturally be changed as well. We will need this equation in the form

$$E_0^2 = E^2 - p^2$$

when we speak about the invariance of rest energy (E_0 = constant) in different inertial frames.

$E = mc^2$

As we have seen in the previous section, Einstein arrived at the conclusion that energy is the same as mass: $E = m$. Let us review his reasoning. Since the mass of a moving body increases as its motion increases, and since motion is a form of energy (kinetic energy), then the increased mass of a moving body comes from its increased energy. Thus kinetic energy and the mass increase must be the same physical quantity. Generalizing this result, it can be said that the concepts of mass and energy are just different ways of looking at the same physical quantity, something that may be called mass-energy. Even though mass and energy are the same, we still tend to think of some forms as mass and other forms as energy. However, this is a distinction without a difference. It is a distinction of tradition, not of essence.

So far in this chapter we have been using a natural system of units in which the speed of light is taken as the basic unit: $c = 1$. The reason is that most of the formulas of relativity theory involve c, and therefore the elimination of this factor considerably simplifies the formulas. Since $c = 1$, it only remains to choose two other basic units. As these units, we can choose the unit of length and the unit of energy. For simplicity, let us choose these as 1 meter (m) and 1 joule (J). We call this system the m, J, $c = 1$ system. In it, time is measured in derived units, namely the light-meter. One light-meter is the time interval during which a light signal (propagating with speed $c = 1$) travels a distance of 1 meter. In SI units, 1 light-meter of time is equal to $0.333 \cdot 10^{-8}$ s. Note: If instead we had chosen the two other basic units as 1 second (the unit of time) and 1 joule (the unit of energy), then we would have the s, J, $c = 1$ system, with the derived unit of distance being the light-second.

Let us now look at dimensions. Since energy is equal to work, and since work is equal to force times distance, energy has dimensions

$$\text{energy} = \text{force} \cdot \text{distance}$$

Since force is mass times acceleration, and since acceleration is distance divided by time squared, force has dimensions

$$\text{force} = \text{mass} \cdot \text{distance}/(\text{time} \cdot \text{time})$$

These two equations give

$$\text{energy} = \text{mass} \left(\frac{\text{distance}}{\text{time}}\right)^2$$

The ratio of distance divided by time is velocity, and the velocity used for conversion between m, J, $c = 1$ units and SI units is the velocity of light c measured in meters per second. The foregoing equation can be written as

$$\text{energy} = \text{mass} \cdot c^2$$

Thus the identity $E = m$ in m, J, $c = 1$ units becomes *Einstein's equation*

$$E = mc^2$$

which holds for SI units: E in joules, m in kilograms, and $c = 3 \cdot 10^8$ m/s. This is the form of Einstein's equation which is most familiar.

This equation illustrates the role of the velocity of light as a conversion factor between quantities E and m which were originally regarded as distinct. If the equivalence of energy and mass had been known at the time of the French Revolution, there would not be separate units for the two (joule and kilogram). Viewed in this context the formula $E = mc^2$ has the same significance as the formula for converting kilometers to miles. The factor c^2 is simply a conversion factor, because the same physical quantity is measured both in joules and kilograms.

Einstein's formula is sometimes referred to as a formula for the conversion of mass into energy. Actually, it is a formula for the conversion of kilograms into joules. For example, suppose we have a 1-kilogram block of matter. The 1 kilogram of mass in the block is energy. In terms of joules, that amount of energy is

$$E = mc^2 = 1 \cdot (3 \cdot 10^8)^2 \text{ joules} = 9 \cdot 10^{16} \text{ joules}$$

Mass and energy are the same thing. A 10-kilometer race is the same as a 6.2-mile race. One kilogram of energy is the same as $9 \cdot 10^{16}$ joules of energy. Alternatively, 1 kilogram of mass is the same as $9 \cdot 10^{16}$ joules of mass. However, from tradition we say that 1 kilogram of mass is the same as $9 \cdot 10^{16}$ joules of energy. Because of the different units (kilogram and joule), we hold on to the artificial distinction between mass and energy.

The equation $E = mc^2$ is often associated with nuclear energy. However, it applies equally well to all forms of energy. From a theoretical point of view, the equation is valuable in comparing a given amount of energy in one of its forms with the same amount in another form. For example, 1 kilogram of coal (about 2.2 lb) in solid form, if converted entirely to electric energy, would be $9 \cdot 10^{16}$ joules, which is 25 billion kilowatt hours. If we heat water from freezing to boiling, the added energy increases its mass by about 1 part in 10^{11}, much too small to be detected. In an ordinary chemical reaction, such as a fire, the amount of mass that becomes heat is about 1 part in 10^9; that is, the combustion products have that much less mass than the fuel and oxygen that were burned. Again this amount is beyond the threshold of measurement. The reason why Einstein's equation is associated with nuclear energy is that in nuclear explosions, mass changes of 1 part in 1000 can take place. This amount can readily be measured, so Einstein's formula can be used to give the energy release.

Let us now review the concept of kinetic energy in terms of SI units. Relativistic energy E depends upon the velocity v of the body with respect to the reference frame. Because the body has different velocities in different reference frames, it follows that the total energy E of a body differs in different reference frames: the greater the velocity v, the greater the relativistic energy E. A body has minimum energy in the reference frame in which it is at rest. This value of the energy of a body is called the *rest energy*, and it is equal to the rest mass m_0 times the square of the velocity of light; that is,

$$E_0 = m_0 c^2$$

Let us now define kinetic energy in SI units. The *relativistic kinetic energy* K of a body is defined as the difference between its total and rest energies. Thus

$$K = E - E_0 = mc^2 - m_0 c^2 = m_0(\gamma - 1)c^2$$

As we have seen, when the magnitude of v is small with respect to c, then $\gamma - 1$ has the expansion

$$\gamma - 1 = \frac{1}{2}\beta^2 + \cdots = \frac{1}{2}\frac{v^2}{c^2} + \cdots$$

Thus we can expand K as

$$K = m_0(\gamma - 1)c^2 = \frac{1}{2}m_0 v^2 + \cdots$$

This equation shows that the relativistic kinetic energy K agrees approximately with the classical expression in the case of small velocity v. Although the classical formula represents an approximation, its accuracy is sufficient for all practical purposes in engineering. Even for velocities of hundreds of kilometers per second, the classical formula $m_0 v^2/2$ differs from the exact expression K by only the order of 1 part in 1 million.

The ratio of kinetic energy to rest energy is

$$\frac{K}{E_0} = \frac{m - m_0}{m_0} = \frac{\gamma m_0 - m_0}{m_0} = \gamma - 1$$

At velocities much lower than the velocity of light (i.e., when β is almost equal to zero in magnitude), the ratio becomes approximately $\beta^2/2$, which means that the kinetic energy of the body is much less than the rest energy. For example, at a velocity of 300 km/s (or $\beta = 0.001$), which is ten times as much as the orbital velocity of the Earth, the kinetic energy is only $10^{-6}/2$ or 0.0000005 times the rest energy.

At relativistic velocities (i.e., β almost equal to one in magnitude), practically all the energy of a body is in its kinetic energy. The rest energy is then substantially less than the kinetic energy. For example, suppose a proton is accelerated to a velocity of $\beta = 0.999923$. Then

$$\frac{K}{E_0} = \frac{1}{\sqrt{1 - \beta^2}} - 1 \approx 79.6 \qquad (7.2)$$

Thus at this velocity the kinetic energy of a proton is almost 80 times its rest energy.

In the previous section we derived the equation

$$E^2 = E_0^2 + p^2$$

relating the total energy E, the rest energy E_0, and the momentum p. This equation is in terms of natural units (for which $c = 1$). In converting this equation to SI units, we must introduce the factor c. In natural units, we know that $E_0 = m_0$, so by the dimension analysis just given, we have $E_0 = m_0 c^2$ in SI units. Let us now make a dimension analysis for momentum. In natural units, momentum $p = m\beta$, where β is a pure number, so momentum has dimensions of mass. In SI units, momentum has dimensions mass times velocity. The velocity used for conversion to SI units is $c = 3 \cdot 10^8$ m/s. Thus in SI units the momentum becomes pc. Thus equation (7.2) becomes

$$E^2 = m_0^2 c^4 + p^2 c^2 \qquad (7.3)$$

This is the required equation for SI or other unnatural units.

Up to now we have considered the energy and momentum of objects having mass. But now let us consider light. Light can only exist in motion traveling at the ultimate velocity. Light carries energy as it travels. Maxwell showed that light also carries momentum. Light momentum causes the phenomenon of radiation pressure. When light strikes a surface, the momentum is transferred to the surface. Under laboratory situations, the radiation pressure is small. However in extreme conditions, as in a hot star, the radiation pressure is so great that it supports the star against collapse.

Maxwell found that a beam of light having energy E carries momentum p given by

$$p = \frac{E}{c} \qquad (7.4)$$

This is *Maxwell's equation*. Einstein's great contribution to quantum physics in 1905 was to show that the photoelectric effect could be explained by considering light as being composed of particles, which are now called *photons*. Since a photon travels at the speed of light ($\beta = c = 1$), its gamma factor is infinite ($\gamma = \infty$). Thus its relativistic mass $m = m_0\gamma$ and its relativistic momentum $p = m_0\gamma\beta$ would be infinite, unless its rest mass m_0 were zero. Hence the rest mass m_0 of a photon must be zero. In the equation

$$E^2 = p^2c^2 + m_0^2c^4$$

we must therefore put $m_0 = 0$. Hence this equation for a photon reduces to

$$E = pc \tag{7.5}$$

As we would expect, this is the same as Maxwell's equation (7.4) given for the momentum of a light beam. In conclusion, light has no rest mass, it carries energy E and momentum $p = E/c$, and it travels at the ultimate speed c.

In addition to the photon, there is another particle called the *neutrino* that also has no rest mass. As a result, a neutrino must travel at the speed of light to carry energy and momentum. Neutrinos show up in the decay of unstable particles and in the beta decay of radioactive nuclei. Like the photon, a neutrino has no electric charge. However, a neutrino differs in its spin angular momentum. A photon has spin one and a neutrino has spin one-half. Neutrinos also differ in the manner and strength that they interact with other particles.

Finally, there is the *graviton*, which is also a particle with no rest mass, travels at the speed of light, and carries energy and momentum. However, gravitons have only been established theoretically, as up to this time none have been observed.

INVARIANTS AND SYMMETRY

The relationships between conservation and symmetry in modern physics has been explored incisively during the past half-century. Eugene Wigner (1902–   ) has pioneered these studies). Symmetry and invariance considerations, and even conservation laws, played an important role in the thinking of Galileo and Newton, as well as in the thinking of other early physicists. However, these ideas were articulated only rarely, and were not given special attention. Newton formulated his laws of motion so that they would apply in every inertial reference frame in such a way that they would be invariant under rotations and displacements (that is, under the Galilean group).

The law of the conservation of energy was instinctively recognized even before Galileo's time. Jordanus de Nemore (ca.1300) recognized essential features of what we now call mechanical energy, and Leonardo da Vinci postulated the impossibility of the perpetuum mobile. The conservation of momentum and conservation of angular momentum in the case of central motion gives one of Kepler's laws. In dealing in Newton's three-body problem, which has occupied the best mathematicians for three centuries, the law of the conservation of angular momentum is always written down as a matter of course. However as a result of relativity theory and atomic physics, symmetry conditions and conservation laws took on new meaning. As a result of the interest in Bohr's atomic model, the conservation of angular momentum became all important. Today there is universal confidence in invariance postulates and the connection between symmetry and conservation laws.

Some kind of invariance is required for there to be symmetry. For example, consider a circle of some definite radius r. To each point on the circle, associate another point on the circle that is a given distance beyond it. Such an association is called a counterclockwise rotation through the angle θ. Even though each point is associated with some new point, the entire set of new points is identical with the original set. In such a case we say that the rotation left the circle the same; that is, the radius r of the circle is invariant under a rotation. The specific value of θ is irrelevant in the argument. Thus the radius is invariant under a whole class of transformations, namely, the rotations about the center. The circle has radial symmetry.

A principle of symmetry states that a certain object is invariant under a given class of transformations. Each symmetry principle is distinguished by the relevant invariant object and the transformation class. Often we do not speak of symmetry unless the class of transformations satisfies the mathematical requirements of a group.

In relativity theory the class of transformations is the Poincaré group (Lorentz transformation). This transformation represents a shear in the space-time coordinates, which means that hyperbolas are left the same. Let us now consider what is invariant.

According to space-time symmetry, the space and time coordinates in one inertial frame are related by the Poincaré group (Lorentz transformation) to those in another frame. As we know, there are also physical quantities which do not change under such a transformation; they are called *invariants*. Let us now revert to natural units ($c = 1$).

Special relativity is based on the fundamental fact that the speed of light is invariant. Another important invariant is the time-space interval τ, whose square is defined as

$$\tau^2 = t^2 - x^2 - y^2 - z^2$$

For a given event, τ remains the same for all inertial frames.

Similarly, the rest mass m_0 is an invariant. If m is the relativistic mass, and

p is the momentum, then the square of the rest mass is

$$m_0^2 = m^2 - p^2$$

For any given body, m_0 remains the same for all inertial frames.

Because rest energy E_0 is the same as rest mass m_0, and relativistic energy E is the same as relativistic mass m, it follows that

$$E_0^2 = E^2 - p^2$$

For any given body, E_0 remains the same for all inertial frames.

In conclusion, the time-space interval τ, the rest mass m_0, and the rest energy E_0 are all invariant in all inertial frames. The invariance of τ shows that space and time shear (i.e., leave hyperbolas invariant) under the Poincaré group. The invariance of m_0 shows that mass m and momentum p shear under the Poincaré group. Finally, the invariance of E_0 shows that energy E and momentum p shear under the Poincaré group. This sums up the essential features of relativity theory.

THE CONSERVATION LAWS

Three principal laws of nature are the law of conservation of energy, the law of conservation of linear momentum, and the law of conservation of angular momentum. These laws are valid only in inertial reference frames. The classic derivation of these laws is based on Newton's second and third laws of motion, which hold only for inertial frames. Each of these *conservation laws* holds only in closed systems, that is, in systems for which the sums of all the external forces and respective moments of force are equal to zero. Furthermore, for the conservation of energy the system must also be adiabatically isolated; that is, it must not participate in heat exchange.

A significant feature of the conservation laws is their generality. The validity of these laws is not influenced by the details of the specific physical process concerned or the features of the various bodies participating in the process. They are applicable to the macroworld of astronomy and to the microworld of molecules, atoms, and the elementary particles making up the atomic nucleus. The conservation laws can be employed in investigating mechanical, chemical, thermal, electrical, and any other processes. During the history of physics, the conservation laws were almost the only ones that retained their validity while various other laws were replaced by newer ones. The conservation laws form a cornerstone of physics.

Relativity theory brought about a hiatus in the concepts of time and space. The classical concept of the simultaneity of events was entirely altered. The dilation of time, the contraction of distance, and the increase in the mass of bodies represented revolutionary changes. Most significantly, two conservation laws of

Newtonian mechanics, namely, conservation of mass and conservation of energy, turned out to be the same law. This follows from Einstein's equation $E = mc^2$ showing the equivalence of mass and energy. That is, these two conservation laws were united in relativity theory into an inclusive law of conservation of mass-energy.

In the microworld the tenets of quantum mechanics make many specific phenomena sharply different from the familiar Newtonian phenomena of large bodies. Because of Heisenberg's uncertainty principle, it is impossible to solve dynamical problems such as the unique determination of the path of an electron. However, the conservation laws remain valid in quantum physics, and often they are the only means of explaining certain quantum phenomena. This universality of the conservation laws makes them particularly valuable.

There is a basic difference between laws such as Newton's second law of motion and the conservation laws. From Newton's second law it is possible to find the detailed course of a body. For example, if we are given the force and the initial conditions, then the path of the body can be found as a function of time. In contrast, the conservation laws give no direct indications of the course of the process under consideration. Instead the conservation laws specify only those processes that are forbidden and as a result cannot and do not occur in nature. Thus the conservation laws may be described as *principles of forbiddenness*. If a phenomenon violates a conservation law, then the phenomenon is forbidden. Any phenomenon that does not violate any conservation law can in principle occur.

As we have seen the conservation laws were originally derived from Newton's laws. However, the conservation laws are still valid in situations such as quantum mechanics where Newton's laws break down. Thus the conservation laws are much more universal than Newton's laws from which they were derived. It would be helpful then to find more general principles upon which the conservation laws could be based. Such principles actually exist. They are the principles of space-time symmetry, that is, the symmetry of the Poincaré group. Specifically, the law of the conservation of energy follows from the *uniformity of time*, the law of conservation of linear momentum follows from the *uniformity of space*, and the law of conservation of angular momentum from the *isotropy of space*. The universal significance of these laws is indicated by the fact that they are derived from such general principles.

As an illustration of the connection between space-time symmetry and the conservation laws, let us indicate the derivation of the law of the conservation of energy from the uniformity of time. The principle that time is uniform means that any experiment, occurring under the same conditions but at different moments of time, takes place in exactly the same way. An experiment performed today, and the same experiment performed tomorrow, give the same results. It follows in particular that the rest mass of a closed and adiabatically isolated system is independent of time. We note that the condition of adiabatic isolation plays no role in the properties of space, so this condition is not required for the conservation of momentum and the conservation of angular momentum. However, adiabatic iso-

The Conservation Laws

lation is essential in the properties of time, because the rest mass would change in the course of time if heat exchange occurs.

In this paragraph calculus is used, so a reader might want just to skim the mathematics. Consider a body moving in an inertial frame with velocity β. Let m_0 be the rest mass and m be the relativistic mass. Then the linear momentum of a body is $p = m\beta$, the total energy is $E = m$, and the rest energy is $E_0 = m_0$. Let a force F act on the body. The kinetic energy is

$$K = E - E_0$$

If we differentiate we obtain

$$dK = dE - dE_0$$

At constant rest energy we have $dE_0 = 0$. Thus the change in the kinetic energy dK of the body equals the change in its total energy dE; that is,

$$dK = dE$$

The invariance of rest energy says that

$$E_0^2 = E^2 - p^2$$

Since $dE_0^2 = 0$, we have

$$0 = dE^2 - dp^2$$

so

$$2EdE = 2pdp$$

Since $E = m$ and $p = m\beta$, we have

$$mdE = m\beta dp$$

so

$$dE = \beta dp$$

Since the system is closed, the external force $F = 0$. We now use Newton's second law $F = dp/dt$, so

$$dp = F\,dt = 0 \cdot dt = 0$$

Thus the equation $dE = \beta dp$ becomes

$$dE = 0$$

We have thus shown that it follows from the uniformity of time that $dE = 0$ for a closed and adiabatically isolated system, or E = constant, which is the law of the conservation of energy.

This derivation is only a sketch, but it illustrates one of the more elegant results in mechanics. Each conservation law can be established from some aspect of symmetry in space or time. Let us now look at the law of conservation of momentum. Consider a system of objects for which the total energy can be written as the sum of the kinetic energy and the potential energy. For example, suppose the objects are moving about in a gravitational field close to the surface of the Earth (in the sense that we can consider the Earth's surface as a flat plane $z = 0$). As a result of these assumptions, the potential energy depends only upon the height z of each object above the Earth's surface. It does not depend upon the x, y coordinates. In this case, it can be shown that the total momentum of these objects in a horizontal direction must remain constant. This is because the potential energy does not depend upon the horizontal location, so as a result no force acts in the horizontal direction. However, the momentum perpendicular to the surface is not constant, because the force of gravity acts in the downward direction. In other words, the underlying assumption in this example is that the space we are considering is uniform in the horizontal plane. The example shows that this symmetry condition is closely connected to the law of the conservation of momentum. In fact, this law follows from the uniformity of space. If we postulate that space is uniform in all directions, it can be shown that momentum is conserved in all directions.

The mathematical principles of invariance and symmetry were recognized first as a fundamental and vital principle in physics by Poincaré. Their meaning and importance were brought out in the special theory of relativity in the symmetry of the Poincaré group of transformations (which include the Lorentz transformation as a special case). In this book we have not considered the full group, but only the Lorentz transformation, and usually only in the two-dimensional x, t case (the space-time hyperbola). The reason is that the x, t case illustrates the essential space-time idea of the full Poincaré group, but allows us to illustrate the mathematics in terms of two-dimensional space-time diagrams.

The Poincaré group contains, first, *displacements in space and time*. This means that connections between events are the same everywhere and at all times. The laws of nature are the same no matter where and when they are established. If this were not so, it would be impossible for physicists to uncover nature's laws. It should be emphasized that it is the physical laws to which the symmetry applies, and not to the events themselves that are governed by these laws. Of course, events vary from place to place. However, the physical relationships inherent in these events are the same everywhere.

The Poincaré group contains, second, a symmetry that postulates the *equivalence of all directions*. This principle can be recognized by realizing that the influence of the Earth's gravity is responsible for the difference between up and down. In the motion of a body, it is not just its three spatial coordinates, but its three coordinates with respect to the Earth.

The Poincaré group contains, third, a symmetry in which the laws of nature are *independent from the state of motion* in which observations are made *as long as the motion is uniform*. This symmetry gives the principle of equivalence of uniformly moving coordinate systems (the inertial frames of reference).

Poincaré introduced group theory into physics in 1905 by showing that the symmetry of the Poincaré group gave the fundamental structure of space-time. However, his mathematics was too involved to be fully appreciated at that time. During the passage of time since then, this work has come to be regarded as fundamental to physics, and in fact group theory is now used universally as a unifying principle in the study of the invariances and the resulting symmetries in nature.

PART II

MATHEMATICS

CHAPTER 8

HYPERBOLIC TRIGONOMETRY

Mock on, Mock on, Voltaire, Rousseau
Mock on, Mock on: 'tis all in vain!
You throw the sand against the wind,
And the wind blows it back again.
And every sand becomes a Gem
Reflected in the beams divine;
Blown back they blind the mocking Eye,
But still in Israel's path they shine.
The Atoms of Democritus
And Newton's Particles of Light
Are sands upon the Red Sea shore
Where Israel's tents do shine so bright.

William Blake (1757–1827)
Mock on Mock on Voltaire Rousseau

Plate 8.1. René Descartes (1596–1650)

EUCLIDEAN AND NON-EUCLIDEAN GEOMETRY

Few persons have such enduring fame as the great Greek geometer Euclid. However, few of the details of Euclid's life are known. We do not know the dates of his birth and death, or even where he was born. We do know that he was active as a teacher in Alexandria, Egypt, about 300 B.C. Although he wrote some other books, parts of which survive, his place in history rests on his great work in geometry, the *Elements*, a treatise in 13 books. Proclus (410–485 A.D.) wrote that Euclid put together the *Elements*, collecting many of Eudoxus' theorems, perfecting many of Theaetetus', and also bringing to irrefragable demonstration the things that were only somewhat loosely proved by his predecessors. Euclid lived in the time of the first King Ptolemy, who once asked him if there were in geometry any shorter way than that of the *Elements*, and he answered "There is no royal road to geometry." Stobaeus wrote that a student once asked Euclid, "What shall I get by learning these things?" Euclid called his slave and said "Give the student a piece of silver, for he must make gain out of what he learns."

The importance of the *Elements* lies not only on its many individual theorems and proofs which form a collection of all known geometric results up to that time, but also on the arrangement of the material in the overall plan of the book. This plan involves, in the first place, the selection of the axioms and postulates. In turn, the theorems are carefully arranged so that each follows logically from its predecessors. So well done is the *Elements* that it has been used as a textbook in its original form for more than 2000 years. Not only is it the most successful textbook ever written, but it has been the model used in the whole development

Plate 8.2. Title page of John Dee's translation of Euclid's *Elements*, 1570.

of mathematics over the centuries. It is the outstanding example of a completely deductive structure and has fascinated great thinkers over the ages. It was a significant factor in the rise of modern science. The influence of Euclid on Sir Isaac Newton is apparent, as Newton wrote his great work *Principia* in terms of geometric arguments in a form similar to that of the *Elements*.

In the course of this book, we will see that the model for the special theory of relativity is constructed out of concepts from Euclidean geometry but with different rules of manipulation. Because these rules are consistent, it follows that the model is just as consistent as Euclidean geometry. If there were any logical paradox in special relativity, then it could be converted into a similar logical paradox involving Euclidean geometry.

As we know, non-Euclidean geometry is an alternative to Euclidean geometry in which the parallel postulate is not held to be true. The *parallel postulate* of Euclidean geometry states: *given a line and a point not on that line, then there is a unique parallel line that passes through the given point.* Over the centuries, there were many attempts to prove this postulate from the others, and if such an attempt succeeded, then a denial of the parallel postulate would be inconsistent. However, none did succeed.

By the end of the nineteenth century, it was clear that the parallel postulate not only was independent of the other postulates, but also that it could be changed in two opposite ways. One way, as proposed by Bolyai, Lobachevski, and Gauss was to assume that there could be an infinite number of parallel lines passing through the given point. Such a geometry is called *hyperbolic geometry* in which all Euclid's other postulates remain valid. A "straight" line is still a geodesic, or shortest line between two points. However, in this hyperbolic space any triangle has the sum of its angles less than 180°. Another way of changing the parallel postulate was proposed by Riemann and Schläfli. It is assumed now that no line can be drawn parallel to the given line. This geometry, called *elliptic geometry*, has the property that the sum of the angles of a triangle is always more than 180°. Every geodesic is finite and closed. The lines in every pair of geodesics cross.

How does one prove consistency for such a new geometry? Poincaré settled the question by constructing a model for non-Euclidean geometry out of Euclidean geometry. Let us now give one of his models of hyperbolic geometry.

This model uses the upper half-plane (that is, points in the x, y plane for which $y > 0$) as the universe. The points of the non-Euclidean geometry to be constructed are taken as just the points in this half-plane. The distinction is that the straight lines defined in the non-Euclidean geometry are not the straight lines of this plane. Instead, a non-Euclidean straight line is defined as any Euclidean circle whose center lies on the x axis. It can be shown that such non-Euclidean straight lines obey all the axioms and postulates as set forth in Euclidean geometry, except for the parallel postulate. For example, many different non-Euclidean straight lines (half-circles A, B, etc.) can be drawn through a point P but do not intersect a given non-Euclidean straight line L. See Figure 8.1.

Thus all lines such as A and B can be called parallel to line L. If the parallel postulate were necessary to a logical statement of geometry (that is, if only one line could pass through P and be parallel to L), then the Poincaré model would have an inconsistency, and this inconsistency would translate into an inconsistency about half-circles in Euclidean geometry. In this way, Poincaré showed that non-Euclidean geometry is just as consistent as Euclidean geometry. Although the consistency of Euclidean geometry has never been established in a mathematical

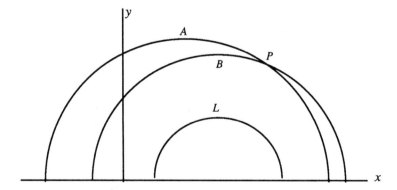

Figure 8.1. Non-Euclidian straight lines.

way, it has never been seriously questioned. As we will see, the model of special relativity is a model which differs from the Euclidean model. However, the difference is not one of having a new definition of straight line, as in the Poincaré non-Euclidean model just discussed. Instead, the difference is in the use of a different curve for measurements, more specifically, in the use of a hyperbola instead of a circle.

Today, we have come to understand that Euclid's geometry is not the only self-consistent geometrical system which can be devised, and during the last 150 years mathematicians have constructed many different non-Euclidean geometries. Indeed, since Einstein's general theory of relativity has been accepted, scientists have realized that Euclidean geometry does not always hold true in the real universe. In the vicinity of black holes and neutron stars, for example, the gravitational field is very intense, and Euclid's geometry does not give an accurate picture. In fact, Euclid's geometry (i.e., geometry with the parallel postulate) does not appear to be a correct description of the universe as a whole. However, in the world in which we live our daily lives as human beings, Euclidean geometry provides an extremely close approximation to reality, and it never has to be questioned in our mundane activities.

We now recognize that all consistent geometric systems are equally valid in the abstract but that the structure of physical space must be determined empirically. It is reported that Gauss actually triangulated three mountain peaks to see if their angles added up to 180°, but because of instrumental errors he obtained inconclusive results. With the advent of Einstein's general theory of relativity in 1915, physicists are no longer disturbed by the notion that physical space has a generalized non-Euclidean structure. But before these concepts can be mastered, it is important to have a good working knowledge of the special theory of relativity, the subject matter of this book.

ANALYTIC GEOMETRY

René Descartes, the great French philosopher, scientist, and mathematician, was born in the village of La Haye in 1596. In 1637, he published his most famous work, *Discourse on the Method for Properly Guiding the Reason and Finding Truth in the Sciences*, which is usually called simply *Discourse on Method*. Appended to the *Discourse* were three essays in which he gives examples of the discoveries he made by the use of his method. In the first appendix, called *Optics*, he presents the law of the refraction of light (which had been discovered earlier by Willebrord Snell). He also discusses lenses and various optical instruments, he describes the functioning of the eye, and he presents a theory of light that is a preliminary version of the wave theory later formulated by Christian Huygens. In his second appendix, he gives the first modern discussion of meteorology. He describes rain, wind, and the clouds and gives a correct explanation for the rainbow. In the third appendix,

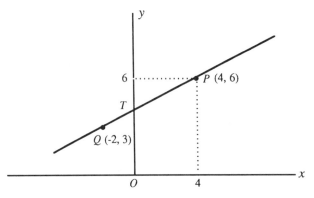

Figure 8.2. Cartesian coordinates.

The Geometry, Descartes gives his most important contribution, analytic geometry. This was a major mathematical advance that prepared the way for Newton's invention of calculus.

Analytic geometry deals with such concepts as points, lines, circles, hyperbolas, and so on, by algebraic or analytic methods, rather than those methods used in demonstrative geometry as put forth by Euclid. Much of the vocabulary of analytic geometry is encountered in other mathematics courses, so we will only give a short review here. See Figure 8.2.

1. Point P is determined by its abscissa $x = 4$ and its ordinate $y = 6$. These values $(4, 6)$ are the coordinates of the point P. Because the axes are perpendicular, this is a system of rectangular coordinates.
2. Negative values of x are measured to the left of the origin O, while negative values of y are measured down from O. Point P in the diagram is in quadrant I, and quadrants II, III, and IV are named in a counterclockwise direction. Point Q is in quadrant II.
3. A first degree, or linear, equation $x - 2y = -8$ is that of a straight line. This equation is satisfied by $(4, 6)$, which is the point P, because $4 - 2 \cdot 6 = -8$. Similarly, it is satisfied by $(-2, 3)$, which is the point Q, since $-2 - 2 \cdot 3 = -8$. In fact, it is satisfied by the coordinates of all points on line QP (and only by such points). These coordinates are all roots of $x - 2y = -8$.
4. The equation $x - 2y = -8$ can be changed to the form $y = (1/2)x + 4$, which corresponds to the form $y = mx + b$, where x and y are variables, while m and b are constants. In this case, $m = 1/2$, $b = 4$.
5. In the next few pages, it will be shown that m is the value of the slope (to be defined later) and b is the intercept on the y axis (line segment OT).

A most important property of a straight-line graph is its inclination (the angle it forms with the x axis). This angle, usually represented by θ (the lowercase

ANALYTIC GEOMETRY

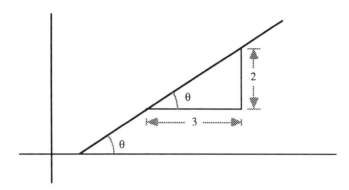

Figure 8.3. Definition of slope.

Greek letter theta), is measured in a counterclockwise direction, the x axis being the initial line. A line parallel to the x axis is considered to have zero inclination.

Figure 8.3 illustrates the term slope, which is represented by m and is defined as the tangent of the angle of inclination θ; that is, $m = \tan \theta$. Thus in the case shown, the slope, the tangent of θ is 2/3 (the ratio of the opposite side to the adjacent side). Hence $m = 2/3$. The same value of m would have been secured if other triangles had been used (such as triangles with sides 4 and 6, or 10 and 15, or 1 and 1.5), since the tangent of a given angle is a constant quantity. The slope of a line parallel to the x axis ($\theta = 0$) is zero (since $\tan 0 = 0$).

Parallel lines have the same slope. We can verify this fact by graphing (1) $y = (1/2) x$ and $y = (1/2)x + 4$, (2) $y = -2x$ and $y = -2x + 4$. The graphs can be sketched by assigning a few values to x and finding the corresponding values of y.

The slopes of perpendicular lines are negative reciprocals of each other. We can verify this fact by graphing (1) $y = (1/2)x + 4$ and $y = -2x + 5$, (2) $y = 2x - 3$ and $y = -(1/2)x + 4$.

From Figure 8.4, we see that $\tan \theta = m = (y - b)/x$. Change this equation to the form $y = mx + b$, where m is the slope, and b is the intercept on the y axis. Conversely, in the graph of $y = mx + b$ show that (1) $b = y$ intercept [let $x = 0$] and (2) $m =$ slope [show that $(y - b)/x = m$].

In summary, the equation of a straight line can be written in the form $y = mx + b$, where m is the slope of the line and b is the y intercept. This form is called the slope-intercept formula and it has many uses.

The slope-intercept formula $y = mx + b$ involves the y intercept, b, which is frequently neither known nor required. We shall now derive a formula for a straight line that can be used when the coordinates of two of its points (x_1, y_1) and (x_2, y_2) are known. It should be noted that x_1 refers to a specific value of x of some definitely fixed point P_1, whereas x (with no subscript) refers to the abscissa of any point P on the graph. The following steps show how the equation of the

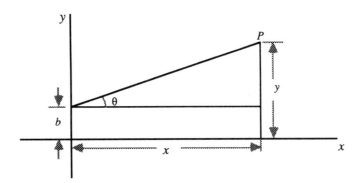

Figure 8.4. Straight line defined by slope and intercept.

line P_1P_2 is found, knowing only the coordinates of the points P_1, given by (x_1, y_1), and P_2, given by (x_2, y_2).

1. Select a point P on the graph, calling its coordinates x and y.
2. Find the expression for the slope (tan θ), using the coordinates of the two given points P_1 and P_2. Thus

$$\tan \theta = \frac{y_2 - y_1}{x_2 - x_1}$$

3. Find another expression for the slope, using the coordinates of P and the coordinates of P_1 (one of the two given points). (Note that either of the fixed points may be named P_1). Thus

$$\tan \theta = \frac{y - y_1}{x - x_1}$$

4. Set these two expressions equal to each other. The result is the two-point formula for the straight line:

$$\frac{y - y_1}{x - x_1} = \frac{y_2 - y_1}{x_2 - x_1}$$

The equation for the straight line in Figure 8.5 is

$$\frac{y - 4}{x - 2} = \frac{6 - 4}{12 - 2} \quad \text{or} \quad x - 5y = -18$$

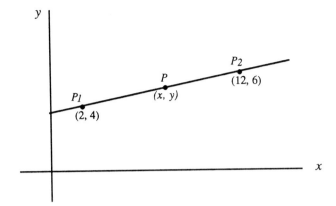

Figure 8.5. Straight line defined by two points.

The length of a line segment is to be sharply distinguished from the equation of the line of which the segment is a part. As an illustrative example, let us find the length of the line segment joining (3, 1) and (7, 4). See Figure 8.6. We have $P_1Q = 7 - 3 = 4$, $QP_2 = 4 - 1 = 3$, and $P_1P_2 = \sqrt{4^2 + 3^2} = 5$. Using the foregoing plan, we have the following formula for the distance between two points:

$$d = \sqrt{(x_2 - x_1)^2 + (y_2 - y_1)^2}$$

This formula holds no matter which point is designated as P_1 or P_2, since the differences are squared.

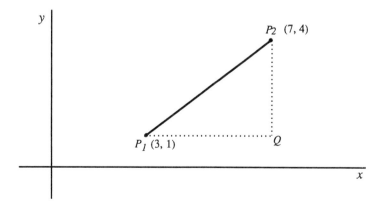

Figure 8.6. Line segment.

Exercises

1. The equation of a straight line parallel to the x axis is of the form $y = b$, where b is a constant. Verify this fact by finding the equation of the line joining (2, 5) and (6, 5).
2. Find the equation of the straight line which has the slope $m = 1/2$ and passes through the point (4, 2).
3. Find the equation of the line passing through the given points (1, 4) and (3, 6).
4. Find the equation of the line with x intercept $x = 2$ and y intercept $y = 3$. [This corresponds to $P_1 = (2, 0)$ and $P_2 = (0, 3)$.]
5. Find the equation of the line that is parallel to the given line $y = -(1/2)x + 2$ and passes through the given point (3, 5). [Use the fact that parallel lines have the same slope, so $m' = m = -1/2$].
6. Find the equation of the line that is perpendicular to the given line $y = (-1/2)x + 2$ and passes through the given point (3, 5). (Use the fact that the slopes of perpendicular lines are negative reciprocals of each other, so $m' = -1/m = -1/(-1/2) = 2$.)
7. Find the equation of the perpendicular bisector of the line segment joining (1, 2) and (7, 4). (Find the slope as the negative reciprocal, and the coordinates of the midpoint M by the formula

$$x_M = (x_1 + x_2)/2, \qquad y_M = (y_1 + y_2)/2$$

8. Find the length of the line joining (2, 1) and (6, 4). Show that (−1, 6) is equidistant from (2, 1) and (4, 3).
9. If (8, 3) is on a circle whose center is (5, −1), find the radius of the circle.

PYTHAGOREAN THEOREM

The formula for the distance between two points is an application of the Pythagorean theorem. This is perhaps the most important theorem in mathematics. The Pythagorean theorem is

> In a right triangle, the square of the hypotenuse is equal to the sum of the squares of the other two sides.

This theorem is one of the striking results achieved in the early days in the study of geometry. The old-Babylonian cuneiform text BM 85196 contains the following problem:

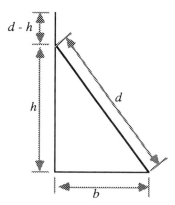

Figure 8.7. Given hypotenuse and one leg, find the other leg.

> A patu (beam?) of length 30 (stands against a wall). The upper end has slipped down a distance 6. How did the lower end move?

The problem amounts to the consideration of a right triangle (Figure 8.7). The hypotenuse is equal to $d = 30$, and one leg is equal to $h = d - 6 = 24$. The other leg is determined by using the Pythagorean theorem and is found to be $b = \sqrt{d^2 - h^2} = 18$.

The faithfulness with which the Babylonians preserved, throughout one and one-half millennia, the tradition of this theorem, is shown by a text from the era of the Seleucids (i.e., after 310 B.C., as the Seleucids were the successors of Alexander the Great). This text, BM 34568, has one of its problems,

> A reed stands against a wall. If I go down 3 yards (at the top), the (lower) end slides away 9 yards. How long is the reed? How high the wall?

The diagram is the same as that shown in Figure 8.7. This time it is given that $b = 9$ and $d - h = 3$, while h and d are required. The solution is

$$d - h = 3$$
$$d^2 = (d - 3)^2 + 9^2 = d^2 - 6d + 9 + 81$$
$$0 = -6d + 90, \quad d = 15, \quad h = d - 3 = 12$$

Other problems of the same text give d and h, or $d + h$ and b, or again $d + h$ and $d + b$, and so on. The last and most complicated problem gives $d + h + b$ and hd. As always, the real difficulty in these problems is algebraic in character. The only geometric property, used over and over again, is the "Theorem of Pythagoras." It would be wrong to suppose that in all these problems the ratio of width, height, and diagonal are 3:4:5. One finds other ratios, such as

$$5:12:13, \quad 8:15:17, \quad 20:21:29$$

These examples of right triangles with rational sides, except the last, are also found in the writings of Heron of Alexandria.

Who does not, when he hears the name Pythagoras, think of the famous theorem about the square of the hypotenuse? (Figure 8.8.) However, the connection between the man and the theorem is in doubt. In his commentary on this theorem, as it appears in *Euclid* (I 47), Proclus says, very indefinitely, "If we listen to those who wish to recount ancient history, we may find some of them referring this theorem to Pythagoras, and saying that he sacrificed an ox in honor of the discovery." But Vitruvius is of the opinion that the ox fell victim to the discovery

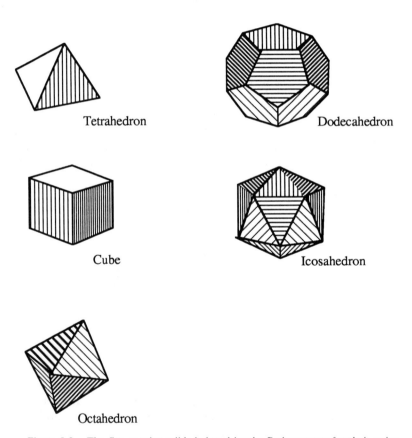

Figure 8.8. The five regular solids beloved by the Pythagoreans for their unity of shape and angle.

of the right triangle whose sides are 3, 4, 5. But the entire story is an impossible one, because the great Pythagoras was strongly opposed to the killing and sacrificing of animals, of cattle especially.

Some have thought that this theorem could not have been known in the days of Pythagoras (ca.600 B.C.). But this objection loses force now that we have found the theorem applied even 1200 years earlier in the cuneiform texts of old Babylon. It is quite possible that Pythagoras became acquainted with the theorem in Babylon. That is about all that can be properly said about it.

CIRCLE AND HYPERBOLA

In plane geometry a *circle* is defined as the locus of a point whose distance from a fixed point is constant. Using this definition we now will derive the equation of a circle, given its center and radius. (See Figure 8.9.) If the center is (a, b), and if (x, y) is an arbitrary point on the circle, it follows by the distance formula that

$$r = \sqrt{(x - a)^2 + (y - b)^2}$$

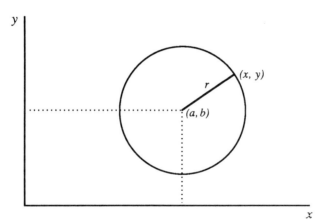

Figure 8.9. Circle.

This yields the general equation of a circle:

$$(x - a)^2 + (y - b)^2 = r^2$$

An important special case is the circle with the center at the origin $(0, 0)$ and radius $r = 1$; the resulting circle, called the unit circle, is given by

$$x^2 + y^2 = 1$$

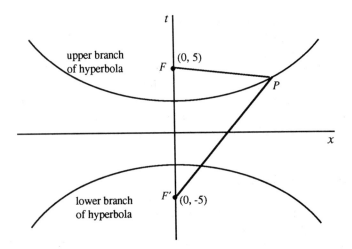

Figure 8.10. Hyperbola.

A *hyperbola* is the locus of a point the difference of whose distances from two fixed points is constant. For example, let us find the equation of a hyperbola whose foci are $(0, -5)$, $(0, 5)$ in the (x, t) plane and whose constant difference is 6. (See Figure 8.10.) See $F'P - FP = 6$, we have

$$\sqrt{x^2 + (t + 5)^2} - \sqrt{x^2 + (t - 5)^2} = 6$$

or

$$3\sqrt{x^2 + (t - 5)^2} = 5t - 9$$

or

$$16t^2 - 9x^2 = 144$$

The reader should reproduce the foregoing derivation, performing all calculations. Also write (by interchanging x and t) the equation of the hyperbola whose foci are $(-5, 0)$ and $(5, 0)$ and whose constant difference is 6.

An important special case is the unit hyperbola whose equation is

$$t^2 - x^2 = 1$$

In Figure 8.11, we see the graph of the unit hyperbola and also its asymptotes $t^2 - x^2 = 0$ (or $t + x = 0$, $t - x = 0$). The hyperbola approaches but never touches its asymptotes.

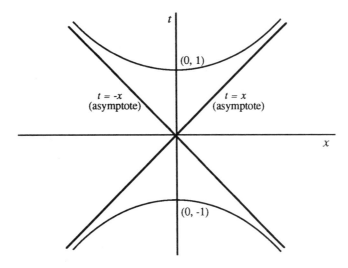

Figure 8.11. Unit hyperbola with its two asymptotes.

TRIGONOMETRY

In the history of mathematics we find that the discipline known as trigonometry was first studied intensively because of its usefulness in surveying and in astronomy. The use of trigonometry in the indirect measurement of angles or distances furnishes ample justification for the study of the subject. However, the numerical side is by no means the only important feature of trigonometry. The theoretical aspects are indispensable not only in mathematics but also in all fields of science.

Let θ represent angle *CAB*. (Figure 8.12.) Right triangles *ABC* and *ADF* are similar and hence

$$\frac{AB}{AC} = \frac{AD}{AF}$$

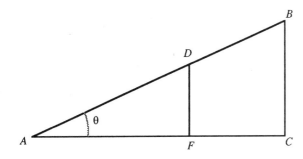

Figure 8.12. Similar triangles.

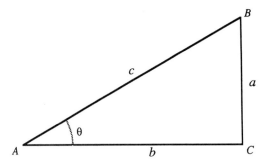

Figure 8.13. Right triangle used to define the circular trigonometric functions.

Thus the difference in size between the two triangles has no effect on the ratio of the hypotenuse to the base. This ratio has the same value for all right triangles with θ as the angle. Similarly we can state that the ratio of any two sides of a right triangle, with θ as one acute angle, depends only upon the size of θ and not on the size of the triangle.

Let θ represent any acute angle. (Figure 8.13.) From any point B on either side of θ, drop a perpendicular BC to the other side, thus forming a right triangle ABC. Let the small letters a, b, c represent the lengths of the sides, respectively, opposite the vertices A, B, and C. Then the values of the ratios which are named below depend only upon the size of θ and not upon the particular point B which was used.

a/b is called the tangent of θ.   Abbreviation: tan θ
a/c is called the sine of θ.   Abbreviation: sin θ
b/c is called the cosine of θ.   Abbreviation: cos θ
c/b is called the secant of θ.   Abbreviation: sec θ

Hence, tan θ, sin θ, cos θ, and sec θ are functions of θ; more explicitly, they are called the *(circular) trigonometric functions*.

The notion of angle can be generalized as follows. (Figure 8.14.) Suppose that a half-line, radiating from a point O, rotates about O, either clockwise or counterclockwise from an initial position OA to a terminal position OP. Then this rotation is said to generate an angle AOP, whose initial side is OA and terminal side is OP. The value of the angle is the *amount of rotation*, where we consider counterclockwise rotation as positive and clockwise rotation as negative. We shall say that an angle θ is in its standard position on a coordinate system if the vertex of θ is at the origin and the initial side of θ lies on the positive part of the horizontal axis. Choose any point P, not at the origin, on the terminal side of θ. (Figure 8.15.) Let (x, y) be the coordinates of P and let r be the radius given by

$$r = \sqrt{x^2 + y^2}$$

TRIGONOMETRY

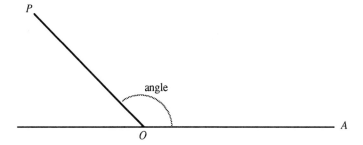

Figure 8.14. Angle AOP is the amount of rotation from initial position OA to the terminal position OP.

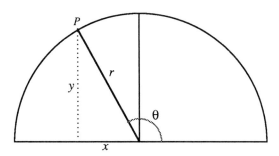

Figure 8.15. Circle used to define the circular trigonometric functions.

Then the trigonometric functions are defined as

$$\tan\theta = \frac{y}{x}, \quad \sin\theta = \frac{y}{r}, \quad \cos\theta = \frac{x}{r}, \quad \sec\theta = \frac{r}{x}$$

From this definition we see that

$$\tan\theta = \frac{\sin\theta}{\cos\theta}$$

The reason that these trigonometric functions are called circular is that they are defined in terms of a circle centered at the origin and of radius r. The equation of the circle is

$$x^2 + y^2 = r^2$$

If we take a circle of radius $r = 1$, then we see that

$$x = \cos\theta, \quad y = \sin\theta$$

Plate 8.3. William Blake's painting *Newton*.

HYPERBOLIC TRIGONOMETRY

Are there other kinds of trigonometric functions? Yes, there are *hyperbolic trigonometric functions*. In this case we take a rectangular hyperbola centered at the origin and with equation

$$t^2 - x^2 = \tau^2$$

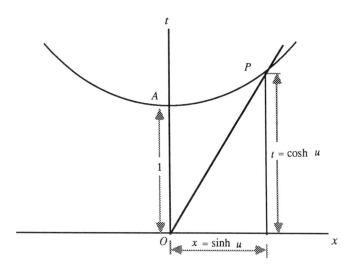

Figure 8.16. Unit hyperbola used to define the hyperbolic trigonometric functions.

HYPERBOLIC TRIGONOMETRY

Now we are calling our coordinate system (x, t) instead of (x, y), as we will always do when we are using hyperbolic functions. The Greek letter τ (tau) appearing in this equation is a constant, even as the letter r appearing in the equation of the circle was a constant. We now take the case where $\tau = 1$. The resulting unit hyperbola, shown in Figure 8.16, is

$$t^2 - x^2 = 1$$

In terms of this diagram, the hyperbolic trigonometric functions are

$$x = \sinh u, \quad t = \cosh u, \quad \tanh u = \frac{\sinh u}{\cosh u}$$

This result is completely analogous to Figure 8.17, which gives the circular trigonometric functions in terms of the unit circle $x^2 + y^2 = 1$, where we now measure the angle θ from the vertical axis, with the positive direction clockwise. This interplay between circular functions and hyperbolic functions is basic. The hyperbolic cosine of u ($\cosh u$) is pronounced to rhyme with "gosh u," the hyperbolic sine of u ($\sinh u$) is pronounced as though it were spelled "sinch u," and the hyperbolic tangent of u ($\tanh u$) is pronounced "tanch u."

We are now in a position to illuminate the meaning of the variable u in the equations

$$x = \sinh u, \quad t = \cosh u$$

as they relate to the point P, given by (x, t), on the unit hyperbola

$$t^2 - x^2 = 1$$

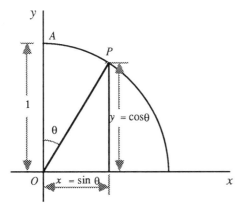

Figure 8.17. Unit circle used to define the circular trigonometric functions.

The variable u is called the *hyperbolic parameter*. Before we do so, however, we shall find the analogous meaning of the variable θ in the equations

$$x = \sin\theta, \quad y = \cos\theta$$

as they relate to the point P, that is, (x, y), on the unit circle

$$x^2 + y^2 = 1$$

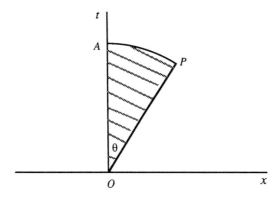

Figure 8.18. For unit circle, the angle θ is equal to twice the shaded area.

The most familiar interpretation of θ is, of course, that it is the radian measure of the angle AOP; that is,

$$\theta = \frac{\text{arc } AP}{\text{radius } OP} = \text{arc } AP$$

because for the unit circle the radius r is 1. But we also recall that the area of a circular sector of radius r and central angle θ (in radians) is given by $(1/2)r^2\theta$. (Figure 8.18). Since we are here dealing with the unit circle, this says that

$$\text{area of sector } AOP = \frac{1}{2}\theta$$

or, if we solve for θ,

$$\theta = \text{twice the area of sector } AOP$$
$$= \text{twice the shaded area}$$

Of course, we must realize that θ is a pure (dimensionless) number and what we mean is that θ is twice the number of square units which the (unit) radius vector OP sweeps out as P moves along the circle to its final position. Thus when the area of the sector AOP is one-half, then $\theta = 1$, and the coordinates of P represent sin 1 and cos 1. Negative values of θ would be interpreted as corresponding to areas swept over in a counterclockwise rotation of OP.

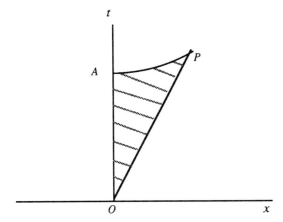

Figure 8.19. For unit hyperbola, the hyperbolic parameter u is equal to twice the shaded area.

Now for the unit hyperbola $t^2 - x^2 = 1$ there is an analogous interpretation for the variable u in the equations $x = \sinh u$, $y = \cosh u$. (Figure 8.19.) As the vector moves from the position OA to the position OP along the unit hyperbola,

$$\text{area of sector } AOP = \frac{1}{2} u$$

or, solving for u,

$$u = \text{twice the area of sector } AOP$$
$$= \text{twice the shaded area}$$

It can be shown that a positive value of u is associated with an area to the right of the t axis, and a negative value of u with an area to the left of the t axis. Areas are measured in terms of the unit square having OA as side. The term *hyperbolic radian* is sometimes used in connection with the variable u, but here again u is just a dimensionless real number. For example, sinh 2 and cosh 2 may be interpreted as the coordinates of P when the area of the sector AOP is just equal to the area of a square having OA as its side.

We can calibrate arc length on the unit circle by marking off the values of the radians θ along its circumference. Because θ is equal to arc length, the scale is uniform. We can also calibrate arc length on the unit hyperbola by marking off the values of the hyperbolic radians u along the hyperbolic curve. However, now the scale will not be uniform. (Figure 8.20.)

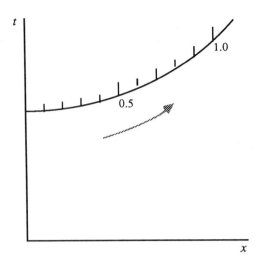

Figure 8.20. Calibration of arc length on the unit hyperbola in terms of the hyperbolic parameter u.

LINEAR TRANSFORMATIONS

In general, a linear transformation has the form

$$x' = ax + by$$
$$y' = cx + dy$$

which can be written in matrix notation as

$$\begin{pmatrix} a & b \\ c & d \end{pmatrix} \begin{pmatrix} x \\ y \end{pmatrix} = \begin{pmatrix} x' \\ y' \end{pmatrix}$$

Here the domain is the set of all ordered pairs of real numbers (x, y), geometrically the points in the Cartesian plane. The range consists of ordered pairs (x', y'), again points in the plane. Thus the mapping will be from the Cartesian plane to itself, or a portion of itself.

As an example, the matrix mapping

$$\begin{pmatrix} 1 & 2 \\ 3 & 4 \end{pmatrix}$$

maps the point $(1, 1)$ to the point

$$x' = (1)(1) + (2)(1) = 3$$
$$y' = (3)(1) + (4)(1) = 7$$

Linear Transformations

We see that the identity mapping

$$\begin{pmatrix} 1 & 0 \\ 0 & 1 \end{pmatrix}$$

maps each point (x, y) to the same point $(x', y') = (x, y)$.

In general, there are two ways in which we can regard a mapping: (1) the point (x, y) moves to a new position (x', y') with the position of the axes fixed; (2) alternatively, we may think that the point (x, y) remains in the same place, but it is assigned new coordinates (x', y') with respect to new axes. Thus a transformation may be regarded as a *change of position* (with fixed axes) or a *change of axes* (with fixed position).

Let us now investigate the transformation

$$\begin{pmatrix} 0 & -1 \\ 1 & 0 \end{pmatrix} \begin{pmatrix} x \\ y \end{pmatrix} = \begin{pmatrix} x' \\ y' \end{pmatrix}$$

which when written out becomes

$$x' = -y, \quad y' = x$$

Figure 8.21 shows fixed axes, with the point $(3, 5)$ in the first quadrant moving to the position $(-5, 3)$ in the second quadrant. Alternatively, Figure 8.22 shows the point fixed, but the axes moving by a 90° rotation. Let us verify that this transformation always produces a 90° rotation on lines through the origin by considering the equation $y = mx$, where m is the slope of the line. The image point of a point (x, y) on the line $y = mx$ is given by

$$x' = -y = -mx, \quad y' = x$$

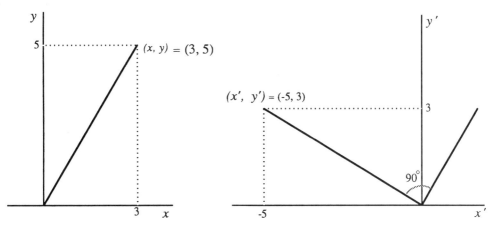

Figure 8.21. Transformation as a change of position with fixed axes.

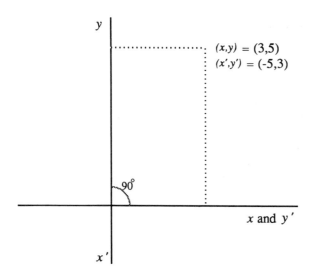

Figure 8.22. Transformation as a change of axes with a fixed position.

Hence, for $m \neq 0$,

$$x' = -my' \quad \text{or} \quad y' = -(1/m)x'$$

Dropping primes, we have found the image of $y = mx$ is $y = -(1/m)x$ (with the point moving and axes fixed). Since their slopes (see the coefficients of x in the two equations) are negative reciprocals of each other, the lines are perpendicular to each other. In this treatment, we have thought of the mapping as moving the points of the plane into new positions, keeping the reference axes in the same fixed position, and we have seen that the rotation is counterclockwise. On the other hand we could have equally as well thought of the points as remaining fixed, but being assigned new coordinates (new names with respect to new axes). In this case the axes are rotated onto each other through a 90° clockwise rotation.

Consider now the four transformations

$$\begin{pmatrix} 0 & -1 \\ 1 & 0 \end{pmatrix}, \quad \begin{pmatrix} -1 & 0 \\ 0 & -1 \end{pmatrix}, \quad \begin{pmatrix} 0 & 1 \\ -1 & 0 \end{pmatrix}, \quad \begin{pmatrix} 1 & 0 \\ 0 & 1 \end{pmatrix}$$

The first, as we have just seen, is a rotation of the plane about the origin through 90° counterclockwise. The second is a rotation through 180° counterclockwise, the third through 270° counterclockwise, and the fourth through 0°, 360°, 720°, ..., which as the identity changes nothing at all. From geometrical considerations the mapping

$$\begin{pmatrix} x' \\ y' \end{pmatrix} = \begin{pmatrix} 0 & -1 \\ 1 & 0 \end{pmatrix} \begin{pmatrix} 0 & 1 \\ -1 & 0 \end{pmatrix} \begin{pmatrix} x \\ y \end{pmatrix}$$

LINEAR TRANSFORMATIONS

rotates a line OP through $360°$ (first through $270°$, then the result through $90°$). Since the line OP is restored to its original position, we have

$$\begin{pmatrix} 0 & -1 \\ 1 & 0 \end{pmatrix} \begin{pmatrix} 0 & 1 \\ -1 & 0 \end{pmatrix} = \begin{pmatrix} 1 & 0 \\ 0 & 1 \end{pmatrix}$$

as can be verified by direct matrix multiplication. Also geometrical arguments (give them) indicate

$$\begin{pmatrix} 0 & 1 \\ -1 & 0 \end{pmatrix} \begin{pmatrix} -1 & 0 \\ 0 & -1 \end{pmatrix} = \begin{pmatrix} 0 & -1 \\ 1 & 0 \end{pmatrix}$$

Let us now discuss reflections. Let L be a line and P a point of the plane. A point P' is called the reflection of P in the line L if L is the perpendicular bisector of PP'. (Figure 8.23.) If P is on L, then $P' = P$.

Three special cases, namely, reflection of P in the x axis, reflection of P in the y axis, and reflection of P in the line $y = x$, are, respectively,

$$\begin{pmatrix} 1 & 0 \\ 0 & -1 \end{pmatrix}, \quad \begin{pmatrix} -1 & 0 \\ 0 & 1 \end{pmatrix}, \quad \begin{pmatrix} 0 & 1 \\ 1 & 0 \end{pmatrix}$$

These transformations convert the point (x, y) to the respective points $(x, -y)$, $(-x, y)$, and (y, x). Let (x, y) be first reflected to (x', y') in the x axis, and then $(x'\ y')$ reflected to (x'', y'') in the y axis. Obtain the single matrix expressing the composite reflection (x, y) to (x'', y''). The answer is

$$\begin{pmatrix} -1 & 0 \\ 0 & 1 \end{pmatrix} \begin{pmatrix} 1 & 0 \\ 0 & -1 \end{pmatrix} = \begin{pmatrix} -1 & 0 \\ 0 & -1 \end{pmatrix}$$

which gives $x'' = -x$, $y'' = -y$. This composite is called a reflection through the

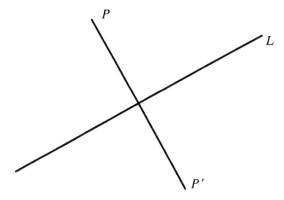

Figure 8.23. P' is the reflection of P with respect to line L, the perpendicular bisector of PP'.

origin because the origin bisects the line segment between (x, y) and (x'', y''). Note that a reflection through the origin is equivalent to a rotation of 180° about the origin. Draw diagrams for the above results.

Let us now consider stretchings and shrinkings. The mapping

$$\begin{pmatrix} 3 & 0 \\ 0 & 1 \end{pmatrix} \begin{pmatrix} x \\ y \end{pmatrix} = \begin{pmatrix} x' \\ y' \end{pmatrix}$$

maps points on the x interval from 0 to 1 to the interval 0 to 3, and maps points on the y interval from 0 to 1 to the interval 0 to 1. Thus it represents a stretch parallel to the x axis. In general, the three matrices

$$\begin{pmatrix} c_1 & 0 \\ 0 & 1 \end{pmatrix}, \quad \begin{pmatrix} 1 & 0 \\ 0 & c_2 \end{pmatrix}, \quad \begin{pmatrix} c_1 & 0 \\ 0 & c_2 \end{pmatrix} \quad \text{with } c_1 > 0, c_2 > 0$$

represent, respectively, a stretch (or shrinking) parallel to the x axis, parallel to the y axis, and parallel to both axes. Note that

$$\begin{pmatrix} 1 & 0 \\ 0 & c_2 \end{pmatrix} \begin{pmatrix} c_1 & 0 \\ 0 & 1 \end{pmatrix} = \begin{pmatrix} c_1 & 0 \\ 0 & c_2 \end{pmatrix}$$

Mappings of the form

$$\begin{pmatrix} c & 0 \\ 0 & c \end{pmatrix}$$

are examples of similarity transformations. The stretch (or shrinking) produced by a similarity transformation is the same along both axes, and this is what produces the similarity.

Let us now consider shears. A shear parallel to the x axis is given by the mapping

$$\begin{pmatrix} 1 & p \\ 0 & 1 \end{pmatrix} \begin{pmatrix} x \\ y \end{pmatrix} = \begin{pmatrix} x' \\ y' \end{pmatrix}$$

where p is a constant. That is,

$$x' = x + py, \quad y' = y$$

If $p = 1$, then Figure 8.24 shows the shear.

Thus every point on the side $P_0 P_1$ remains fixed under this shear. Points on the y axis map to points on the line $y = x$. In particular the point $P_3(0, 1)$ has

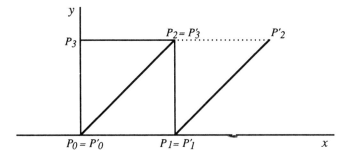

Figure 8.24. A shear parallel to the x axis.

the image $P'_3(1, 1)$. Thus a shear parallel to the x axis is a mapping of the plane in which each point is moved parallel to the x axis through a distance that is proportional to its distance from the x axis (this is the py of the mapping).

A shear parallel to the y axis is given by the mapping

$$\begin{pmatrix} 1 & 0 \\ q & 1 \end{pmatrix} \begin{pmatrix} x \\ y \end{pmatrix} = \begin{pmatrix} x' \\ y' \end{pmatrix}$$

where q is a constant. Analysis similar to that performed for a shear parallel to the x axis should be carried out by the reader.

ROTATIONS AND SHEARS

In the foregoing section we considered special cases of rotations and shears. Now we want to consider more general cases. Consider the transformation

$$\begin{pmatrix} x' \\ y' \end{pmatrix} = \begin{pmatrix} 1 & -\beta \\ \beta & 1 \end{pmatrix} \begin{pmatrix} x \\ y \end{pmatrix}$$

which is

$$x' = x - \beta y, \qquad y' = \beta x + y$$

If we consider this mapping as a change of axes with the points fixed, then the new axes are

$$x' = 0, \quad \text{or } x - \beta y = 0, \quad \text{or } x = \beta y, \quad \text{or } y = (1/\beta)x$$
$$y' = 0, \quad \text{or } \beta x + y = 0, \quad \text{or } y = -\beta x$$

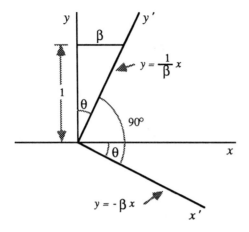

Figure 8.25. Rotation of axes by the rotation angle θ. The slopes of the new axes x', y' are negative reciprocals of each other.

From Figure 8.25 we see that the x' and y' axes are perpendicular (as the slopes of $y = (1/\beta)x$ and $y = -\beta x$ are negative reciprocals). Also we see that the y' axis has been rotated clockwise from the y axis by an angle θ, where $\beta = \tan \theta$. Thus we can write this rotation of axes as

$$\begin{pmatrix} x' \\ y' \end{pmatrix} = \begin{pmatrix} 1 & -\tan \theta \\ \tan \theta & 1 \end{pmatrix} \begin{pmatrix} x \\ y \end{pmatrix}$$

The quantity $x^2 + y^2$ is the squared distance from the origin to the point (x, y) in the old coordinate system, whereas $x'^2 + y'^2$ is the similar quantity in the new coordinate system. Since

$$x' = x - y \tan \theta, \qquad y' = x \tan \theta + y$$

we have

$$x'^2 + y'^2 = (x - y \tan \theta)^2 + (x \tan \theta + y)^2 = (1 + \tan^2 \theta)(x^2 + y^2)$$

Carry out the missing calculations. In general, we are interested in rotations that preserve distance. We can obtain such a transformation if we divide the coefficients of the foregoing transformation matrix by

$$\sqrt{1 + \tan^2 \theta} = \frac{1}{\cos \theta}$$

In other words, we must multiply the transformation matrix by $\cos \theta$. The result is

$$\begin{pmatrix} \cos \theta & -\tan \theta \cos \theta \\ \tan \theta \cos \theta & \cos \theta \end{pmatrix} = \begin{pmatrix} \cos \theta & -\sin \theta \\ \sin \theta & \cos \theta \end{pmatrix}$$

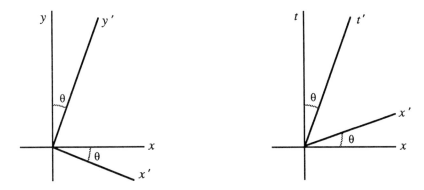

Figure 8.26. The given axes are x, y. The new axes are x', y'. Left: Rotation of axes by the rotation angle θ. Right: Shear of axes by the shear angle θ.

This matrix thus represents a *rotation of axes* clockwise through an angle θ such that distance is preserved. (Figure 8.26.) Verify that if

$$x' = x \cos \theta - y \sin \theta, \qquad y' = x \sin \theta + y \cos \theta$$

then

$$x'^2 + y'^2 = x^2 + y^2$$

In a similar vein to the arguments just given, consider the transformation

$$\begin{pmatrix} x' \\ t' \end{pmatrix} = \begin{pmatrix} 1 & -\beta \\ -\beta & 1 \end{pmatrix} \begin{pmatrix} x \\ t \end{pmatrix}$$

Note now that we are using the variables t and t' instead of y and y', as we generally will do when we are considering shears instead of rotations. This mapping is

$$x' = x - \beta t, \quad t' = -\beta x + t$$

If we consider this mapping as a change of axes with the points fixed, then the new axes are

$$x' = 0, \quad \text{or } x - \beta t = 0, \quad \text{or } x = \beta t, \quad \text{or } t = (1/\beta)x$$
$$t' = 0, \quad \text{or } -\beta x + t = 0, \quad \text{or } t = \beta x$$

From Figure 8.27, we see that the new axes (the x' and t' axes) are not perpendicular to each other.

We would expect this situation, because of the shearing action of the transformation. We see that the t' axis has been rotated clockwise from the t axis by

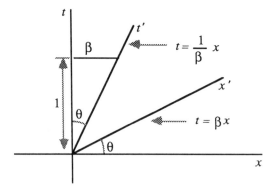

Figure 8.27. Shear of axes by the shear angle θ. The slopes of the new axes x', y' are positive reciprocals of each other.

an angle θ, where tan θ = β. However, the x' axis has been rotated counterclockwise from the x axis by the same angle θ. These conflicting rotations of the separate axes are characteristic of the shear of the two axes taken as an entity.

In the case of a rotation we saw that slope of the y' axis is 1/β and that the slope of the x' axis is −β. The fact that these two slopes are *negative reciprocals* indicated that these two axes are *orthogonal in the Euclidean geometry of rotation*. Now we have the case of a shear. In this case the slope of the t' axis is 1/β, and the slope of the x' axis is β. The fact that these two slopes are *positive reciprocals* indicates that these two axes are *orthogonal in the affine geometry of shears*.

Replacing β by tan θ, the matrix equation becomes

$$\begin{pmatrix} x' \\ t' \end{pmatrix} = \begin{pmatrix} 1 & -\tan\theta \\ -\tan\theta & 1 \end{pmatrix} \begin{pmatrix} x \\ t \end{pmatrix}$$

In dealing with shears, we are interested in squared time-space intervals defined as $t^2 - x^2$ before the shear, and as $t'^2 - x'^2$ after the shear. Since

$$x' = x - t\tan\theta, \qquad t' = -x\tan\theta + t$$

we have

$$t'^2 - x'^2 = (-x\tan\theta + t)^2 - (x - t\tan\theta)^2 = (1 - \tan^2\theta)(t^2 - x^2)$$

Carry out the missing calculations.

In general, we are interested in shears that preserve the so-called time-space interval. We can obtain such a transformation if we divide the coefficients of the transformation matrix by

$$\sqrt{1 - \tan^2\theta}$$

The result is

$$\begin{bmatrix} \dfrac{1}{\sqrt{1-\tan^2\theta}} & -\dfrac{\tan\theta}{\sqrt{1-\tan^2\theta}} \\ -\dfrac{\tan\theta}{\sqrt{1-\tan^2\theta}} & \dfrac{1}{\sqrt{1-\tan^2\theta}} \end{bmatrix}$$

Verify that the determinant of this matrix is equal to one. In terms of $\beta = \tan\theta$, this matrix is

$$\begin{bmatrix} \dfrac{1}{\sqrt{1-\beta^2}} & -\dfrac{\beta}{\sqrt{1-\beta^2}} \\ -\dfrac{\beta}{\sqrt{1-\beta^2}} & \dfrac{1}{\sqrt{1-\beta^2}} \end{bmatrix}$$

so the shearing transformation is in fact the Lorentz transformation

$$x' = \frac{x}{\sqrt{1-\beta^2}} - \frac{\beta t}{\sqrt{1-\beta^2}}$$

$$t' = -\frac{\beta x}{\sqrt{1-\beta^2}} + \frac{t}{\sqrt{1-\beta^2}}$$

If we define

$$\gamma = \frac{1}{\sqrt{1-\beta^2}}$$

then we obtain the standard form of the Lorentz transformation

$$x' = \gamma(x - \beta t)$$

$$t' = \gamma(-\beta x + t)$$

Figure 8.28 shows the unit hyperbola $t^2 - x^2 = 1$ with its asymptotes $t = x$ and $t = -x$. It also shows the axes x', t' which have undergone a shear of θ. Point P, which marks the intersection of the hyperbola with the t' axis, has coordinates $x = \sinh u$, $t = \cosh u$. These coordinates are the sides of the right triangle OQP, and since θ is the angle opposite the side $\sinh u$, we have

$$\tan\theta = \frac{\sinh u}{\cosh u} \quad \text{or} \quad \tan\theta = \tanh u$$

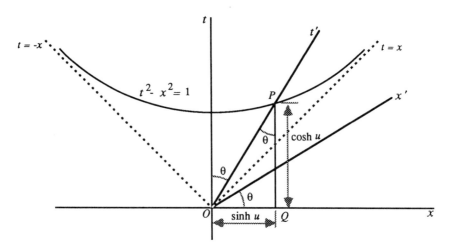

Figure 8.28. Unit hyperbola.

This equation gives the relationship between the shear angle θ (in radians) and the hyperbolic parameter u (in hyperbolic radians). Since $\beta = \tan \theta$, we can write this equation as

$$\beta = \tanh u$$

Let us now use the result

$$\sqrt{1 - \beta^2} = \sqrt{1 - \tanh^2 u} = \frac{1}{\cosh u}$$

Thus the shear (or Lorentz) transformation

$$x' = (1 - \beta^2)^{-1/2} x - (1 - \beta^2)^{-1/2} \beta t$$
$$t' = -(1 - \beta^2)^{-1/2} \beta x + (1 - \beta^2)^{-1/2} t$$

becomes

$$x' = (\cosh u)x - (\sinh u)t$$
$$t' = -(\sinh u)x + (\cosh u)t$$

The matrix of the shear (or Lorentz) transformation is

$$\begin{pmatrix} \cosh u & -\sinh u \\ -\sinh u & \cosh u \end{pmatrix}$$

The determinant of this matrix is

$$\cosh^2 u - \sinh^2 u = 1$$

Compare the foregoing shear transformation matrix with the rotation transformation matrix

$$\begin{pmatrix} \cos\theta & -\sin\theta \\ \sin\theta & \cos\theta \end{pmatrix}$$

which has determinant

$$\cos^2\theta + \sin^2\theta = 1$$

ROTATION OF SPACE COORDINATES

We now want to tell the story of two surveyors, the first who takes his directions by the North Star, and the second who uses a compass. That is, the first surveyor uses true north and the second surveyor magnetic north. The result would be two different sets of records for the same points:

	True North Survey	Magnetic North Survey
Same Point	(x, y)	(x', y')

The distance between the point and the origin is given in the true north survey as

$$\sqrt{x^2 + y^2} \qquad (8.1)$$

and in the magnetic north survey as

$$\sqrt{x'^2 + y'^2} \qquad (8.2)$$

In Euclidean geometry, both of these quantities are the same. Thus we have discovered the principle of the *invariance of distance*. That is, one get exactly the same distances from the magnetic north coordinates as from the true north coordinates, despite the fact that the two sets of surveyors numbers are quite different. This invariance of the distance means that the two frames of reference are related by a rotation transformation. We can verify this by substituting the rotation transform

$$x' = x\cos\theta - y\sin\theta$$
$$y' = x\sin\theta + y\cos\theta$$

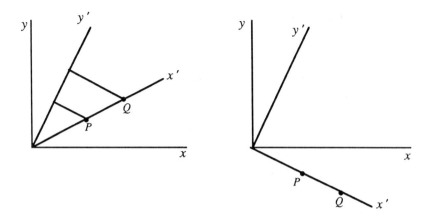

Figure 8.29. Left: Points P and Q on the x' axis correspond to different distances on the y' axis, so these two axes are not orthogonal in Euclidean geometry. Right: Points P and Q on the x' axis correspond to the same distance on the y' axis, so these two axes are orthogonal in Euclidean geometry. The slopes of these two axes are negative reciprocals of each other.

into expression (8.2) for the distance. The result will be expression (8.1) for the distance.

Consider now a true north Cartesian system of coordinates x, y for a true north surveyor. By construction the x axis is perpendicular (i.e., orthogonal) to the y axis. Now consider two other straight lines which intersect at the origin, and let these two lines be the coordinate system of another surveyor. Call him the primed surveyor, and let these two lines be his axes x', y'. If *for him*, each point on his x' axis corresponds to the same y' distance, his two axes are called *orthogonal in Euclidean geometry*. (See Figure 8.29.)

The notion of *orthogonality* is a keystone in geometry, and there is a simple algebraic formula, known as the inner product, which characterizes it. Suppose that x' and y' are two lines, and we want to determine whether they are orthogonal or not. We suppose that they go through the origin. Take the points $a_1 = (x_1, y_1)$ on x', and $a_2 = (x_2, y_2)$ on y', to serve as direction vectors for x' and y'. See Figure 8.30.

Now define the *inner product* (or dot product) of the two vectors a_1 and a_2 to be

$$a_1 \cdot a_2 = x_1 x_2 + y_1 y_2$$

In three-dimensions, the inner product of the vectors $a_1 = (x_1, y_1, z_1)$ and $a_2 = (x_2, y_2, z_2)$ would be

$$a_1 \cdot a_2 = x_1 x_2 + y_1 y_2 + z_1 z_2$$

ROTATION OF SPACE COORDINATES

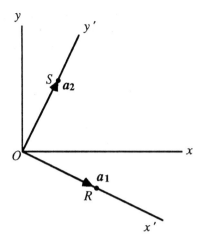

Figure 8.30. Direction vectors OR for x' and OS for y'.

That is, the inner product is the sum of the cross-products of the components. In the language of matrix multiplication, the inner product (in two dimensions) is

$$a_1 \cdot a_2 = (x_1, y_1) \begin{pmatrix} 1 & 0 \\ 0 & 1 \end{pmatrix} \begin{pmatrix} x_2 \\ y_2 \end{pmatrix}$$

By the analytic geometry of the Cartesian plane, we know that two lines are orthogonal if their slopes are negative reciprocals of each other. Because the origin $(0, 0)$ and the point (x_1, y_1) each lie on line x', its equation is

$$\frac{y - 0}{x - 0} = \frac{y_1 - 0}{x_1 - 0}$$

or

$$y = \frac{y_1}{x_1} x$$

so its slope is y_1/x_1. Similarly, the slope of line y' is y_2/x_2. The condition that these two slopes are negative reciprocals of each other is

$$\frac{y_1}{x_1} = -\frac{x_2}{y_2}$$

which is

$$x_1 x_2 + y_1 y_2 = 0$$

But this is exactly the same condition that the inner product of the direction vectors of the two lines be zero. Thus we conclude that the two lines x' and y' are orthogonal in Euclidean geometry if and only if the inner product of any two direction vectors a_1 and a_2 for these lines vanishes. That is,

$$x' \perp y' \quad \text{if and only if} \quad a_1 \cdot a_1 = 0$$

Suppose now that a_1 and a_2 represent the same point; that is, $a_1 = a_2 = a = (x, y)$. Then the inner product becomes the sum of squares

$$a \cdot a = x^2 + y^2$$

and we recognize this expression as the square of the distance d from the origin to the point; that is,

$$a \cdot a = d^2$$

The quantity d^2 being a sum of squares is always positive unless the point a is the origin, in which case d^2 is zero. Thus the quantity d^2 is never negative so the Euclidean distance d can be defined as the positive square root of d^2.

In this section we have considered Euclidean spatial geometry. If two different orthogonal coordinate systems are used, one being a rotation of the other, the distance d between two points is an invariant. Also we have introduced the concept of Euclidean inner product. The condition that two lines be orthogonal is the same as the condition that the inner product of their direction vectors vanish. Finally, the inner product of a direction vector with itself must be nonnegative, so the nonnegative distance d can be defined as the positive square root of this inner product.

SHEAR OF SPACE-TIME COORDINATES

We now want to tell the story of two people in different inertial frames, the first who measures spatial distance as x and time as t and the second who measures distance as x' and time as t'. An event is defined as a point in space-time. Thus the two different sets of records for the same event are

	First Person	Second Person
Same Event	(x, t)	(x', t')

The time-space interval between the event and the origin for the first person is defined as

$$\tau = \sqrt{t^2 - x^2} \tag{8.3}$$

SHEAR OF SPACE-TIME COORDINATES

and for the second person the time-space interval is defined as

$$\tau = \sqrt{t'^2 - x'^2} \qquad (8.4)$$

We require that these two time-space intervals be the same; that is, both are equal to the same τ. The two people have different space and time coordinates for the same two events, but when they calculate their time-space intervals between these two events, their results will agree. This is the principle of the *invariance of time-space interval*. Invariance of interval means that the two frames of reference are related by a shear transformation. We can verify this by substituting the shear transformation

$$x' = x \cosh u - t \sinh u$$
$$t' = -x \sinh u + t \cosh u$$

into expression (8.4) for τ. The result will be expression (8.3) for τ.

Consider now a space-time system of coordinates x, t for a given inertial frame. By construction the x axis is perpendicular (i.e., orthogonal) to the t axis. Now consider two other straight lines x', t' which intersect at the origin. Let line t' represent the path of a person moving at constant speed in the given inertial frame. If line x' represents a set of events that are simultaneous for him, then lines x' and t' are said to be *orthogonal in space-time geometry*. See Figure 8.31.

Let us now find the algebraic formula for the inner product that characterizes orthogonality in space-time geometry. Suppose that x' and t' are two lines, and we want to determine whether they are orthogonal or not. Each line goes through the origin. Take the events $\boldsymbol{a}_1 = (x_1, t_1)$ on x', and $\boldsymbol{a}_2 = (x_2, t_2)$ on t', to serve as direction vectors for x' and t'. Now define the *time-space inner product*

$$\boldsymbol{a}_1 \cdot \boldsymbol{a}_2 = t_1 t_2 - x_1 x_2$$

That is, the inner product is the difference of the cross-products of the components.

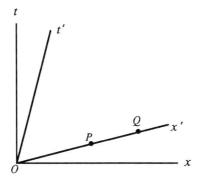

Figure 8.31. Events P and Q on the x' axis correspond to the same time on the t' axis, so these two axes are orthogonal in space-time geometry. The slopes of these two axes are positive reciprocals of each other.

The inner product can also be written as

$$a_1 \cdot a_2 = (x_1, t_1) \begin{pmatrix} -1 & 0 \\ 0 & 1 \end{pmatrix} \begin{pmatrix} x_2 \\ t_2 \end{pmatrix}$$

In space-time geometry, we know that two lines are orthogonal if their slopes are positive reciprocals of each other. The slope of the x' axis is t_1/x_1, and the slope of the t' axis is t_2/x_2. The condition that these two slopes are positive reciprocals of each other is

$$\frac{t_1}{x_1} = \frac{x_2}{t_2}$$

which is

$$t_1 t_2 - x_1 x_2 = 0$$

But this is exactly the same condition that the inner product of the direction vectors of the two lines be zero. Thus we conclude that the two lines x' and t' are orthogonal in space-time geometry if and only if the time-space inner product of any two direction vectors a_1 and a_2 of these lines vanishes. That is,

$$x' \perp t' \quad \text{if and only if} \quad a_1 \cdot a_2 = 0$$

Suppose now that a_1 and a_2 are the same vector; that is, $a_1 = a_2 = a = (x, t)$. Then the inner product becomes the difference of squares

$$a \cdot a = t^2 - x^2$$

and we recognize this expression as the square of the time-space interval τ; that is,

$$a \cdot a = \tau^2$$

The quantity τ^2 being a difference of squares can be positive, zero, or negative. Thus we can characterize three kinds of vector a, according to the table

$a = (t, x)$ is	$a \cdot a = \tau^2 = t^2 - x^2$ is
Timelike	Positive
Lightlike	Zero
Spacelike	Negative

Note that as an alternative we could define the *space-time inner product* of $a_1 = (x_1, t_1)$ and $a_2 = (x_2, t_2)$ as

$$a_1 \cdot a_2 = x_2 x_1 - t_1 t_2$$

This inner product is the negative of the time-space inner product just defined. In case a_1 and a_2 are the same vector $a = (x, t)$, the space-time inner product becomes

$$a \cdot a = x^2 - t^2$$

which is equal to the square of the space-time interval σ; that is,

$$a \cdot a = \sigma^2$$

The square of the time-space interval τ is thus the negative of the square of the space-time interval σ; that is,

$$\tau^2 = -\sigma^2 = t^2 - x^2$$

In this book we generally use the time-space interval τ instead of the space-time interval σ, but this is a matter of preference only.

In this section we considered space-time geometry. If two different orthogonal coordinate systems are used, one being the shear of the other, the time-space interval τ between two points is an invariant. Also we have introduced the concept of time-space inner product. The condition that two lines be orthogonal is the same as the condition that the inner product of their direction vectors vanish. Finally, the inner product of a direction vector with itself can be positive, zero, or negative, in which cases the line is called timelike, lightlike, or spacelike, respectively.

Timelike straight lines can be called *inertia lines* as they represent the paths of unaccelerated bodies. Lightlike lines are simply called *lightlines*, and they are the paths of photons, neutrinos, and gravitons. Spacelike straight lines are sometimes called *separation lines*, since particles that occupy distinct locations on a spacelike straight line must be separate and distinct particles. The reason is that two events on a spacelike straight line cannot both lie on the path of a single particle.

Now we come to the most difficult aspect of relativity theory, the one that can never fully be understood in a physical sense. We consider one particle of light, a photon, traveling at speed $c = 1$ in vacuum. Suppose that it travels along the lightlike line given by $t = x$ in our frame of reference. Now we want to try to explain what we mean when we say that a photon has no inertial frame of reference. Let us try to construct such a frame mathematically and see what results. The lightlike line in question would serve both as the distance axis x' and the time axis t' of the inertial frame of the photon, if such as inertial frame existed. Because $\tau^2 = 0$ for a lightlike line, it follows that a lightlike line is orthogonal to itself. Thus the x' and t' axes, which coincide, are orthogonal as required for an inertial frame. Thus mathematically the frame of reference for a photon consists of two orthogonal axes x' and t' that coincide with the light line. Because this frame is one dimensional, it can only reach events that lie along the given lightlike line of the given two-dimensional x, t frame. Thus this constructed photon frame is not a bona fide inertial frame, and this is what we mean when we say that there can be no inertial frame for a photon.

Because the x' and t' axes of the constructed photon frame coincide, we can say that all distances exist at the same time, and also that all times exist at the same distance. That is, a photon is at all places at the same time, and at all times at the same place. In other words, when we perceive that a photon has traveled a distance x in time t, the photon perceives that it has spent no time ($\tau = 0$) for this journey. Thus for a photon all locations in its path are simultaneous, so the photon covers an infinite distance (to us) in zero time (to it). This means that a photon is infinite because its presence in space goes from negative infinity to infinity. This is the space aspect of a photon.

Now let us look at the time aspect. When we perceive that a photon has traveled a distance x in time t, the photon perceives that it has covered no distance ($\sigma = 0$) for this journey. Thus for a photon all times in its path are at the same location, so the photon spends an eternity of time (to us) and yet remains at the same place (to it). This means that a photon is eternal since its presence in time goes from negative eternity to eternity.

A photon is not duration and space, but a photon endures and is present. For a photon, all times t (for us) are the same time ($\tau = 0$) for it and all places x (for us) are the same place ($\sigma = 0$) for it. Time does not pass for a photon (because $\tau = 0$), so a photon is ageless. Space does not extend for a photon (because $\sigma = 0$), so a photon is spaceless. A photon endures forever and is everywhere present.

Time and space are frozen (i.e., null and void) for a photon which to us appears to be traveling at speed c in vacuum. By its clock, the photon does not age. It takes no time (to it) to travel any distance (to us). By its rod, a photon does not move. It requires no distance (to it) to travel any length of time (to us). A photon is at the same time at all places, and at the same place at all times. Time and space (in our human sense) do not exist for a photon. A photon is infinite, eternal, perfect.

ROTATION VERSUS SHEAR

Let us compare the rotation transform with the shear (or Lorentz) transform. We will first think of these transformations as a change of position with fixed axes. The rotation transform maps the circle $x^2 + y^2 = r^2$ into the circle $x'^2 + y'^2 =$

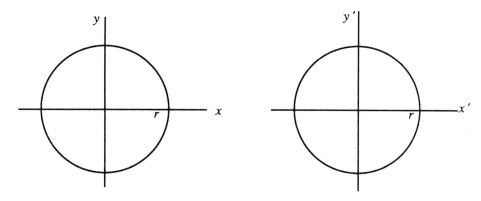

Figure 8.32. Invariance of the radius r in transformation results in circular symmetry.

r^2. To verify this, we write (fill in the missing steps)

$$x'^2 + y'^2 = (x \cos \theta - y \sin \theta)^2 + (x \sin \theta + y \cos \theta)^2 = x^2 + y^2 = r^2$$

Thus the radius r is invariant under the rotation transform. From Figure 8.32, we see that both circles have exactly the same shape.

The shear transform maps the hyperbola $t^2 - x^2 = \tau^2$ into the hyperbola $t'^2 - x'^2 = \tau^2$. To verify this, we have (fill in missing steps)

$$\tau^2 = t'^2 - x'^2 = (-x \sinh u + t \cosh u)^2$$
$$- (x \cosh u - t \sinh u)^2 = t^2 - x^2 = \tau^2$$

Thus the time-space interval τ is invariant under the shear transformation. From Figure 8.33, we see that both hyperbolas have exactly the same shape.

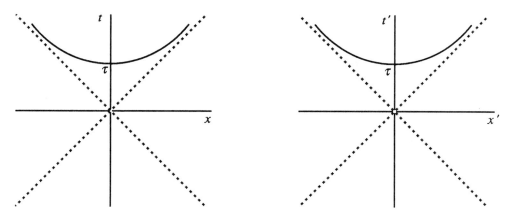

Figure 8.33. Invariance of the time-space interval τ in transformation results in hyperbolic symmetry.

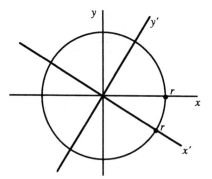

Figure 8.34. Invariance of *r* means that circle is left intact under rotation of axes.

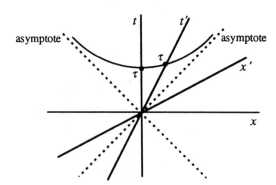

Figure 8.35. Invariance of τ means that hyperbola is left intact under shear of axes.

Next let us think of these transformations as a change of axes with the same positions. The circle remains fixed under the rotation transform, as seen by Figure 8.34. In either frame of reference, a circle is a circle. Under the shear transform, the hyperbola remains fixed. (Figure 8.35.) Certainly, the curve is a hyperbola with respect to the x, t frame. But is it a hyperbola with respect to the sheared axes x', t', as seen in the same diagram with the x, t axes removed? (Figure 8.36.) The curve is indeed a hyperbola in the sheared frame, with the apex at the point $x' = 0$, $t' = \tau$ on the t' axis. On sheared axes, the grid system consists of parallelograms instead of squares, as seen in Figure 8.37.

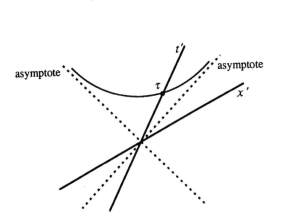

Figure 8.36. Hyperbola plotted on sheared axes.

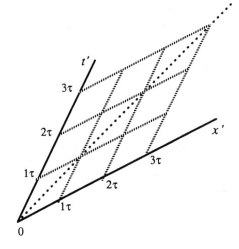

Figure 8.37. Grid system of parallelograms.

GROUP THEORY

A group may be described as a set of operations to be performed on a given object so chosen that, when several operations are performed in succession, the result is equivalent to just one operation of the group. Also any sequence of operations can always be undone by some operation within the group. As a single example, let a group contain the following two operations as elements: (1) doing nothing and (2) moving a switch for an electric light bulb. Originally, the light is out. It can be seen that any succession of element 1 reproduces element 1 (leaving the room dark). An even number of applications of element 2 is equivalent to element 1. However, an odd number of applications of element 2 is equivalent to element 2 (the room lights up). In this section, we will use these ideas.

Group theory is a powerful and elegant branch of algebra that emerges from the mathematical study of invariance and symmetry. Much of the basic mathematical structure of group theory, especially in its geometrical and topological aspects, is due to Poincaré. In 1905, he introduced group theory into physics by devising the fundamental group which describes the space-time structure of the universe. This is the Poincaré group, which includes the Lorentz transformation as a subgroup. In this book we deal only with this subgroup, and especially only with its two-dimensional x, t form. However, this simplified approach has the advantage of letting us see the underlying hyperbolic structure of space-time without the hindrance that usually comes from introducing an excessive amount of formal

Plate 8.4. John Dee (1527–1608) at age 67.

mathematics. But some abstract mathematics is useful, so we will now proceed to the definition of a group.

The formal definition is given in this paragraph. A set G of elements, satisfying the following four conditions, is called a *group:*

1. A law of composition (which for use of a word will be called *multiplication*) is defined for the set, such that the multiplication of each pair of elements f and g gives an element h of the set. This can be written as $fg = h$. Element h is called the product of the elements f and g, which are called factors. In general, the product of two factors depends upon the order of multiplication, so that the elements fg and gf can be different.

2. The multiplication is *associative.* If f, g, and h are any three elements, the product of the element f with the element gh is equal to the product of the element fg with h:

$$f(gh) = (fg)h$$

3. The set G contains a unit element e (called the *identity*), giving for every element f in the set G the relation

$$ef = fe = f$$

4. For any element f in the set G there is an element f^{-1} (called the *inverse*) such that

$$f^{-1}f = ff^{-1} = e$$

Let us make this additional note. If the multiplication is commutative, that is, if for any pair of elements f, g we have $fg = gf$, then the group is said to be *commutative* or *Abelian.*

Having given the formal definition, let us now try to recount the central ideas. We have a collection of objects and a rule for multiplying them. Suppose that the set is made up of just two numbers, 1 and -1, and the set operation of multiplication is just ordinary multiplication. The multiplication of any two numbers in this set, say, -1 and -1, is also a member of the set, in this case 1. The number 1 acts as the identity element, because it does not change what it multiplies. The inverse of an element undoes the effect of that element. Because $(-1) \cdot (-1) = 1$, we see that the inverse of -1 is -1 itself.

An essential property of quantum physics is that experiments affect one another and their sequential order is essential. Quantum theory is inherently noncommutative, and its natural language is group theory.

Let us examine the x, y coordinate plane of analytic geometry. A rotation of this plane about the origin through an angle θ can be considered as a symmetry of the plane, since it preserves the relations of distance and angle. The set of all these rotations form a group. Indeed, if we perform a rotation through an angle θ_1 and follow this with a rotation through an angle θ_2, then the result is also a rotation, namely, one through the angle $\theta_1 + \theta_2$. This represents the rule for the multiplication (in the group theory sense) of any two rotations. It is not hard to check that the remaining three properties required for a group are satisfied. For example, a rotation through the angle θ_1 followed by a rotation through the angle $\theta_2 + \theta_3$ is the same as a rotation through the angle $\theta_1 + \theta_2$ followed by a rotation through the angle θ_3. A rotation through the angle zero can be used for the identity element e. The inverse of the rotation θ would be the rotation $-\theta$.

Instead of dealing with rotations in two-dimensions, we can do the same in three dimensions, or in higher-dimensional spaces. In each case the rotations about the origin form a group. Also, for the plane, we can consider the totality of all rigid motions of the plane. These also form a group, which includes the rotation group as a subgroup. In addition to the rotations, this larger group includes translations of the plane. Similarly, groups of rigid motions can be formed for spaces of higher dimensions. Group theory can be made the foundation of geometry. Poincaré was instrumental in this kind of study. He showed the way in which non-Euclidean geometries are generated by an application of group theory to nonrigid bodies.

Let us now be more specific and show that the set of rotations about the origin in the x, y plane makes up a group. As we know from the section on rotations and shears, a rotation through an angle θ such that distance is preserved is represented by the matrix

$$\begin{pmatrix} \cos\theta & -\sin\theta \\ \sin\theta & \cos\theta \end{pmatrix}$$

This matrix is an element of the set. The set has an infinite number of elements, one for every value of θ. The law of composition is matrix multiplication. The multiplication of two rotations, one with angle θ_1 and the other with angle θ_2, is

$$\begin{pmatrix} \cos\theta_1 & -\sin\theta_1 \\ \sin\theta_1 & \cos\theta_1 \end{pmatrix} \begin{pmatrix} \cos\theta_2 & -\sin\theta_2 \\ \sin\theta_2 & \cos\theta_2 \end{pmatrix}$$

Upon multiplication we find that this product is equal to

$$\begin{bmatrix} \cos(\theta_1 + \theta_2) & -\sin(\theta_1 + \theta_2) \\ \sin(\theta_1 + \theta_2) & \cos(\theta_1 + \theta_2) \end{bmatrix}$$

which is the element representing a rotation of $\theta_1 + \theta_2$. Since matrix multiplication is associative, that condition for a group is satisfied. The unit element is obtained by letting $\theta = 0$; the result is the unit matrix

$$\begin{pmatrix} 1 & 0 \\ 0 & 1 \end{pmatrix}$$

Finally, the inverse element for the rotation θ is the element for rotation $-\theta$, namely,

$$\begin{pmatrix} \cos \theta & \sin \theta \\ -\sin \theta & \cos \theta \end{pmatrix}$$

This group is called the *rotation group*.

Next let us show that the set of shears about the origin in the x, t plane makes up a group. As we know, a shear with angle θ has hyperbolic parameter u given by

$$\beta = \tan \theta = \tanh u$$

If we require that the time-space interval be preserved, then the shear is represented by the matrix

$$\begin{pmatrix} \cosh u & -\sinh u \\ -\sinh u & \cosh u \end{pmatrix}$$

This matrix is an element of an infinite set. The law of composition is matrix multiplication. A shear with hyperbolic parameter u_1 followed by a shear with hyperbolic parameter u_2 gives a shear with hyperbolic parameter $u_1 + u_2$:

$$\begin{pmatrix} \cosh u_1 & -\sinh u_1 \\ -\sinh u_1 & \cosh u_1 \end{pmatrix} \begin{pmatrix} \cosh u_2 & -\sinh u_2 \\ -\sinh u_2 & \cosh u_2 \end{pmatrix}$$
$$= \begin{pmatrix} \cosh (u_1 + u_2) & -\sinh (u_1 + u_2) \\ -\sinh (u_1 + u_2) & \cosh (u_1 + u_2) \end{pmatrix}$$

The associative condition is satisfied because matrix multiplication is associative. The unit element is obtained by letting $u = 0$; the result is the unit matrix. Finally, the inverse element for hyperbolic parameter u is the element for hyperbolic parameter $-u$, namely,

$$\begin{pmatrix} \cosh u & \sinh u \\ \sinh u & \cosh u \end{pmatrix}$$

This group is the *shear group*, or two-dimensional Lorentz transformation. It is a subgroup of the Poincaré group, which describes the fundamental space-time symmetry of nature. Henry Wadsworth Longfellow writes[9]

> Slowly the hour-hand of the clock moves round;
>   So slowly that no human eye hath power
>   To see it move! Slowly in shine or shower
>   The painted ship above it, homeward bound,
> Sails, but seems motionless, as if aground;
>   Yet both arrive at last; and in his tower
>   The slumberous watchman wakes and strikes the hour,
>   A mellow, measured, melancholy sound.
> Midnight! The outpost of advancing day!
>   The frontier town and citadel of night!
>   The watershed of Time, from which the streams
> Of Yesterday and Tomorrow take their way,
>   One to the land of promise and of light,
>   One to the land of darkness and of dreams! . . .
> It is the mystery of the unknown
>   That fascinates us; we are children still,
>   Wayward and wistful; with one hand we cling
> To the familiar things we call our own,
>   And with the other, resolute of will,
>   Grope in the dark for what the day will bring.

---

[9] Longfellow, Henry Wadsworth. "The Two Rivers." From *The Complete Poetical Works of Henry Wadsworth Longfellow*. Houghton Mifflin Company, Boston, 1882.

CHAPTER 9

CATENARY AND RELATIVITY

When to the sessions of sweet silent thought
I summon up remembrance of things past
I sigh the lack of many a thing I sought
And with old woes new wail my dear time's waste.

William Shakespeare (1564–1616)
Sonnet 30

SCIENCE AS FABLE

Science is a set of relevant, consistent statements of general validity that also contain an element of surprise. In particular, relativity theory is almost always surprising, startling, an affront to conventional perceptions. This element of surprise gives

Plate 9.1. Galileo Galilei (1564–1642)

rise to difficulties, but it is the essence of the excitement. When first encountered, the statements of special relativity are not obvious. They become acceptable only after considerable study in relation to many other facts and explanations. The novelty inherent in the theory has always produced an initial feeling of disbelief and consternation. The feeling is that the theory is too complicated to understand, especially because it deals with ideas that cannot be discovered or verified in a simple, direct way. Yet one of the most basic drives is the desire to understand. For most of humankind's existence, the response to human curiosity has been the invention of myths and fables. They are fascinating and often profound. Yet many of the historical ones lack consistency and simplicity. Science is a fable that has been made consistent. In this chapter, we wish to introduce the main concepts of special relativity by means of a fable, the metaphor of the ants.

METAPHOR OF ANTS

An old man has set up a colony of ants. The ants live at the lowest point of a long hanging rope suspended at both ends. (Figure 9.1.) Galileo believed that the shape of a rope of uniform linear density hanging from its end points was part of a parabola. This is not the case. The curve it forms, called a catenary (from *catena*, Latin for "chain"), has an equation given by the hyperbolic cosine. The ants live at the lowest point of the catenary.

Soldier ants can venture out from the center of the catenary (i.e., the lowest point) and climb up the rope. By forming a chain of ants, they can measure the distance along the rope (i.e., the arc length) from the center to any point on the rope. This distance we call x.

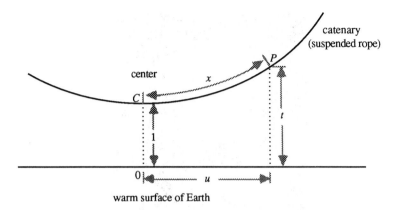

Figure 9.1. The ants live at point C on the catenary CP.

The surface of the ground is warm, and for this reason the ants live at the center of the catenary. It takes one day for an egg to hatch at the center. However, if a new egg is taken to a distance x, the egg takes a longer time to hatch. This time we call t. In fact, this time t is given by the vertical distance to the warm surface. The reason is that the time of incubation depends upon how far away the egg is from the warm body.

We take the point (on the level ground) directly under the center of the catenary as our origin of coordinates. The abscissa represents the hyperbolic parameter u and the ordinate the time t. Point P on the catenary is represented by the coordinates (u, t). The center C is $(0, 1)$.

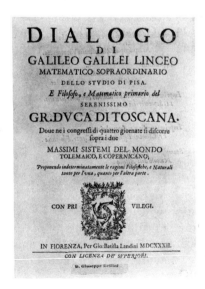

Plate 9.2. Title page of Galileo's *Dialogo*, 1632.

Plate 9.3. Engraved title page of Galileo's *Dialogo*, 1632, showing Aristotle, Ptolemy, and Copernicus discussing the structure of the Universe.

The equation of the catenary is

$$t = \cosh u$$

and this equation gives the time for an egg to hatch. At the center ($u = 0$) the time is

$$t = \cosh 0 = 1$$

which agrees with what we said.

The arc length along the catenary is

$$x = \sinh u$$

At the center ($u = 0$) the arc length is

$$x = \sinh 0 = 0$$

as we would expect.

TIME-SPACE INTERVAL

So much for the mathematics. However, the ants are quite clever. They make a table of incubation time t as a function of distance x. They perform various numerical operations, and finally they discover that if they subtract x^2 from t^2, they

always obtain 1; that is

$$t^2 - x^2 = 1$$

Because square units are confusing, they take the positive square root and call the result the time-space interval which characterizes every point on the catenary to which they are confined:

$$\text{time-space interval} = \sqrt{t^2 - x^2} = 1$$

The old man confirms their result, for

$$\sqrt{t^2 - x^2} = \sqrt{\cosh^2 u - \sinh^2 u} = \sqrt{1} = 1$$

LORENTZ TRANSFORM

Close by the old man has set up another colony of ants (called the primed colony) on a hanging rope identical to that of the original colony (called the unprimed colony) but displaced by the horizontal distance u_0 along the ground. (Figure 9.2.) The origin of coordinates of the primed colony is directly under the center of their catenary, and they measure time and distance in the same way as the unprimed colony. For the primed colony, time is denoted by t', distance along their catenary by x', and horizontal distance by u'.

Because the ants have tunnel vision, neither colony is aware of the other colony. However, humans have conquered isolation by means of television, and in this spirit the old man sets up a laser holographic system for the ants. Each point on the unprimed catenary is projected vertically to the point directly above

Plate 9.4. The Golden Gate Bridge, with two catenaries.

Lorentz Transform

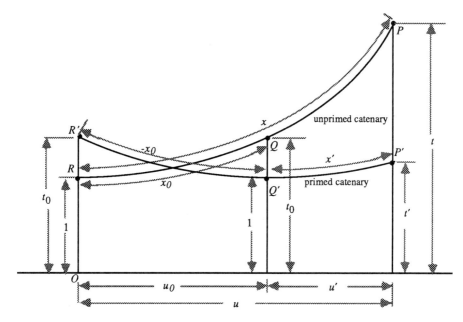

Figure 9.2. The unprimed colony lives at the lowest point R on catenary RQP, and the primed colony lives at the lowest point Q' on catenary $R'Q'P'$.

or below it on the primed catenary. Similarly, each point on the primed catenary is projected vertically to the point directly above it or below it on the unprimed catenary. In this way a faithful three-dimensional image of everything on the unprimed catenary is projected to the primed catenary, and vice versa. As a result, each colony of ants thinks that the other colony lives on its own catenary. Neither colony is aware of the existence of the other catenary at all. As far as the ants know, point Q is the same point as Q', point P is the same point as P', and so on.

The space-time coordinates of the joint point P, P' are (x, t) according to the unprimed colony and (x', t') according to the primed colony. Thus there is an honest difference of opinion as to the coordinates of what the two colonies perceive as the exact same point. The unprimed colony is located at R and the primed colony at Q'. The unprimed colony perceives point Q' as point Q, which in their coordinates is given by (x_0, t_0). Similarly, the primed colony perceives point R as R', which in their coordinates is $(-x_0, t_0)$.

After much scientific labor, a worker ant named Hendrik Lorentz discovered the relationship among the coordinates (x', t'), (x_0, t_0), and (x, t). It is

$$x' = t_0 x - x_0 t$$

$$t' = -x_0 x + t_0 t$$

In matrix form the transformation is

$$\begin{pmatrix} x' \\ t' \end{pmatrix} = \begin{pmatrix} t_0 & -x_0 \\ -x_0 & t_0 \end{pmatrix} \begin{pmatrix} x \\ t \end{pmatrix}$$

They call this transformation from the (x, t) coordinates of the point P to the x', t' coordinates of the point P' the *Lorentz transformation*. Both colonies, of course, perceive P and P' as the same point.

The old man knows that the Lorentz transformation is nothing more than the relationship between the horizontal positions, namely,

$$u' = u - u_0$$

For, by taking the hyperbolic sine, he obtains

$$\sinh u' = \sinh (u - u_0)$$

or

$$\sinh u' = \sinh u \cosh u_0 - \cosh u \sinh u_0$$

or

$$x' = x t_0 - t x_0$$

which is the first Lorentz equation. Similarly, by taking the hyperbolic cosine he obtains

$$\cosh u' = \cosh (u - u_0)$$

or

$$\cosh u' = \cosh u \cosh u_0 - \sinh u \sinh u_0$$

or

$$t' = t t_0 - x x_0$$

which is the second Lorentz equation.

VELOCITY

The ants realize that the offset between their two colonies is important. For the first colony, the second colony is offset by x_0, t_0. The ratio of these coordinates is

$$\beta = \frac{x_0}{t_0}$$

Velocity

Because β is the ratio of distance over time, the quantity β is called the velocity of primed colony with respect to the unprimed colony. Similarly, $-\beta$ is the velocity of the unprimed colony with respect to the primed colony. That is, the quantity β is the relative velocity between the two colonies. Also, the incubation time t_0 is now denoted by γ. In terms of these new quantities, the Lorentz transformation

$$x' = t_0\left(x - \frac{x_0}{t_0}t\right)$$

$$t' = t_0\left(-\frac{x_0}{t_0}x + t\right)$$

becomes

$$x' = \gamma(x - \beta t)$$

$$t' = \gamma(-\beta x + t)$$

The old man admires the ants in a small way, but chides them because they did not realize that the gamma factor can be expressed in terms of the velocity by the equation

$$\gamma = \frac{1}{\sqrt{1 - \beta^2}}$$

Of course, the old man knows that

$$\beta = \frac{x_0}{t_0} = \frac{\sinh u_0}{\cosh u_0} = \tanh u_0$$

and

$$\gamma = t_0 = \cosh u_0 = \frac{1}{\operatorname{sech} u_0} = \frac{1}{\sqrt{1 - \tanh^2 u_0}} = \frac{1}{\sqrt{1 - \beta^2}}$$

A young ant named Albert Einstein recognizes that the old man is subtle (he had constructed the holograph) but not malicious (he had projected vertically, so the u axis remained intact, thereby making possible the Wright transform $u' = u - u_0$). The Lorentz transform was so precious to Einstein that he proclaimed that any two colonies with relative velocity β between them must have coordinates that are related by the Lorentz transform. There was an ant named Sir Isaac Newton whose earlier work on mechanics had to be revised accordingly. However, the ant named James Clerk Maxwell did not have to modify his electromagnetic equations, as they were already in fact invariant under the Lorentz transform.

PHYSICAL VELOCITY ADDITION FORMULA

After much work, an ant named Albert Michelson found that in the coordinates (x, t) of any colony, the velocity $\beta = x/t$ could never be greater than one (in magnitude). For the first colony, the ultimate speed was one, and likewise for the second colony, even though they had a relative speed of β. After much intellectual labor, Poincaré and Einstein produced the *addition equation for physical velocities*

$$v' = \frac{-\beta + v}{1 - \beta v}$$

The old man knows that this equation is nothing more than the formula for the hyperbolic tangent of the difference of two hyperbolic parameters; that is,

$$u' = -u_0 + u$$

so

$$\tanh u' = \frac{-\tanh u_0 + \tanh u}{1 - \tanh u_0 \tanh u}$$

Because $v' = \tanh u'$, $\beta = \tanh u_0$, and $v = \tanh u$, we see that this is the physical velocity addition formula.

LARMOR TIME DILATION

The first colony had an egg named "moving muon" that an ant named Larmor took to point Q, with velocity $v_0 = x_0/t_0$. The other colony had an egg named "rest muon" that was at the center of the colony Q', which in their coordinates had velocity $v' = x'/t' = 0$. According to the second colony (at rest), the incubation time of the "rest muon" is one. However, according to the first colony, the "moving muon" traveled a distance of x_0 and had an incubation time of t_0. Clearly t_0 is greater than one, so the incubation time t_0 of the "moving muon" is greater than the incubation time 1 of the "rest muon." We say that the incubation time of a moving muon is "dilated" or "increased" with respect to the incubation time of the rest muon. This is the Larmor dilation of time.

Appendix: Derivation of the Catenary

In this section calculus is used, so the reader might want to omit or skim this material. The *catenary* is an infinitely long curve that is the idealized shape of a uniform rope hanging freely under the action of gravity. The curve lies in a plane,

Larmor Time Dilation

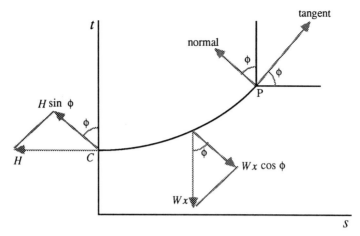

Figure 9.3. Catenary *CP*.

which we take to be the *s*, *t* plane, with the *s* axis horizontal and the *t* axis vertical. Let *W* denote the weight of a unit length of the rope. We consider the forces acting on arc *CP*, where *C* is at the lowest point and *P* is an arbitrary point on the curve. See Figure 9.3.

Let *x* be the *arc length* from *C* to *P* measured along the curve. At point *C* we have $x = 0$. The tangent at *P* makes a certain *inclination angle* ϕ with the *s* axis. By considering various points *P* on the rope, we may regard the inclination ϕ as a function of the arc length *x*, and vice versa. The three forces that act on arc *CP* are

1. The tension *T* at *P* acting along the tangent
2. The tension *H* along the horizontal tangent at *C*
3. The weight *Wx* in the vertical direction

Because the conditions at *C* remain constant regardless of where we pick the point *P*, the horizontal tension *H* at *C* is constant for a given rope. Since the foregoing three forces are in equilibrium, we can set the sum of their components in any given direction equal to zero. Let us pick that direction to be the direction of the normal to the curve at *P*.

Because the tension *T* acts along the tangent, and the normal is perpendicular to the tangent, it follows that the component of the tension *T* in the direction of the normal is zero. This is why we picked the direction of the normal. The unknown and uninteresting *T* is eliminated. The component of the horizontal tension *H* in the direction of the normal is $H \sin \phi$. The component of the weight *Wx* in the direction of the normal is $-Wx \cos \phi$. The reason for the minus sign is that this component acts in the opposite direction to the normal. The sum of these components is equal to zero; that is,

$$H \sin \phi - Wx \cos \phi = 0$$

This equation gives

$$x = \frac{H}{W} \tan \phi$$

Denote H/W by the constant a, so

$$x = a \tan \phi$$

This equation is the so-called *intrinsic equation of the catenary*. It gives the manner in which the arc length x depends on the inclination angle ϕ.

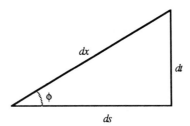

Figure 9.4. Element of arc dx.

Let us now deduce the Cartesian equation, that is, the equation that gives ordinate t as a function of abscissa s. From Figure 9.4, we see that

$$ds = dx \cos \phi, \qquad dt = dx \sin \phi$$

In Chapter 12, the Gudermannian function gd u is defined as

$$\text{gd } u = \sin^{-1} (\tanh u)$$

Define the *hyperbolic parameter* u such that gd u is equal to the inclination angle ϕ; that is, gd $u = \phi$, or

$$\sin^{-1} (\tanh u) = \phi$$

This relationship $\tanh u = \sin \phi$ is illustrated by the triangle in Figure 9.5.
From Figure 9.5, we see that

$$\cosh u = \sec \phi$$

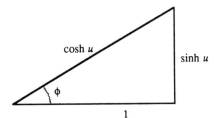

Figure 9.5. Relationship of inclination angle ϕ and hyperbolic parameter u.

If we differentiate this equation, we obtain

$$\sinh u\, du = \sec \phi \tan \phi\, d\phi$$

Also from Figure 9.5, we see that

$$\sinh u = \tan \phi$$

If we divide the two equations, we get

$$du = \sec \phi\, d\phi$$

Now let us differentiate $x = a \tan \phi$. We obtain $dx = a \sec^2 \phi\, d\phi$. Thus

$$ds = dx \cos \phi = a \sec^2 \phi \cos \phi\, d\phi = a \sec \phi\, d\phi = a\, du$$

$$dt = dx \sin \phi = a \sec^2 \phi \sin \phi\, d\phi = a \sec \phi \tan \phi\, d\phi$$

$$= a\, d(\sec \phi) = a\, d(\cosh u)$$

Taking the lowest point C (where $x = 0$, $\phi = 0$, and $u = 0$) to be $s = 0$, $t = a$, we find

$$s = a u, \quad t = a \cosh u$$

In the fable, a is taken to be one. Thus,

$$s = u, \quad t = \cosh u$$

which is the equation for the catenary as used in the fable.

CHAPTER 10

THE MINKOWSKI WORLD

Go, wond'rous creature, mount where Science guides,
Go, measure earth, weigh air, and state the tides;
Instruct the planets in what orbs to run,
Correct old Time, and regulate the Sun.

Alexander Pope (1688–1744)
An Essay on Man, II, 19-22

TIME-SPACE INTERVAL AS THE METRIC

Herman Minkowski was born in Alexota, Russia, on June 22, 1864, and died in Göttingen, Germany, on January 12, 1909, at age 44. He earned his Ph.D. at the University of Köningsberg in 1885 where he taught a few years, before going to

Plate 10.1. Herman Minkowski (1864–1909)

the University of Zurich and finally to Göttingen. At Zurich, Einstein was one of his students. At Göttingen, Max Born who later was to win the Nobel Prize in Physics, was his research assistant. Born in his book *Physics in My Generation* (Springer-Verlag, New York, 1969) recalls that Minkowski became interested in the electrodynamics of moving bodies at Göttingen in 1904. Born writes that in 1905 Minkowski conducted a seminar in which "we studied papers by Hertz, FitzGerald, Larmor, Poincaré, and others, but also got an inkling of Minkowski's own ideas." By 1907 Minkowski developed a formal geometric interpretation of the symmetries of the Poincaré group, and wrote that his goal was to exhibit those symmetries which "had not occurred to any of the previously mentioned authors, not even to Poincaré himself." More specifically, Minkowski developed the idea previously put forth by Poincaré to treat four-dimensional space-time as a purely geometric notion. In this chapter, we will treat this concept, but for simplicity, we will not use the representation of time with an imaginary coordinate as done by Poincaré and Minkowski, but will retain the representation of time with a real coordinate.

Minkowski's geometric intuition had been displayed at age 17 when he won a prize from the Académie des Sciences for his essay on the geometrical analysis of quadratic forms. His mathematical training and ability prepared him to understand and appreciate the work of Poincaré. From Minkowski's papers of 1907 and 1908, it is clear that he was impressed by Poincaré's discovery of the fundamental space-time (Poincaré) group, and by Poincaré's proposal that the invariance of the time-space interval τ could be utilized to seek other invariances and the resulting symmetries of the corresponding quantities that transformed under this group. In his 1907 lecture *Des Relativitätsprinzip*, Minkowski said that the mathematician's familiarity with certain necessary concepts from the theory of invariants placed him in a particularly good position to understand new ideas in physics, whereas the physicists were currently making their way painfully through a primeval forest of obscurities. Minkowski's plan was to continue Poincaré's exploitation of the theory of invariants and the symmetries of the space-time group.

In his 1908 address *Space and Time* delivered at the 80th Assembly of German Natural Scientists and Physicians, at Cologne, Minkowski said that relativity made it necessary to take time into account as a fourth dimension (treated mathematically in a different way from the three spatial dimensions because of the imaginary coordinate). Einstein adopted the ideas of Poincaré and Minkowski and went on to develop them to still greater heights in his general theory of relativity published seven years later in 1915. Unfortunately, Minkowski could not participate in this development, having died of appendicitis in 1909. Nor could Poincaré, who died from a medical operation in 1912 at age 58, at the height of his mathematical ability.

Minkowski is best remembered in elementary physics books by the statement in his address *Space and Time*: "The views of space and time which I wish to lay before you have sprung from the soil of experimental physics, and therein lies their strength. They are radical. Henceforth, space by itself, and time by itself, are doomed to fade away into mere shadows, and only a kind of union of the two will preserve an independent reality." Minkowski was a great mathematician, and in writing this, he showed that he understood the depth of the problem as well as the desire for mathematical beauty which most scientific people seek. The mathematics is elegant, and as we will see, Minkowski geometry involves working with the shears of affine geometry instead of the rotations of Euclidean geometry.

The *Minkowski world* is defined as a four-dimensional manifold made up of points. Each such point has coordinates constituted by four numbers, x, y, z, t defining the four-dimensional point. The coordinates x, y, z give the location, and the coordinate t gives the time of the event represented by this point. (Recall that an event is given by a particular location at a particular time.) For our purposes, we suppose that x, y, z, t are measured in natural units in which the velocity of light is one ($c = 1$). The four-dimensional space-time, or Minkowski world, is purely a geometric notation.

Geometrical properties of the Minkowski world can be established after some invariant relationship between coordinates of points is found, which can be interpreted as the *metric* between two points of the manifold. When such a metric between points can be defined, we pass from the geometric concept of *manifold* (a collection of points) to the geometric concept of *space* (collection of points with a metric defined between any two points). But how can the necessary invariant relation (the metric) be found? It should not be forgotten that the coordinates of the world points are defined with physically different quantities, that is, distances x, y, z and time t, so it is impossible to presume in advance that the metric in this world can be defined by the usual Euclidean distance

$$[(t_2 - t_1)^2 + (x_2 - x_1)^2 + (y_2 - y_1)^2 + (z_2 - z_1)^2]^{1/2}$$

between the points (x_1, y_1, z_1, t_1) and (x_2, y_2, z_2, t_2). But, as we have seen, the theory of relativity answers the question as to metric unambiguously. If we con-

Plate 10.2. King Akhenaten as a sphinx before the sun god Aten. New Kingdom, Eighteenth Dynasty, Ancient Egypt.

sider only inertial frames of reference, then the time-space interval between events, defined as

$$[(t_2 - t_1)^2 - (x_2 - x_1)^2 - (y_2 - y_1)^2 - (z_2 - z_1)^2]^{1/2}$$

remains invariant for any pair of events, or, in terms of geometry, for any pair of points in the Minkowski world. The transformation from one inertial frame to another is described by a shear (the Lorentz transformation), and no other transformation is needed in the framework of the special theory of relativity. Consequently, we can take the *time-space interval* between events as the basic *invariant defining the metric* in the Minkowski world. Thus, the two principles of special relativity (the relativity principle and the light principle), from which the invariance of the interval between events follows, signify that the geometry of the Minkowski world is determined by this metric. It is seen from the appearance of this metric [plus sign in front of $(t_2 - t_1)^2$ and minus signs in front of $(x_2 - x_1)^2$, $(y_2 - y_1)^2$, and $(z_2 - z_1)^2$] that space coordinates and the time coordinates are not equivalent, but treated quite differently. There is a fundamental difference between space and time. A person has the option of moving at will in any of the three spatial coordinates, but he must flow with time. Thus, the space of the four variables x, y, z, t of the special theory of relativity is a four-dimensional space, but it is non-Euclidean. The reason is that the Minkowski space is not originated by just adding the fourth (time) coordinate t to the three spatial ones x, y, z, but through the peculiar definition of the invariant metric (the time-space interval) between the points of this space.

WORLD LINES

Characteristic features of the Minkowski world can be illustrated by means of the Minkowski plane. One of the two coordinate axes must necessarily represent the time axis, since in the special theory of relativity purely spatial geometry (the

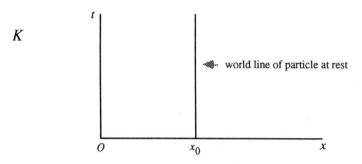

Figure 10.1. World line of particle at rest.

geometry of x, y, z) remains Euclidean. Only space-time geometry is described by the Minkowski construction. Thus, in our choice of reference frames, it is most convenient to consider the x, t plane and suppose that the relative velocity between two frames is in the direction of the x axis.

Every event in our real physical world occurs at a definite world point of the Minkowski world. A particle at a given location at a given time represents an *event*. No matter how this particle moves or not, the continuous sequence of events represented by this particle yield a certain curve in space-time called the *world line* of the particle. Let us draw the x, t axes of the frame K at right angles to each other. Let a particle at rest be located at the location $x = x_0$ in the frame K; its world line in the x, t plane of the Minkowski world is the straight line parallel to the t axis, as shown in Figure 10.1.

Let another particle move uniformly along the x axis in the frame K at the velocity β. Its world line in this frame will be a straight line inclined at the angle θ to the t axis, as seen in Figure 10.2. The inclination angle θ is determined by the equation

$$\beta = \tan \theta$$

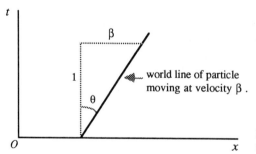

Figure 10.2. World line of moving particle.

Because $\beta < 1$ (since β is in natural units, i.e., $\beta = V/c$, where V is in meters per second and $c = 300,000,000$ meters per second, the ultimate velocity), the angle θ cannot exceed 45° for any moving object (i.e., if $\beta < 1$, then $\theta = \tan^{-1} \beta <$

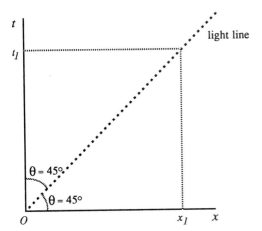

Figure 10.3a. Light line in orthogonal coordinates.

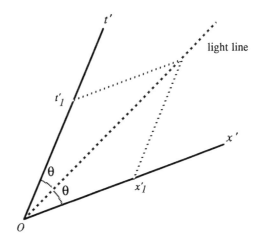

Figure 10.3b. Light line in oblique coordinates.

45°). As we know, light plays a unique role. For a light particle (a photon), $\beta = V/c = c/c = 1$, so the angle $\theta = \tan^{-1} 1 = 45°$. Hence, the world line of a light ray is always at 45° to the coordinate axes. The *light line* is defined as the world line of light that goes through the origin and is thus the bisecting line of the angle between the coordinates. Figure 10.3a shows the case when the coordinates are drawn at right angles, and Figure 10.3b the case when the coordinates are drawn not at right angles.

Let us now consider two inertial frames K and K', with frame K' moving to the right at a velocity β relative to frame K. As w know, the x', t' axes are related to the x, t axes by the Lorentz transformation

$$x' = \gamma x - \gamma \beta t$$

$$t' = -\gamma \beta x + \gamma t$$

where the gamma factor is defined as

$$\gamma = (1 - \beta^2)^{-1/2}$$

The two sets of axes can be plotted on the same diagram, with the x, t axes at right angles and with the x', t' axes drawn together in a scissorslike manner to the bisecting light line (Figure 10.4.)

The Lorentz transformation shears the x and t axes through the angle $\theta = \tan^{-1} \beta$ around the origin, both axes being sheared in the direction of the light line (the bisector of the angle between each set of coordinate axes) to their final positions x' and t'. The straight lines $x' = $ constant are now parallel to the t' axis, and the straight lines $t' = $ constant are parallel to the x' axis. Thus, in the diagram,

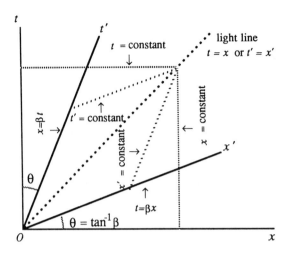

Figure 10.4. Orthogonal axes x, t sheared to oblique axes x', t'.

the x, t coordinate system is a rectilinear right-angled system, whereas the x', t' coordinates are a rectilinear oblique-angled system.

Alternatively, we may represent the Lorentz transformation as the shear of the x and t axes (plotted as an oblique-angled system) through an angle $\theta = \tan^{-1}\beta$ to the final positions as the x' and t' axes (plotted as a right-angled system), as shown in Figure 10.5.

The light line bisects the angle of each set of axes. The straight lines $x =$

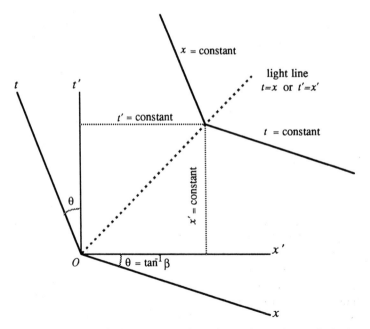

Figure 10.5. Oblique axes x, t sheared to orthogonal axes x', t'.

constant are parallel to the t axis, and the straight lines $t =$ constant are parallel to the x axis. Finally, we could plot the two sets of axes on a diagram in which both sets are oblique angled. Such a diagram is that of Figure 10.6 in the next section.

RELATIVITY OF SIMULTANEITY

So much discussion occurs on the relativity of simultaneity that it is important enough to state it in the form of a theorem: *two events E and F simultaneous in a given inertial frame are not simultaneous in any other inertial frame.*

Let us now prove this theorem. First, use the two events to define the coordinates' axes of the given inertial frame as follows. Connect E and F by a straight line, call this line the x axis, and let E be the origin. Let ψ be the angle the x axis makes with either light line and reflect the x axis about the light line by the angle ψ. Call the resulting line the t axis. See Figure 10.6.

Any other reference frame will have a different set of axes, say, x' and t'. These axes will each, respectively, make an angle ψ' to the light line, as shown in the figure. In the x', t' frame, simultaneous events lie along lines parallel to the x' axis. Thus, we draw a line through F parallel to the x' axis, and we let F' be the intersection of this line with the t' axis. By construction, events F and F' are simultaneous in the x', t' frame. Because F' lies above E on the t' axis, event F' occurs after event E. Therefore, event F occurs after E in the x', t' frame. We conclude that events E and F, which are simultaneous in the given (unprimed) frame, are not simultaneous in the other (primed) frame.

Events that are simultaneous with reference to one inertial frame are not simultaneous with respect to another inertial frame, and vice versa. Every inertial frame has its own particular time; unless we are told the inertial frame to which

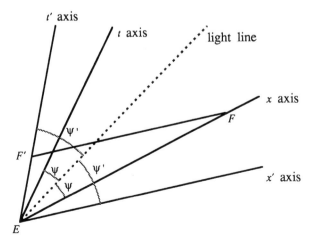

Figure 10.6. Relativity of simultaneity.

the statement of time refers, there is no meaning in the statement of the time of an event. Before the advent of the special theory of relativity, it had always tacitly been assumed in physics that the statement of time had an absolute significance, that is, that it is independent of the inertial frame in question and so time was the same in all inertial frames. But this assumption of absolute time was incompatible with the tenets of relativity theory, and so had to be abandoned.

EINSTEIN'S TRAIN

In the textbook treatments of relativity theory, the *relativity of simultaneity* is not presented as a mathematical theorem as we have done, but instead it is given in the form of examples. The premium example of simultaneity, from which all the others are modeled, was given by Einstein in his book *Relativity, The Special and the General Theory* (Crown Publishers, New York, 1961). We will now discuss this example, called the train example.

We suppose that a very long train is traveling along the rails with constant velocity β to the right. People traveling in this train use the train as an inertial frame in which they are at rest, and they regard all events in reference to the train. Then every event that takes place on the embankment also takes place at a particular

Plate 10.3. Max Planck (1858–1947), father of quantum physics, and Albert Einstein in 1929.

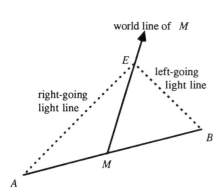
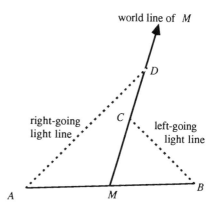

Figure 10.7. Events A and B are simultaneous.

Figure 10.8. Events A and B are not simultaneous.

point of the train. The railway embankment is also an inertial frame, and people on the embankment are at rest in this frame.

Lightning has struck the rails on the railway embankment at two places A and B far distant from each other. Suppose we make the additional assertion that these two lightning flashes occurred simultaneously. How do we determine whether there is sense in this statement? We offer the following method with which to test the simultaneity. By measuring along the rails, the length of the connecting line AB should be determined. Place an observer at the midpoint M of the distance AB. If the observer perceives the two flashes of the lightning at the same time, then they are simultaneous.

In Figure 10.7, the flash from A travels on the right-going light line and the flash from B travels along the left-going light line. The observer at M perceives the two flashes reaching her at the same time if and only if the two light lines intersect the world line of M at the same event E, as depicted in the figure. Thus, the diagram shows the simultaneity of events A and B. In such a case, the world line of M can serve as the t axis, and line AB as the x axis, of the inertial frame.

If the two light lines did not intersect the world line of M at the same event, then the two flashes would not be simultaneous, as shown in Figure 10.8. In the diagram, event D would occur after event C. Thus, the diagram shows the nonsimultaneity of events A and B. Whereas the world line of M can serve as the t axis of the inertial frame, the line AB cannot serve as the x axis.

It is clear that this definition can be used to give an exact meaning not only to two events, but to as many events as we care to choose. By the foregoing definition, it is easy to show that when three events A, B, C occur in different places in such a manner that A is simultaneous with B, and B is simultaneous with C, then the criterion for the simultaneity of the pair of events A, C is also satisfied.

Let us now deal with the main issue of the train example. The issue is: Are two events (e.g., the two strokes of lightning A and B) which are simultaneous

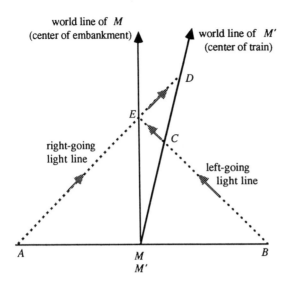

Figure 10.9. Simultaneous events A and B in the embankment frame are not simultaneous in the train frame.

with reference to the embankment frame also simultaneous with reference to the train frame? By use of the theorem given in the previous section, the answer is no. However, let us not use the theorem but follow the analysis given by Einstein. He shows that the answer must be in the negative in the following way.

When we say that the lightning strokes A and B are simultaneous with respect to the embankment, we mean that the rays of light emitted at the places A and B, where the lightning occurs, meet each other at the midpoint M of the length AB of the embankment. But the events A and B also correspond to positions A and B on the train. Let M' be the midpoint of the distance AB on the traveling train. Just when the flashes of lightning occur (as judged from the embankment), this point M' naturally coincides with the point M, but it moves toward the right with the velocity β of the train. An observer sitting in the position M on the embankment does not possess this velocity, so he remains permanently at M, and the light flashes emitted by the lightning A and B would reach him simultaneously; that is, they would meet just where he is situated (event E). See Figure 10.9.

A person at the midpoint M' of the train (considered with reference to the railway embankment) is hastening toward the flash of light coming from B while he is riding ahead of the flash coming from A. As seen in the diagram, the flash from B hits the world line of M' at C, and the flash from A hits the same world line at D. Because C is below D on this world line, the observer at M' sees the beam of light emitted from B earlier than he will see that emitted from A. Thus, observers in the train frame, knowing that M' is the midpoint of the train and seeing the flash from B arrive before the flash from A, must conclude that the lightning occurred at B before it occurred at A. In conclusion, observers in the train frame say that A and B are not simultaneous, despite the fact that observers in the embankment frame say that A and B are simultaneous. Thus ends the celebrated train example of Einstein, an example that appears in nearly every book on relativity, but in words only, and not with the Minkowski diagrams, which we have added for clarity.

RELATIVITY OF THE SAME PLACE

The counterpart of the relativity of simultaneity is the relativity of the same place, which we can state as the theorem: *two events E and F that are at the same place in a given inertial frame are not at the same place in any other inertial frame.*

Let us prove this theorem as follows (Figure 10.10). Let the two events E and F that occur at the same place define the coordinate axes in the given inertial frame. Connect E and F by a straight line. Call this line the t axis and let E be the origin. Let ψ be the angle the t axis makes with either light line and reflect the t axis about the light line by the angle ψ. Call the resulting line the x axis.

Any other inertial frame will have a different set of axes, say, x' and t'. These axes will each make an angle ψ' to the light line, as seen in Figure 10.10. In the x', t' frame, events at the same place lie along lines parallel to the t' axes. Thus, we draw a line through F parallel to the t' axis, and we let F' be the intersection of this line with the x' axis. By construction, events F and F' occur at the same place in the x', t' frame. Because F' lies to the right of E on the x' axis, event F' occurs to the right of E. Thus, event F occurs to the right of E in the x', t' frame. Therefore, we have shown that events E and F that occur at the same place in the given (unprimed) frame are not at the same place in the other (primed) frame.

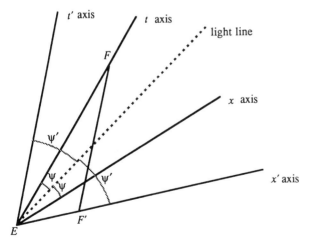

Figure 10.10. Relativity of the same place.

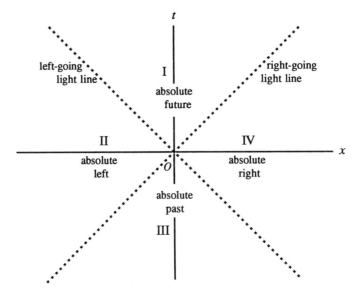

Figure 10.11. The four quadrants of the Minkowski plane defined by the two light lines.

TIMELIKE AND SPACELIKE EVENTS

Let us consider the x, t axes of frame K, as shown in Figure 10.11. The square of the time-space interval between two world points is defined by the expression

$$\tau^2 = (t - t_0)^2 - (x - x_0)^2$$

Let us suppose that event O occurred at the point $x_0 = 0$ at the time $t_0 = 0$, that is, at the origin O. Any events that occurred in space (represented by the x axis) before and after event O are depicted by points in the x, t plane. Thus the square of the time-space interval (i.e., the square of the metric) from event O to any other event (x, t) is equal to

$$\tau^2 = t^2 - x^2$$

We see that the plane can be divided into four quadrants, labeled in the diagram as I, II, III, IV. These quadrants are separated by the two light lines $t = x$ and $t = -x$. For a point on either of these two light lines, the squared metric from event O is $t^2 - x^2 = t^2 - (\pm t)^2$ which is zero. The light line $x = t$ depicts the continuous sequence of events consisting of the emission of a right-going photon (particle of light) from the location $x = 0$ at time $t = 0$ and its arrivals at the points x at time t. Similarly, the light line $t = -x$ depicts the corresponding events for a left-going photon. Because the squared metric between any two events

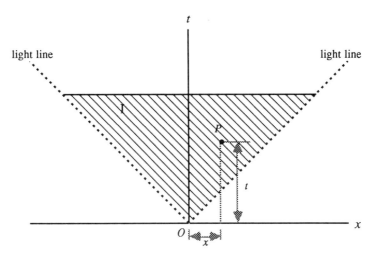

Figure 10.12. Quadrant I in the x, t frame. The coordinates of point P are x, t.

located on either of the light lines is zero, that is, because for $t = x$,

$$\tau^2 = t^2 - x^2 = 0$$

and similarly for $t = -x$, we say that such events are lightlike.

Consider now an event x, t in quadrant I. (Figure 10.12.) We see that $\tau^2 = t^2 - x^2 > 0$. Consequently, we say that the time-space interval (metric) between any event of quadrant I and event O is positive-timelike. The reason for this designation is that for all events in quadrant I, the time is positive ($t > 0$); consequently all of them will occur after event O, and no choice of reference frame can alter this situation. That is, if we shear the x, t axes into the x', t' axes, we still have $t' > 0$ for the points in the new quadrant I, labeled I' in Figure 10.13. Note from Figure 10.12 and Figure 10.13, that quadrant I and quadrant I' are indeed identical, both being the wedge in the upper half plane between the light lines (which do not change when we shear the axes). As a result, we can say that quadrant I is the region of the absolute future with respect to event O.

In quadrant III, we also have $\tau^2 > 0$, but here for all events $t < 0$. Hence, quadrant III is the region of the absolute past with respect to event O.

In quadrants II and IV, the squared metric is negative, $\tau^2 < 0$. We say that events in quadrant IV are positive-spacelike. The reason for this designation is that for all events in quadrant IV, the location is positive ($x > 0$); consequently, all of them occur to the right of the location of event O, and no choice of reference frame can alter this situation. Thus, if we shear the x, t axes into the x', t' axes, we still have $x' > 0$ for the points in the new quadrant IV, labeled IV' in Figure 10.14. Quadrant IV and quadrant IV' are indeed identical, because as we know,

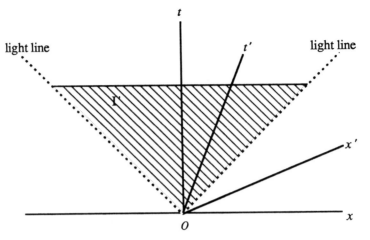

Figure 10.13. Quadrant I' in the x', t' frame is the same as Quadrant I in the x, t frame.

light lines do not shear. As a result, we can say that quadrant IV is the region of the absolute right with respect to event O. Similarly, quadrant II is the region of the absolute left, and points in this quadrant are said to be negative-spacelike.

In summary, we have

Quadrant	Events
I (absolute future)	Positive-timelike ($t > 0$, $t' > 0$)
II (absolute left)	Negative-spacelike ($x < 0$, $x' < 0$)
III (absolute past)	Negative-timelike ($t < 0$, $t' < 0$)
IV (absolute right)	Positive-spacelike ($x > 0$, $x' > 0$)

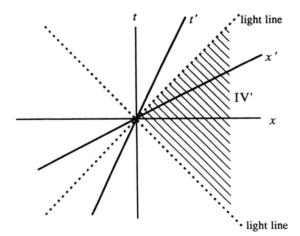

Figure 10.14. Quadrant IV' in the x', t' frame is the same as Quadrant IV in the x, t frame.

PROPER TIME AND PROPER LENGTH

Let us now show that if the event P with coordinates (x_1, t_1) is timelike with respect to event O, then we can find an inertial frame in which the two events occur in the same position of space. That is, we can find a Lorentz transformation which converts the point (x_1, t_1) into the point $(x_1', t_1') = (0, \tau)$. The required situation is shown in Figure 10.15.

We merely let $\beta_1 = x_1/t_1$. The resulting shear gives the reference frames x', t' in which the event (x_1, t_1) becomes the event $(0, t_1')$. As confirmation, let us substitute (x_1, t_1) into the Lorentz equations:

$$x_1' = \gamma_1(x_1 - \beta_1 t_1) = \gamma_1[x_1 - (x_1/t_1)t_1] = 0$$

$$t_1' = \gamma_1(-\beta_1 x_1 + t_1) = \gamma_1[-(x_1/t_1)x_1 + t_1]$$
$$= [1 - (x_1^2/t_1^2)]^{-1/2}[(-x_1^2/t_1) + t_1] = (t_1^2 - x_1^2)^{1/2} = \tau$$

Instead of using the Lorentz transformation, we could have arrived at the same result for t_1' by using the fact that the time-space interval is invariant; thus

$$t_1^2 - x_1^2 = t_1'^2 - x_1'^2 = \tau^2 - 0 = \tau^2$$

The time $t_1' = \tau$ is called the proper time of the event. We will usually denote proper time by the symbol τ. Thus, for any timelike event (x_1, t_1) the proper time is defined as $\tau = \sqrt{t_1^2 - x_1^2}$ if (x_1, t_1) is positive-timelike (in quadrant I) and $\tau = -\sqrt{t_1^2 - x_1^2}$ if (x_1, t_1) is negative-timelike (in quadrant III). The proper time is the metric (time-space interval) between a timelike point (x_1, t_1) and the origin O.

Next let us show that if the event Q with coordinates (x_2, t_2) is spacelike with respect to the event O, then we can find an inertial frame in which the two events

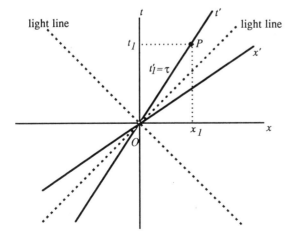

Figure 10.15. Given the timelike event P, the frame x', t' is chosen so that event P occurs at the same location as event O.

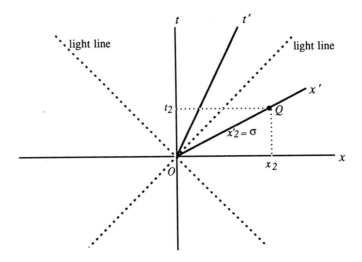

Figure 10.16. Given the spacelike event Q, the frame x', t' is chosen so that event Q occurs at the same time as event O.

occur at the same moment of time. That is, we can find a Lorentz transformation which converts the point (x_2, t_2) to the point $(x_2', t_2') = (\sigma, 0)$. The required situation is shown in Figure 10.16.

We merely let $\beta_2 = t_2/x_2$. The resulting shear gives the reference frame x', t' in which the event (x_2, t_2) becomes the event with coordinates

$$x_2' = \gamma_2(x_2 - \beta_2 t_2) = \gamma_2[x_2 - (t_2/x_2)t_2]$$
$$= [1 - (t_2^2/x_2^2)]^{-1/2}[x_2 - (t_2^2/x_2)] = (x_2^2 - t_2^2)^{1/2} = \sigma$$
$$t_2' = \gamma_2(-\beta_2 x_2 + t_2) = \gamma_2[-(t_2/x_2)x_2 + t_2] = 0$$

Instead of using the Lorentz transformation, we could have arrived at the same result for x_2' by using the fact that the space-time interval is invariant. Thus, for any spacelike point, the proper length is defined as

$$\sigma = \sqrt{x_2^2 - t_2^2}$$

if (x_2, t_2) is positive-spacelike (in quadrant IV) and

$$\sigma = -\sqrt{x_2^2 - t_2^2}$$

if (x_2, t_2) is negative-spacelike (in quadrant II)

In terms of $\beta_1 = x_1/t_1$, the proper time for the timelike point (x_1, t_1) can be written as

$$\tau = \pm \sqrt{t_1^2 - x_1^2} = t_1\sqrt{1 - \beta_1^2} = t_1/\gamma_1$$

and thus

$$t_1 = \frac{\tau}{\sqrt{1 - \beta_1^2}} = \gamma_1 \tau$$

Note that t_1 and τ have the same sign, because we assume the radical is positive. Since $\gamma_1 \geq 1$, this equation shows that the proper time τ is the smallest measure of time between the two events O and (x_1, t_1); the time t_1 between these two events relative to any other inertial observer is equal to or greater than τ in magnitude.

In terms of $\beta_2 = t_2/x_2$, the proper length for the spacelike point (x_2, t_2) can be written as

$$\sigma = \pm \sqrt{x_2^2 - t_2^2} = x_2\sqrt{1 - \beta_2^2} = \frac{x_2}{\gamma_2}$$

and thus

$$x_2 = \frac{\sigma}{\sqrt{1 - \beta_2^2}} = \gamma_2 \sigma$$

Since $\gamma_2 \geq 1$, this equation shows that the proper length σ is the smallest measure of distance between two events O and (x_2, t_2); the distance x_2 between these two events relative to any other inertial observer is equal to or greater than σ in magnitude.

MINKOWSKI GEOMETRY

For those readers who like to think in terms of geometry, we will look at the special theory of relativity in terms of the geometry of the Minkowski plane. When depicting the Minkowski plane on a sheet of paper, we should remember that the Minkowski geometry is non-Euclidean, but that we are used to such relations between lengths of rulers which are customary in the Euclidean plane. As a result, we must completely reorient our thinking in the Minkowski plane. Although this may seem a little difficult at first, it actually leads to quite a bit of enjoyment.

In Figure 10.17, we see a right triangle in the Minkowski plane. Side AC is equal to $x_2 - x_1$ and side BC is $t_2 - t_1$. But in the Minkowski plane we have the relation

$$AB^2 = BC^2 - AC^2$$

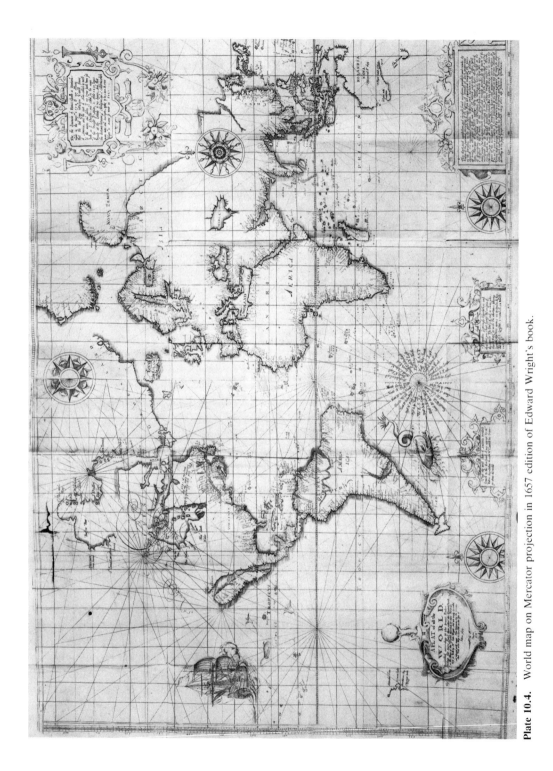

Plate 10.4. World map on Mercator projection in 1657 edition of Edward Wright's book.

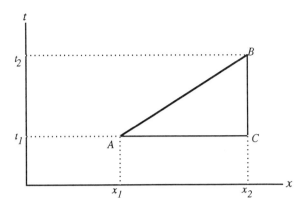

Figure 10.17. Right triangle in the Minkowski plane. Side BC is the time duration, side AC is the space length, and hypotenuse AB is the time-space interval.

which follows from the definition of the square of the time-space interval

$$(\text{time-space interval})^2 = (t_2 - t_1)^2 - (x_2 - x_1)^2$$

The foregoing relation is contrary to the Pythagorean theorem

$$AB^2 = BC^2 + AC^2$$

which holds in the Euclidean plane but which does not hold on the Minkowski plane. Thus, the relation

$$(\text{hypotenuse})^2 = (\text{vertical side})^2 - (\text{horizontal side})^2$$

which holds in the Minkowski plane is called the *pseudo-Pythagorean theorem*. Because we are not used to such a relationship in ordinary life, it follows that we should cautiously perform all comparisons of lengths in the Minkowski plane.

The most outstanding difference is this: in the Euclidean plane the Pythagorean theorem says that the square of the hypotenuse is always a nonnegative number; in the Minkowski plane AB^2 is positive if AB is timelike, zero if AB is lightlike, and negative if AB is spacelike. This is the case of the metric defined by the time-space interval. For the metric defined by the space-time interval, AB^2 would be negative if AB is timelike, zero if AB is lightlike, and positive if AB is spacelike.

In the Euclidean plane (x, y) the locus of points equidistant from the origin of coordinates (at a fixed distance r) is given by the circumference of the circle (where r^2 must be positive):

$$x^2 + y^2 = r^2$$

However, for a point (x, t) in the Minkowski plane, the square of the *time-space interval* τ from the origin O is defined by the relationship

$$\tau^2 = t^2 - x^2$$

and the square of the *space-time interval* σ from the origin O is defined by

$$\sigma^2 = x^2 - t^2$$

It follows that the locus of points "equidistant," or better equiinterval, or equimetric, from O is made up by the two equilateral hyperbolas given by the equations

$$t^2 - x^2 = \tau_1^2 \quad \text{(for timelike events: } t^2 > x^2, \text{ so } \tau_1^2 > 0\text{)}$$

$$x^2 - t^2 = \sigma_2^2 \quad \text{(for spacelike events: } x^2 > t^2, \text{ so } \sigma_2^2 > 0\text{)}$$

where we set $\tau_1^2 = \sigma_2^2$. Each hyperbola is made up of two semihyperbolas, so altogether there are four semihyperbolas. If we let $\tau_1^2 = \sigma_2^2 = 1$ in the preceding equations, we obtain the four unit equilateral semihyperbolas $t^2 - x^2 = 1$, $x^2 - t^2 = 1$, which are plotted in the (x, t) coordinate system in Figure 10.18.

Since the Lorentz transform leaves the square of the time-space interval τ and the square of the space-time interval σ each invariant, we also obtain the unit equilateral hyperbolas $t'^2 - x'^2 = 1$ and $x'^2 - t'^2 = 1$ in the new oblique-angled coordinate system (x', t'). (In the Figure 10.18, these new hyperbolas coincide with the old ones, so no additional curves need to be graphed.) Because the new

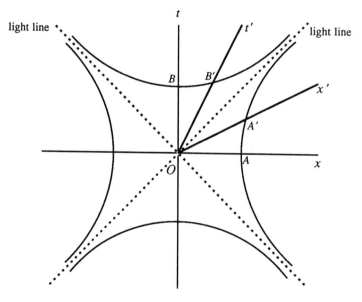

Figure 10.18. Four unit calibration semihyperbolas in the Minkowski plane.

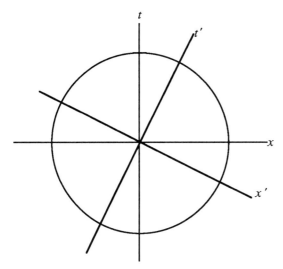

Figure 10.19. Unit calibration circle in the Euclidean plane.

hyperbolas in their x', t' frame have the same Minkowski properties as the old hyperbolas have in their x, t frame, it follows that these four unit equilateral semihyperbolas cross the axes x, t, x', t' at metrics (or intervals) from the origin each equal to unity. The unit hyperbolas plotted are referred to as *scale*, or *calibration*, *hyperbolas*. Thus, Figure 10.18 is the Minkowski counterpart to the familiar unit circle diagram shown in Figure 10.19.

The rotation transform leaves the distance $x^2 + y^2$ invariant. Thus, from the old circle $x^2 + y^2 = 1$, we obtain by the rotation transform the new circle $x'^2 + y'^2 = 1$ in the new right-angled coordinate system (x', y'). (In Figure 10.19, the new circle coincides with the old circle, so no additional curve needs to be graphed.) Because the new circle in the x', y' frame has the same Euclidean properties as the old circle has in the x, y frame, it follows that the circle crosses the axes x, y, x', y' at distances from the origin each equal to unity. In conclusion, when we go to the Minkowski plane, we must wipe out our concept of distance as measured by the unit circle and in its place use the concepts of time-space interval and space-time interval as measured by the unit hyperbolas.

We have now seen that if we choose the unit timelike hyperbola $t^2 - x^2 = 1$, and the unit spacelike hyperbola $x^2 - t^2 = 1$, and if we draw rays from the origin O until they intersect with these hyperbolas, the section of each such ray determines the unit time-space interval, or the unit space-time interval, in the given direction.

It is possible to give a physical interpretation for the hyperbola $t^2 - x^2 = 1$. Let nuclear particles having various velocities, but with identical lifetime (proper time) $\tau = 1$, be generated at the Minkowski world point $x = 0, t = 0$ (the origin O). Then the locus of the world point (x, t) at which these particles decay will be the hyperbola $t^2 - x^2 = 1$. More generally, nuclear particles having various

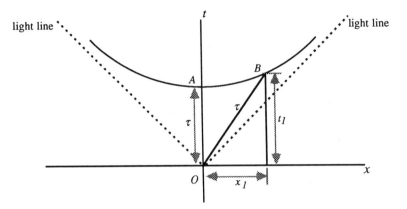

Figure 10.20. World line OB of moving particle with proper lifetime τ.

velocities, but with identical lifetime (proper time τ), generated at the origin O decay on the hyperbola $t^2 - x^2 = \tau^2$. The world lines of these particles are represented by the rays outgoing from the origin and reaching this hyperbola, as shown in Figure 10.20. The world line OA represents a particle at rest, whereas the world line OB represents a particle with velocity $\beta_1 = x_1/t_1$, where (x_1, t_1) are the coordinates of B.

The four unit semihyperbolas $t^2 - x^2 = 1$, $x^2 - t^2 = 1$ subdivide the Minkowski plane into four quadrants, each one containing one semihyperbola. The dividing lines of these quadrants are the asymptotes $t = x$, $t = -x$ of these hyperbolas, and as we know, these asymptotes are the light lines. The two quadrants associated with the hyperbola $t^2 - x^2 = 1$ are quadrant I (absolute future) and quadrant III (absolute past), whereas the two quadrants associated with the hyperbola $x^2 - t^2 = 1$ are quadrant II (absolute left) and quadrant IV (absolute right). See Figure 10.21.

Each of these semihyperbolas intersects only one of the axes, x or t. The points of intersection of the hyperbola $x^2 - t^2 = 1$ with the x axis are determined by the condition $t = 0$, so they are $x = 1$ and $x = -1$. In a similar way, we find that the hyperbola $t^2 - x^2 = 1$ intersects the t axis at $t = 1$ and $t = -1$. We thus say that these hyperbolas cut off unitary sections on the coordinate axes. Since the expression $t^2 - x^2$ is invariant under the Lorentz (or shear) transformation, the equations $t'^2 - x'^2 = 1$, $x'^2 - t'^2 = 1$ will be valid in frame K'. It follows directly that the same hyperbolas cut off unitary sections on the new oblique-angled axes as well. Let us explain.

If we look at the x axis and the x' axis, we see the label "1" on each of the points of intersection with the semihyperbola in quadrant IV. That is, two different points in quadrant IV are labeled as 1. However, the Euclidean distance from O to the point of intersection of the semihyperbola with the x axis is not equal to the Euclidean distance from O to the point of intersection of the semihyperbola with the x' axis. But this is the Minkowski plane, so we must discard the notion of

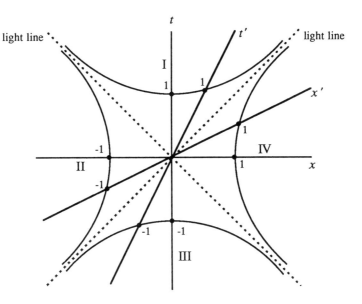

Figure 10.21. One unit semihyperbola in each of the four quadrants (I absolute future, II absolute left, III absolute past, IV absolute right).

Euclidean distance. The Minkowski space-time interval from O to the point of intersection of the semihyperbola with the x axis is equal to 1, and also the Minkowski space-time interval from O to the point of intersection of the semihyperbola with the x' axis is equal to 1. This is why in the diagram we have marked all the points of intersection as either 1 or -1.

FITZGERALD LENGTH CONTRACTION AND LARMOR TIME DILATION

The word *contract* means to reduce in length, to shorten, to shrink, and so a contraction represents a shortening. The word *dilate* means to enlarge, to extend, and so a dilation is an enlargement.

Let us think of two inertial frames: frame K with coordinates x, t and frame K' with coordinates x', t'. The two frames are moving uniformly with respect to each other with relative velocity β. The Minkowski space is shown in Figure 10.22. We recall that in any given frame, the *proper time* is the time duration between two events at the same place, and the *proper length* is the distance between two events at the same time.

Let the frame be K and let one of the events in each case be the origin O. Let A be a point on the x axis and let B be a point on the t axis. Then OA is equal to the proper length σ and OB is equal to the proper time τ. The reason is that O with coordinates $(0, 0)$ occurs at the same time as event A with coordinates

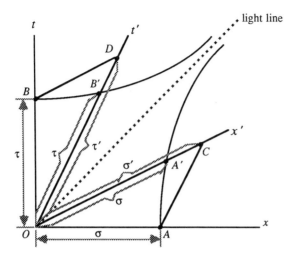

Figure 10.22. Two inertial frames in the Minkowski plane.

$(\sigma, 0)$, and O occurs at the same place as event B with coordinates $(0, \tau)$. In the diagram, we have drawn the semihyperbolas through A and B, and A' and B' are respectively the points of intersection of these semihyperbolas with the x' and t' axes of frame K'. By the Minkowski geometry, line segment OA' is numerically equal to σ, and line segment OB' is equal to τ. This result merely says that σ is a proper length in frame K' (because O and A' occur at the same time in K') and τ is proper time in K' (because O and B' occur at the same place in K'). So much for points A' and B'.

Let us now focus on frame K'. In K', let point C be the oblique projection of A on the x' axis. (We construct a line through A parallel to the t' axis, and C is defined as the point of intersection of this constructed line with the x' axis.) Thus, the x' coordinate of A, and the x' coordinate of C, are the same number, which we call σ'. Thus, A and C occur at the same place in the K' frame.

In frame K', let D be the oblique projection of B on the t' axis. (We construct a line through B parallel to the x' axis, and D is defined as the point of intersection of this constructed line with the t' axis). Thus, the t' coordinate of B, and the t' coordinate of D, are the same number which we call τ'. Thus, B and D occur at the same time in the K' frame.

If we look at the x' axis, we see that $OC > OA'$, which says that $\sigma' > \sigma$. If we look at the t' axis, we see that $OD > OB'$, which says that $\tau' > \tau$.

Let us now explain in geometrical terms how the *contraction of a moving ruler* comes about. The line segment OC represents a ruler of length σ' at rest in K'. Its world lines in the frame K' are encompassed by two world lines, one at each end. At the left end, the world line is the t' axis, for all points on this axis have coordinate $x' = 0$. The other world line is the segment AC (and its continuation), for all points on this line have coordinate $x' = \sigma'$. But in terms of frame K the same-time position of the ends of the ruler OC at the moment $t = 0$ corresponds to the intersection of these two world lines with the x axis, that is, to the points O and A. Thus, the length of the ruler in frame K is OA, which is equal to σ. Since we considered the ruler to be at rest in K', the ruler is moving in K. Thus, the length $OA = \sigma$ of the moving ruler is less than the length $OC = \sigma'$ of the ruler at rest, because as we have shown, $\sigma < \sigma'$. We conclude that there is a contraction of the ruler's length (from σ' to σ) when the length is measured in a reference frame K in which the ruler moves uniformly and rectilinearly. This result is called the *FitzGerald length contraction* of a moving ruler.

Let us next explain in geometric terms how the *dilation of the time kept by a moving clock* comes about. The line segment OB represents a clock with period τ at rest in frame K. (Note: The period is the time between the ticks of the clock.) The separation lines of this period in frame K' are encompassed by two separation lines, one at each end. At the bottom end, the separation line is the x' axis, for all points on this axis have coordinate $t' = 0$. The other separation line is the segment BD (and its continuation), for all points on this line have coordinate $t' = \tau'$. But in terms of frame K' the same-place position of the ends of the period OB at the location $x' = 0$ correspond to the intersection of these two separation lines with the t' axis, that is, to the points O and D. Thus, the period of the clock in frame K' is OD, which is equal to τ'. Since we considered the clock to be at rest in K, the clock is moving in K'. Thus, the period $OD = \tau'$ of the moving clock is greater than the period $OB = \tau$ of the clock at rest, because as we have shown, $\tau' > \tau$. Thus, we conclude that there is a dilation of the clock's period (from τ to τ') when the period is measured in a reference frame K' in which the clock moves uniformly and rectilinearly. This result is called the *Larmor time dilation* of a moving clock.

Note that for the ruler contraction, we let σ' be the ruler's length in the rest frame K', and let σ be the contracted length of the ruler which is seen moving in frame K, where $\sigma' > \sigma$. For the clock's dilatation, we let τ be the clock's period in the rest frame K, and let τ' be the dilated period of the clock which is seen moving in frame K', where $\tau < \tau'$. As we have presented them, we see that both arguments are essentially the same, but they are so arranged that lengths contract (shorten) for a moving ruler, and times dilate (lengthen) for a moving clock, both cases being with respect to a frame at rest. In the words of Macbeth: "Tomorrow, and tomorrow, and tomorrow, creeps in this petty pace from day to day, to the last syllable of recorded time."

CHAPTER 11

LEONHARD EULER AND AFFINE GEOMETRY

Così vedi le cose contingenti
    anzi che sieno in sé, mirando il punto
    a cui tutti li tempi son presenti.
[So you do see contingent things
    before they exist in themselves, gazing at the point
    to which all times are present.]

Dante Alighieri (1265–1321)
Paradiso, XVII, 16-18

AFFINE GEOMETRY

Leonhard Euler, Swiss mathematician, was born in Basel on April 15, 1707, and died in St. Petersburg, Russia, on September 18, 1783. Euler was the most prolific mathematician of all time, writing on every branch of mathematics. He was always

Plate 11.1. Leonhard Euler (1707–1783)

careful to describe his reasoning and to list any false paths he had followed. Few have equaled or approached Euler in the mass of sound and original work of the first importance which he put out.

Euler developed the most compact and famous of all mathematical formulas:

$$e^{\pi i} + 1 = 0$$

It relates zero 0, the unit 1, Euler's imaginary unit i, the irrational number $\pi = 3.14159\ldots$ (the ratio of circumference to diameter of a circle), and Euler's irrational number $e = 2.71828\ldots$ (the base of the natural logarithms). The American mathematician and astronomer Benjamin Peirce (1809–1880) regarded it a revelation, saying to his students "Gentlemen, that is surely true, it is absolutely paradoxical; we cannot understand it, and we don't know what it means, but we have proved it, and therefore we know it must be the truth." So too with relativity theory.

The axiomatic geometer sets up a system of axioms, and then from these axioms he derives theorems of many kinds. The prime example of this approach is that of Euclid. Actually, the axiomatic statements in Euclid were divided into two classes, postulates (αιτηματα) and axioms (κοιναι εννοιαι, common notions). His postulates are geometric, whereas his axioms in general refer to magnitudes. What Euclid did, and what made him great, was to take all the knowledge accumulated in geometry up to his time (ca. 350 B.C.) and codify it into an axiomatic system. His axioms and postulates are marvels for their beauty, elegance, and brevity. He then arranged the theorems in a manner so logical that no improvements were made in his geometry for over 2000 years.

Because Euclid collected his geometric concepts from various ancient sources, it is not surprising that we can extract from the *Elements* two self-contained geometries that differ in their logical foundation. These are known as *absolute geometry* and *affine geometry*. Absolute geometry, first recognized by the Hungarian mathematician Janos Bolyai (1802–1860), is the part of Euclidean geometry that depends upon Euclid's first four postulates,

1. A straight line may be drawn between two points.
2. A finite straight line can be extended in a straight line.
3. A circle may be drawn with any center and radius.
4. All right angles are equal.

but not on the fifth,

5. (The parallel postulate.) If a straight line cuts two other straight lines so that the sum of the two interior angles on one side of it is less than two right angles, the two other lines if extended indefinitely will meet on that side.

The study of absolute geometry is motivated by the fact that it gives theorems that are valid not only in Euclidean geometry, but also in non-Euclidean geometry. Absolute geometry is geometry without the assumption of a unique parallel line through a given point to a given line. In other words, the axioms of absolute geometry do not make any decisions about parallelism. Euler recognized affine geometry as being a separate branch of mathematics. This was a major achievement in mathematics. In affine geometry, the unique parallel (Euclid's fifth postulate) plays a leading role. Euclid's third and fourth postulates are not used, as circles are not mentioned and angles are never measured. Affine geometry is what remains after practically all ability to measure length, area, angles, and so on has been removed from Euclidean geometry. However, there still remains the concept of parallelism. The importance of affine geometry lies in the fact that its theorems hold not only in Euclidean geometry, but also in Minkowski's geometry of time and space.

Another important line of geometric thought had its origin in some geometric problems in art and military engineering in the Renaissance. Mathematicians, extending the investigations that were begun by artists and engineers, constructed out of their findings a new geometry known as *projective geometry*. A major achievement was the nineteenth-century discovery that Euclidean geometry and the two important non-Euclidean geometries (elliptical geometry and hyperbolic geometry) are all branches of projective geometry.

Renaissance artists who were trying to paint realistically had to grapple with this problem of technique: How do you make a two-dimensional drawing of a three-dimensional object so that the drawing really looks like the object? To solve this problem, they first simplified it by assuming that the viewer uses only one eye. The rays of light that travel from the object to the eye may be represented by straight lines (called the projection) converging from the object to the eye. If a

Plate 11.2. Perspective drawing in 1531. The artist in this illustration is using the gridlike panes of a window for his perspective.

glass window is interposed between the eye and the object, the plane of the window intersects each line at a point. The set of all these points is called the section of the projection. This section is a two-dimensional representation of a three-dimensional object, and the eye sees the object as if it were a picture on the glass window. Thus, the artists' problem is reduced to the purely geometric problem of determining the points on the section of the projection of the object.

Most graphs that we see drawn in everyday life fall within the branch of geometry known as *metric geometry*, the best example of which is Euclidean geometry. In metric geometry, angles and distances have constant, clear meanings. In general, length and angle measurements are simple and direct in a metric graph. However, not all geometries are metric, and the best example of a nonmetric geometry is projective geometry. In projective geometry, although the connections and intersections of points and lines have their usual meanings, parallel lines do not always look like parallel lines (for example, parallel railroad tracks appear to converge to a point), two equal angles do not always look like equal angles, and two equal lengths do not always look equal.

Affine geometry may be described as lying in between metric geometry and projective geometry. Affine geometry is like projective geometry, but with one additional condition, namely, parallel lines look parallel. But affine geometry is not metric geometry, because in affine geometry, lengths and angles do not have fixed properties. However, as we will now see, affine geometry can be turned into a metric geometry by the introduction of a suitable metric. The result is *metric affine geometry*. The importance of metric affine geometry is that, with the time-space interval as the metric, it is the geometry that describes the space-time of Poincaré and Minkowski.

For many centuries the postulates of Euclid were considered the best for geometry, but not any more. The modern source for axiomatic systems of geometry is linear algebra. *Linear algebra* is the algebraic theory of vector spaces. The axioms for a vector space give rise to axiomatic systems for all affine and projective, nonmetric and metric geometries. Such unity and simplicity cannot be claimed by the classical axiom system of Euclid, nor by the improved axiom system of David Hilbert (1862–1943), nor even by the famous *Erlangen program* (so called because it originated from Felix Klein's inaugural address at the University of Erlangen in 1872). However, the axiomatic systems that come from linear algebra are eminently suited to deal with specific problems in geometry, a virtue that the axioms of Euclid and Hilbert do not have. The geometries have not changed, only the choice of axioms.

A *vector space* is a set of objects or elements that can be added together and multiplied by numbers, the result being an element of the set, in such a way that the usual rules of calculation hold. Linear algebra can be regarded as the theory of vector spaces. The elements of a vector space are called vectors, and the numbers by which they can be multiplied are called scalars. For our purposes, the (x, t) coordinates of a point may be regarded as the vector from the origin to the point. Then we can add two such vectors as

$$(x_1, t_1) + (x_2, t_2) = (x_1 + x_2, t_1 + t_2)$$

and multiply a vector by a scalar a as

$$a(x, t) = (ax, at)$$

Affine geometry is the study of parallelism. To make affine geometry metric, we must introduce a measuring rod (a metric) into affine space. In this way we can study such metric concepts as distance, orthogonality, similarity, congruence, and so on. As we will see, there are many measuring rods that can be imposed on the same affine plane. Although such a measuring rod allows new concepts to be introduced into affine geometry, the underlying affine structure of the space is in no way affected.

According to linear algebra, the measuring rod for an affine space is defined by choosing a specific bilinear form for the inner product. We have already discussed inner products in Chapter 8. The choice of a suitable inner product changes the affine geometry into a *metric affine geometry*, that is, an affine space upon which a metric has been induced.

We can now make the following definition. Two-dimensional *Minkowski space* (x, t) is a metric affine space for which the inner product of the vectors (x_1, t_1) and (x_2, t_2) is given by

$$(x_1, t_1) \begin{pmatrix} -1 & 0 \\ 0 & 1 \end{pmatrix} \begin{pmatrix} x_2 \\ t_2 \end{pmatrix}$$

AFFINE GEOMETRY

By carrying out the matrix multiplication, we find that the inner product becomes

$$t_1 t_2 - x_1 x_2$$

This inner product defines the concept of orthogonality in the Minkowski plane, namely, the two vectors are *orthogonal* (or *perpendicular*) provided that their inner product vanishes, that is,

$$t_1 t_2 - x_1 x_2 = 0$$

Suppose now that the two vectors are the same, that is, $(x_1, t_1) = (x_2, t_2) = (x, t)$. Then the inner product becomes the square of the *time-space interval* τ; that is, it becomes

$$t^2 - x^2 = \tau^2$$

If τ^2 is positive, we say that the point (x, t) is timelike, in which case τ can be obtained as the square root of $t^2 - x^2$. On the other hand, if τ^2 is negative, we say the point (x, t) is spacelike. Then $\sigma^2 = x^2 - t^2$ is positive, which means that the *space-time interval* σ can be defined as the square root of $x^2 - t^2$. The quantities τ and σ act as the metric, or measuring stick, for the Minkowski plane.

In contrast, the two-dimensional *Euclidean space* (x, t) is a metric affine space defined by the inner product

$$(x_1, t_1) \begin{pmatrix} 1 & 0 \\ 0 & 1 \end{pmatrix} \begin{pmatrix} x_2 \\ t_2 \end{pmatrix} = x_1 x_2 + t_1 t_2$$

The metric, or measuring stick, for the Euclidean plane is thus the familiar distance

$$d = \sqrt{x^2 + t^2}$$

The Minkowski plane and the Euclidean plane provide us with a good example of how different metrics can be imposed on the same vector space. These two geometries are certainly distinct. Since every nonzero vector (x, t) on the Euclidean plane has a positive squared metric d^2, it is not possible to find a coordinate system for the Euclidean plane such that the matrix of the Euclidean metric becomes that of the Minkowski metric, that is, such that

$$\begin{pmatrix} 1 & 0 \\ 0 & 1 \end{pmatrix} \text{ becomes } \begin{pmatrix} -1 & 0 \\ 0 & 1 \end{pmatrix}$$

However, the mathematician Gudermann found a way around the inconvenience. Gudermann's construction is not general because it only works with two coordinate systems at a time, but it is adequate for most purposes. Gudermann's method allows us to look at two different sets of coordinates for the Minkowski plane, but at the same time treat distances by the familiar distance formula d of the Euclidean plane. Gudermann's discovery makes dealing with the Minkowski plane much simpler for those who feel uncomfortable with the different scales of measurement imposed by the calibration hyperbolas. The purpose of this chapter is to lead up to and present the Gudermann result.

MINKOWSKI PLANE

Figure 11.1 illustrates the affine geometry known as the Minkowski world. The horizontal axis is the space axis x, on which the unit length of light second is marked off. The vertical axis is the time axis t, marked off in seconds as the time unit. Any point such as E represents an event. This event occurred at location x_1 and time t_1. The lines branching out from E also have their meanings in space-time. Line L_5 lies parallel to the space axis x. To an observer in the x, t frame, all events lying along L_5 are simultaneous, taking place at different positions, all at time t_1. The x axis itself also represents simultaneous events in the x, t frame, all taking place at different positions at time $t = 0$.

The line L_1 lies parallel to the time axis t. To an observer in the x, t frame, all events lying along L_1 occur at the same position x_1, but at different times. Thus, L_1 represents a stationary particle that stays at the single position x_1, but undergoes a passage of time. The t axis also represents events occurring at a single position ($x = 0$), but at different times. Like line L_1, the t axis also gives the time history of a point at rest in the x, t frame.

The line L_2 is slanted, so it does not lie parallel to either the x or t axes.

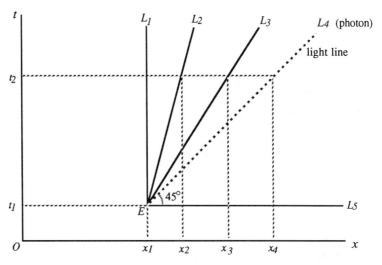

Figure 11.1. The Minkowski space-time plane with coordinates (x, t).

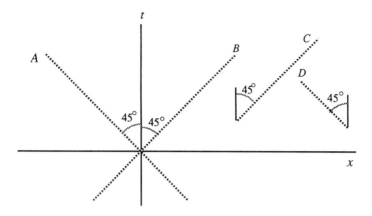

Figure 11.2. The Minkowski plane with light lines A, B, C, D.

Thus, L_2 represents the history of an object in motion in the x, t frame, the object passing through different locations at different times. The line L_3 also represents a moving point, but L_3 is not so steep as L_2. Thus, it represents a point moving faster in the x, t frame than does L_2. The reason is that between time t_1 and time t_2 an object moving along L_2 goes from location x_1 to location x_2, but an object moving along L_3 goes from x_1 all the way to x_3 in the same time interval. In general, the flatter a line is drawn, the greater the speed it represents in the x, t frame. But, there is a limiting speed, the speed of light. The scale of our graphs is chosen such that the speed of light is unity ($c = 1$). Line L_4 has slope unity, so it represents a particle (photon) traveling at the speed of light. Only a photon could travel all the way from x_1 to x_4 in the time increment $t_2 - t_1$.

A history in space-time is called a *world line*. The t axis and L_1 are world lines of objects at rest in the x, t frame, and L_2 and L_3 are world lines of objects in uniform motion in the x, t frame. The line L_4 is the world line of a photon, and so is called a *light line*. No world line of an object can be drawn flatter than the light line, because it would represent an object moving at a speed greater than the speed of light, the ultimate speed. Thus, L_5 and the x axis are not world lines of objects, because they cannot represent the histories of objects or photons.

Figure 11.2 shows Minkowski space with the light lines A and B through the origin. These two light lines make a right angle (45° + 45° = 90°) with each other. They are the world lines of photons moving to the left (A) and right (B). The world lines of objects must be steeper than the 45° world lines of photons. World lines have meaning no matter where they are placed in Minkowski space. For example, the two segments C and D are also light lines, even though they do not pass through the origin of the x, t frame.

In Figure 11.3, t' is the world line of an observer who is in constant motion in the x, t frame. Now, since he is moving in the x, t frame, this observer has his own frame of reference. To him, his own world line is his time axis t'. His space

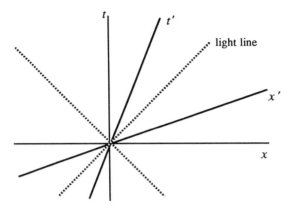

Figure 11.3. The Minkowski plane with two orthogonal coordinate systems (unprimed and primed) shown, each set of axes being symmetric with respect to the same light line.

axis x' is represented by the reflection of his time axis about the light line. This observer does not think that events along the x axis are simultaneous. In his x', t' frame, simultaneous events fall on the x' axis, or on lines parallel to the x' axis.

The effects of metric affine geometry now enter. Although the angle between t' and x' seems smaller than the angle between t and x, the two frames are actually structured the same way. The angle between t' and x' is a right angle (like the angle between t and x), but it does not look like a right angle in Minkowski space.

The world lines of a traveler are shown in Figure 11.4 as he makes a round trip in space. His world line on the trip out is t', and his world line on the trip back is t''. Now another effect of affine geometry enters the situation. In Euclidean geometry, a straight line is the shortest path between two points. In the metric affine geometry of Minkowski space, a straight line is the longest path between two events. Between the events E_1 and E_2, an observer at rest in the x,

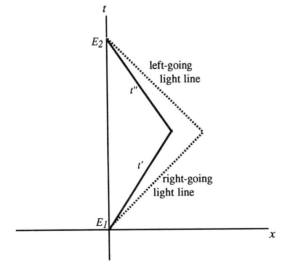

Figure 11.4. World lines of a round trip (there via t' and back via t'') of a traveler.

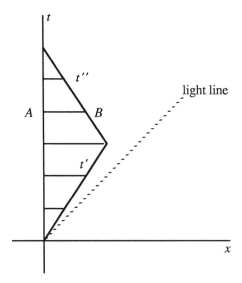

Figure 11.5. Regular time intervals along the t axis are shown by the horizontal lines (i.e., lines parallel to the x axis).

t frame with world line t follows the longest possible path, a straight line. The length of his path is his proper time. The proper time of the traveler is less since he follows a shorter path between E_1 and E_2, namely, the nonstraight path t' and t''. Thus, less time has passed for the traveler than the sedentary person. The shortest possible path between E_1 and E_2 is a path of a light signal sent into space and reflected back again, as shown by the dashed lines in the figure. Since a photon does not age, the travel time of this light signal is zero.

Let us note that if the traveler keeps going indefinitely along t' (and does not turn back along t''), his world line is also straight, so he and the sedentary observer in the x, t frame will never have the chance to compare proper time passages, because they will never meet again.

Regular time intervals are shown from the viewpoint of the sedentary observer in the x, t frame in Figure 11.5. The lines parallel to the x axis make up his time marks. Thus, he regards the event A in this frame and the event B in the traveler's frame to be simultaneous because they both lie on one of these time marks.

In Figure 11.6, regular time intervals are shown from the viewpoint of the traveling observer, first while he is in the x', t' frame, and then while he is in the x'', t'' frame. His time marks are parallel first to the x' axis and then to the x'' axis. The traveler thinks that events along these lines are simultaneous, such as events C and D. His pairing of simultaneous events is obviously very different from the sedentary observer. We also see that as a result of the traveler changing his frame from x', t' to x'', t'', there is a large gap in the sedentary observer's experience that does not appear to be simultaneous to any of the traveler's experiences. This gap is a result of the traveler's sudden change of reference frame and, therefore, in his methods of calling events simultaneous.

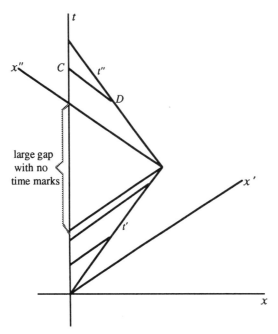

Figure 11.6. Regular time intervals along the t' axis are shown by the lines sloping to the right (i.e., lines parallel to the x' axis). Regular time intervals along the t'' axis are shown by the lines sloping to the left (i.e., lines parallel to the x'' axis).

TIME AND DISTANCE BY RADAR MEASUREMENTS

Suppose that we have a traveler A and some event Q at which A is not present. The event Q could occur either to the right or left of A, but not on his path. What sort of on-the-spot observations are permitted? The observer A can send out a light signal which is reflected from an object, and then the signal returns to A. This method is a description of radar, which consists of sending a microwave signal to a target and then recording the echo. The radar apparatus records the sending time t_s (where the subscript s stands for sending) and the receiving time t_r (where the subscript r stands for receiving). Radar signals, like light signals, are electromagnetic waves and so travel at the speed of light.

Figure 11.7 shows A's world line and event Q. The zero of A's clock is marked O. The observer A sends a light signal at S which is reflected at Q and received by A at R. Thus, A has two hard facts, time t_s of sending a signal to Q and time t_r of receiving the reflected one from Q. This is the only firm information that A can have about the event Q. Now we seek rules by which we can calculate the time and distance of event Q, according to A, from the times t_s and t_r.

With reference to Figure 11.7, there are two distances to consider: the spatial distance between events S and Q and the spatial distance between events Q and R. Are these two distances equal or unequal according to observer A? The answer is, equal. Let us now carefully work out and formulate the objections to this answer. We see that the horizontal distance between S and Q is not equal to

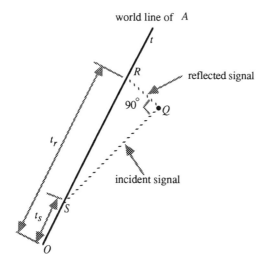

Figure 11.7. A light signal is sent by observer A along space-time path SQ. Then it returns by reflection back to A along space-time path QR.

the horizontal distance between Q and R. But we have not shown that horizontal distances in the diagram mean anything, so we cannot use them to deduce anything. Besides, we know that A's world line can be given any slope (steeper than $45°$). Hence, we cannot deduce anything from a property of the diagram that depends upon this arbitrary slope.

Another objection might be that it all depends upon A's motion relative to Q. If A is moving toward Q, then the signal has farther to go on the outward journey than on the return journey. The answer to this objection is that A is not moving, for he is at rest in his inertial frame. For example, suppose A is standing on one step of an escalator. He is sedentary with respect to his inertial frame (the escalator) even though the escalator is moving with respect to the building. A young girl runs up the escalator from A, turns immediately when she reaches the top, and runs down to A again. The young girl takes the same number of steps each way (the spatial distances are equal according to A). If A uses light signals instead of the girl, the same would be true.

More generally, distances according to any observer must be measured in a frame of reference that is stationary relative to him. All the measuring apparatus must be stationary relative to him. Then it follows that the spatial distance from him to any event is equal to the distance from the event back to him. Thus we have answered the question, that is, the spatial distances (according to A) covered by the incident and reflected signals are the same.

Let us now return to our main line of reasoning. We have the sending and receiving times t_s and t_r of the radar (light) signal. We now know that the radar pulse (or photon) has the same spatial distance to go on its outward journey as on its return journey, and it travels at the same speed ($c = 1$) both ways. Therefore, it takes the same time from event S to event Q as from Q to R (still according to A). Thus we conclude that the time t_q of event Q is halfway between the sending time t_s of event S and the receiving time t_r of event R by A's clock. In other words, t_q is the average:

$$t_q = \frac{t_s + t_r}{2}$$

In common language, t_q is the *midpoint* of the sending time t_s and receiving time t_r.

Now let us consider distances. Again we use the radar approach. Observer A measures the distances in terms of time. In natural units, the speed of light is $c = 1$. According to A, in the time between events S and R, light traveled from him to Q and back; that is, the light covered twice the distance between himself and Q. By using natural units, this is the same as saying that twice the distance of Q from A is equal to the time between S and R. Thus, we conclude that the distance of event Q from A, according to A, expressed in natural units, is half the time by A's clock between events S and R. In other words, x_q is given by

$$x_q = \frac{t_r - t_s}{2}$$

In common language, x_q is the *half-offset* between the sending time t_s and receiving time t_r, the full offset being defined as $t_r - t_s$.

Let the midpoint M be the tick of A's clock halfway in time between S and R. Then t_q, the time of event Q, is the time $(t_r + t_s)/2$ of the event M on A's world line. And the distance of Q from A (in natural units) is the time interval from S to M (or equivalently from M to R) on A's world line. See Figure 11.8.

When we speak of the time of Q according to A, we actually refer to the time of an event (M) on the world line of A, namely, the tick of A's clock midway

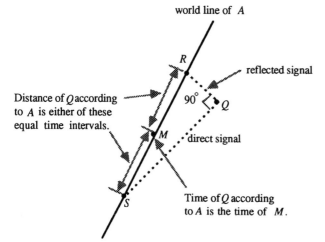

Figure 11.8. The time of event Q according to A is the time of event M. The distance of event Q according to A is equal to either of the two equal time intervals SM or MR.

in time between sending and receiving the radar signal. What we call the distance of Q according to A is actually a time interval on A's world line, namely, half the time that elapses between sending and receiving the radar signal.

MINKOWSKI COORDINATE SYSTEMS

Using the radar method described in the foregoing section, we now want to construct the coordinate system for Minkowski space. It is easier to get a good drawing if we first put in signal lines at 45° slope before putting in the world line of A. Thus we have the diagram shown in Figure 11.9.

If a lamp is flashed at S, then light signals travel both ways in the one-dimensional space of a single spatial axis. The light line for the right-going signal is SQ and for the left-going signal is SP. See Figure 11.10.

The signal QR represents the left-going echo from Q, and the signal PR represents the right-going echo from P, where each echo is received by A at event R. Note that events Q and P are related to A in the same way but on opposite sides. Next draw the line joining P and Q, and let M be the point where this line crosses the world line of A. The figure SPRQ is a rectangle, since its sides are all at 45° to the horizontal. Hence the intervals SM, MQ, PM, MR are all equal. Also, M is the same midpoint as described in the foregoing section.

In the foregoing section, we say that the distance (in natural units) of Q from A according to A is equal to the time interval from S to M (i.e., the interval SM). But to represent the distance from A to Q, it would be better to have some interval stretching from A's world line to the event Q. Interval MQ, being equal to interval SM (see the preceding paragraph), will do nicely. Thus we conclude that the distance (in natural units) of event Q from observer A according to A is represented (on A's distance scale) by the interval MQ in Figure 11.10.

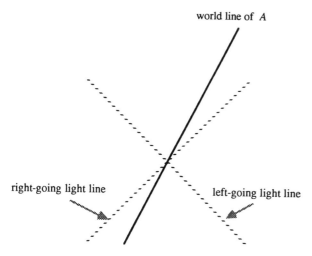

Figure 11.9. First, draw the two light lines (the dotted lines with the 45° slopes); then, draw the world line of A (the sloping solid line).

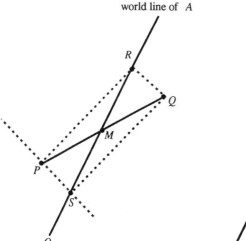 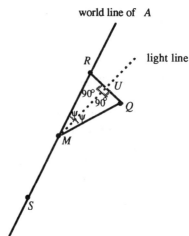

Figure 11.10. A light pulse emitted at event S produces two light signals from observer A. One light signal has outgoing path SQ and, by reflection, return path QR back to A. The other light signal has outgoing path SP and, by reflection, return path PR back to A.

Figure 11.11. Space-time interval MQ equals time-space interval MR. Both MQ and MR make the same angle ψ with the light line, but they lie on opposite sides of the light line.

To avoid confusion, we use the convention that the word *distance* refers to the one-dimensional space axis (x axis, or x' axis, etc.). Also, we agree that the word *direction* always refers to one of the two directions (left or right) on the one-dimensional spatial axis. The word *time* or *duration* refers to the time axis (t axis or t' axis, etc.). When we want to describe the way a line in a diagram points, we will speak of its slope. The world line of a right-going particle has slope greater than 45° upward from left to right, and the world line of a left-going particle has slope greater than 45° upward from right to left. Thus particles going in opposite directions have world lines that slope in opposite directions.

The word *interval* refers to the segment between two events in the Minkowski space-time plane (x, t, or x', t', etc.). There are two kinds of intervals: the *time-space interval* (usually used in this book) and the *space-time interval*. The time-space interval is real if the segment has a slope greater than 45°. The space-time interval is real if the segment has a slope less than 45°. In this discussion, whenever we say "interval," we always take the one (time-space or space-time) that is real for the case at hand.

The interval MQ (representing the distance from A to Q according to A) is not measured horizontally. The slope of MQ depends upon the slope of the world line of A. The relationship is as follows. (Figure 11.11.) Draw the light line

Minkowski Coordinate Systems

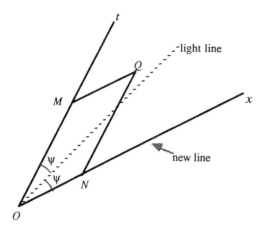

Figure 11.12. The time axis t is chosen as the world line of A. This world line makes an angle ψ with the light line. The space axis x is chosen as the line that is on the other side of the light line and that makes the same angle ψ with the light line.

(45° slope to the right) through M. Since interval MQ equals interval MR, and since the constructed light line MU and light line RQ are at right angles, it follows that right triangle RMU is congruent to right triangle QMU. Thus angles RMU and QMU are equal. We therefore conclude that line MQ makes the same angle ψ with the light line as line MR (the world line of A) makes with the light line, but on the other side. In other words, line MQ is the reflection of line MR about the light line.

No matter what point we take for Q, the foregoing rule will always give the same slope for MQ. The conventional way to draw the axes on the Minkowski diagram thus is to draw a line ON through O that makes the same angle ψ with the light line as does the world line of A, but on the other side, as shown in Figure 11.12.

The world line OM of A becomes the time axis t of A, and the new line ON becomes the distance axis x of A. Distances like MQ are parallel to this new line, the x axis. Thus the x axis specifies the slope to be used when measuring intervals that represent distances according to A. Thus, we have the following recipe for reading A's distances from the diagram. If Q is any event, draw two lines through Q, one parallel to the x axis and meeting the t axis at M and the other parallel to the t axis and meeting the x axis at N. Then, in natural units, the distance of event Q is given by MQ and the time of event Q is given by NQ.

We can therefore state the following propositions. *All equidistant events for A (in A's frame) are represented by the points of a line parallel to his time t axis. All equitime (simultaneous) events for A (in A's frame) are represented by the points of a line parallel to his distance x axis.*

The whole system that we have developed for laying out the diagram and making measurements is called the coordinate system of A. This discussion, of course, illustrates the absolute necessity of the concept of parallel lines and the concept of metric as given by the interval (time-space and space-time). These two concepts are inherent in metric affine geometry.

Plate 11.3. Relief carving of Ahura-Mazda, the ancient Persian god of light. Persepolis, Iran. Achaemenid period, reign of Darius I–Xerxes I, ca. 522–465 B.C.

GUDERMANNIAN SPACE-TIME DIAGRAM

We now want to discover again why the interval between two events on a Minkowski space-time diagram is an invariant. To do this, we need to learn one more trick about space-time diagrams. An observer's distance and time measurements are represented by scales on his x and t axes. Another observer in a differential inertial frame measures distances and times on his axes, which we call x' and t'. As we know, the scales of these two observers are generally different, and it is for that reason we introduced the unit hyperbolas in Chapter 10 to calibrate the various axes. Thus, each set of axes must be separately calibrated, and we cannot directly compare them without taking these calibration marks into effect. But when we have only two observers to deal with, it would be convenient if both their distance scales and their time scales had the same calibration. We now want to show that we can draw their world lines in such a way as to arrange this.

The calibration scale used for the axes for any observer must depend upon the slope of his world line. We recall that his world line is his time axis, and we recall that his distance axis is found by reflecting his world line about the light line. (That is, both his time axis and his distance axis make the angle ψ with the light line, but the two axes are on opposite sides of the light line.) Now we make use

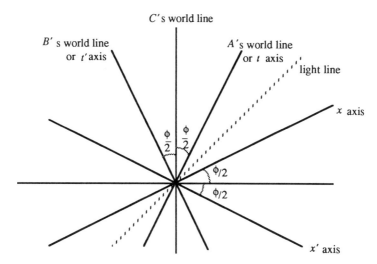

Figure 11.13. Construction of the Gudermannian space-time diagram.

of this fact: *world lines with equal slopes have equal scales.* Let us draw the world lines of the two observers with slopes of the same magnitude but opposite algebraic sign. Thus the two world lines have the same angle $\phi/2$ on either side of the vertical, as shown in Figure 11.13.

Let us now construct the special type of Minkowski space-time diagram which is known as the Gudermannian space-time diagram. The two world lines in Figure 11.13 are labeled the t axis and t' axis. We then reflect them about the light line to obtain the x axis and the x' axis. With this construction, the scales used for representing time and distance measurements in Figure 11.13 are the same for both observers. For example, we can use 1 inch on each of the four axes to represent one unit of time or distance. As usual, we are working in natural units, with the velocity of light equal to unity ($c = 1$).

There is nothing special about any inertial frame, according to the principle of relativity. Thus, if we pick any two observers, calling them A and B, we can construct a diagram for them like the preceding one. Let us now consider a third observer C whose motion is such that A and B are moving, relative to C, at the same speed but in opposite directions. We give C a vertical world line. Then the world lines of A and B are equally inclined to this vertical, but on opposite sides. Both A and B will have the same scale, but it will be different from the scale used for C.

Let us now look carefully at the diagram in Figure 11.13. We see that the t' axis is at right angles to the x axis and that the t axis is at right angles to the x' axis. That is, *the distance axis of each observer (A or B) is perpendicular to the other's time axis.* This is the special feature of the Gudermannian diagram. The reason is as follows. By construction, the time axes have slopes of equal magnitude, so each makes an angle $\phi/2$ with the vertical. The angle ϕ is known as the

Gudermannian angle. Because the x axis is a reflection of the t axis about the 45° light line, it follows that the x axis makes the same angle $\phi/2$ with the horizontal. Thus the t' axis represents a rotation of $\phi/2$ counterclockwise from the vertical, whereas the x axis represents a counterclockwise rotation by the same angle $\phi/2$ from the horizontal. Because the vertical and horizontal are perpendicular, it follows that the t' axis and the x axis are perpendicular. A similar statement holds for the t axis and x' axis.

Let us now search for an invariant in terms of this construction. Given any two events O and Q, we can represent them each as a point on the space-time diagram. However, observers A and B have two different sets of axes, so A will say that the distance and time between O and Q are x and t, respectively, whereas B will say that they are x' and t', respectively. Even though A and B have the same scales on the space-time diagram, as done in the preceding construction, $x \neq x'$ and $t \neq t'$. This fact is evident in the diagram of Figure 11.14.

Let us now find an equation that connects x, t, x', t'. In Figure 11.14, draw the line NM. Consider triangle NMQ. It is a right triangle because two of its sides are parallel to axes that meet at right angles. Also, triangle NMO is a right triangle. These two right triangles have side NM in common. Thus by the double use of the Pythagorean theorem we have

$$OM^2 + ON^2 = NM^2 = QM^2 + QN^2$$

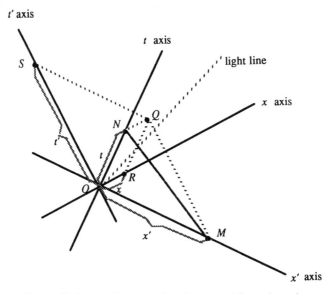

Figure 11.14. Gudermannian space-time diagram. All axes have the same scale. For example, 1 inch would be 1 second (or light-second) on each of the four axes. The four dotted lines give the corrdinates of Q: NQ is x, RQ is t, SQ is x', and MQ is t'.

which is

$$x'^2 + t^2 = NM^2 = t'^2 + x^2$$

To find an equation on which the two observers agree, we need an equation that contains only A's measurements x, t on one side and B's measurements x', t' on the other side. The preceding equation gives us the required equation as

$$t^2 - x^2 = t'^2 - x'^2$$

Observers A and B disagree about the time between O and Q and about the distance between them. However, if each uses his own version about time and distance to calculate

(time from O to $Q)^2$ − (distance from O to $Q)^2$

then they agree about this quantity, which is the square of the time-space interval between O and Q. *Thus we see that the time-space interval between two events is invariant*, for if any two observers agree, then all observers agree.

Plate 11.4. Front row: Albert Abraham Michelson, Albert Einstein, and Robert Andrews Millikan (1868–1953) in California in 1930.

PROPER AND IMPROPER TIMES

Let us consider two inertial observers, A and B, who pass through the origin O with relative speed β (expressed as a fraction of the velocity of light). Let the frame of A be x, t and the frame of B be x', t'. The Minkowski space-time diagram is shown in Figure 11.15.

Let the B clock tick with a constant period between ticks, with one tick at O and the next tick at Q. Since the B clock is fixed in the B reference frame, the

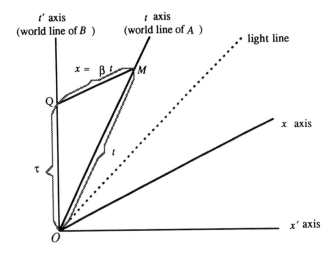

Figure 11.15. Minkowski space-time diagram with perpendicular axes x', t' and oblique axes x, t.

distance x' between O and Q according to B is zero. If we let τ stand for the time period between ticks for the B clock, then τ is the time between O and Q according to B. Thus, the squared time-space interval between O and Q according to B is

$$\tau^2 - 0^2 = \tau^2$$

Now B is moving with a velocity β relative to A. Let A say that the time period between the ticks of the B clock is equal to t. Because B has traveled away from A at velocity β, it follows that, at t, B has covered a distance $x = \beta t$. Thus, event Q has coordinates $x = \beta t$, $t = t$ in A's reference frame. Thus, the squared time-space interval between O and Q according to A is

$$t^2 - x^2 = t^2 - (t\beta)^2 = t^2(1 - \beta^2)$$

The key is the invariance of the time-space interval. From this invariance, we see that the two versions must be the same.

$$\tau^2 = t^2(1 - \beta^2)$$

We have thus found a connection between the times of A and B, namely,

$$t = \frac{\tau}{\sqrt{1 - \beta^2}}$$

which is

$$(\text{time from } O \text{ to } Q \text{ by the } A \text{ clock}) = \frac{(\text{time from } O \text{ to } Q \text{ by the } B \text{ clock})}{\sqrt{1 - \beta^2}}$$

PROPER AND IMPROPER TIMES   345

In the present case, which time is proper and which improper? Observe that B is present at both events O and Q, so τ is the proper time betwen the events. But A is present at O but not present at Q, so t is improper. Thus the above equation states

$$\text{(improper time of } A) = \frac{\text{(proper time of } B)}{\sqrt{1-\beta^2}}$$

We recognize this equation as the equation of the *dilation of time*. Because the factor $\sqrt{1-\beta^2}$ is less than one, this equation states that the proper time is less than the improper.

Now let us verify the dilation of time by making use of a space-time diagram in which the world lines of A and B have slopes with equal magnitude but opposite sign. (As we have just seen, such a construction with symmetric world lines about the vertical is called the Gudermann world.) See Figure 11.16.

The axes of A are x and t, and the axes of B are x' and t', where the t' axis is perpendicular to the x axis and the t axis is perpendicular to the x' axis. Let both A and B have the same proper time τ (the ticks between each of their clocks are the same). Because all the axes have the same scale, we mark off τ as 1 inch on both the t' and t axes, thus defining points Q and q, respectively. Draw line QM parallel to the x axis and line qm parallel to the x' axis, as shown in Figure 11.16. Then OM is the improper time t and Om is the improper time t'. Since OQ is less than OM, and since both scales are the same, we see that $\tau < t$. Also,

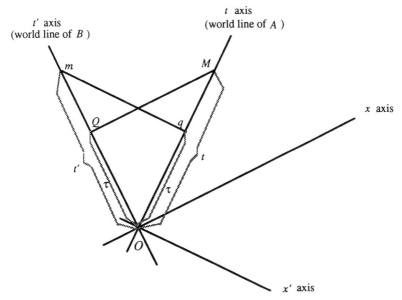

Figure 11.16. Gudermannian space-time diagram corresponding to the Minkowski space-time diagram of Figure 11.15.

since Oq is less than Om, and again both scales are the same, we see that $\tau < t'$. Thus in each case, the proper time is less than the improper.

In summary, A thinks that the B clock is running slow, because the time period $\tau = OQ$ representing the period actually produced by the B clock is less than the time duration $t = OM$ representing what A thinks the period is. On the other hand, the time period $\tau = Oq$ of the A clock is less than the time duration $t' = Om$ as perceived by B. Thus, B thinks that the A clock is running slow. Hence by exactly parallel arguments, each observer thinks the other's clock is running slow.

FITZGERALD LENGTH CONTRACTION

The space-time diagram in the symmetric slopes geometry (Gudermann world) helps us understand the contraction of a ruler. Let observer B carry a ruler, holding it by its left end. See Figure 11.17.

We now want to talk about the length of the ruler. Since B holds the left end, the right end is always at the same distance λ from him. Thus, the world line of the left end is the same as the world line B (the t' axis), and the world line of the right end is a line parallel to the t' axes (the *dashed line* in Figure 11.17). The distance OP on the x' axis is equal to the length λ of the ruler according to the person (B) holding the ruler.

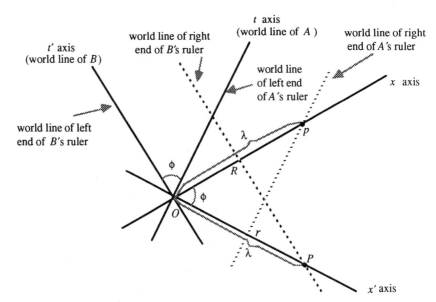

Figure 11.17. The t' axis and the dashed line PR are the world lines of the left and the right ends, respectively, of a ruler held by B. The t axis and the dotted line rp are the world lines of the left and right ends of a ruler held by A.

Although B is carrying the ruler, it still has a space-time existence, independently of what B thinks of it. The ruler as a space-time entity is represented by the whole area between the t' axis and the dashed parallel line. Every point in that area represents the event of a bit of a rod being instantaneously at that location. However, observer A splits space-time differently from observer B. In other words, whereas B uses coordinates x', t' for an event, observer A uses x, t for the same event. Thus, we must ask what represents the ruler at one instant ($t = 0$) according to A? The answer is the segment of the x axis from O to the dashed line, that is, OR in the diagram. The length of the ruler is the distance between its ends at the same time, namely, $\lambda = OP$ for B and OR for A. And we see now that the disagreement about length is a simple consequence of the different coordinates used by A and B. Since OR is perpendicular to the t' axis, it is shorter than $\lambda = OP$. Thus A thinks that the ruler that B holds has contracted; that is,

$$\lambda > OR$$

Now let A carry the ruler. Since A holds the left end, the right end is always the same distance λ from him. Thus the world line of the left end is the same as the world line of A (the t axis), and the world line of the right end is a line parallel to the t axis (the *dotted line* in Figure 11.17). The distance Op on the x axis is equal to the length of the ruler according to the person (A) holding the ruler. However, to B, the ruler appears as the segment of the x' axis from O to the dotted line, that is, Or in Figure 11.17, because $t' = 0$ along the x' axis. The length of the ruler is the distance between its ends at the same time, namely, $\lambda = Op$ for A and Or for B. Since Or is perpendicular to the t axis, it is shorter than $\lambda = Op$. Thus, B thinks that the ruler that A holds has contracted; that is,

$$\lambda > Or$$

In summary, if an observer in an inertial frame holds a ruler of length λ, then an observer in another inertial frame will think that the ruler contracts to a length less than λ. The amount of contraction can be found as follows. Let ϕ be the angle between the t and t' axes, or equivalently let ϕ be the angle between the x and x' axes. As we know, the angle ϕ is known as the *Gudermannian angle*. Then from Figure 11.17 we have

$$OR = \lambda \cos \phi, \quad Or = \lambda \cos \phi$$

Let us now use the diagram in Figure 11.18 to find $\cos \phi$. In the time span $t = 1$, observer B moves away from observer A by an amount $x = \beta t = \beta(1) = \beta$. We have labeled β and $t = 1$ in Figure 11.18, which shows that $\sin \phi = \beta$. Therefore, it follows that

$$\cos \phi = \sqrt{1 - \sin^2 \phi} = \sqrt{1 - \beta^2}$$

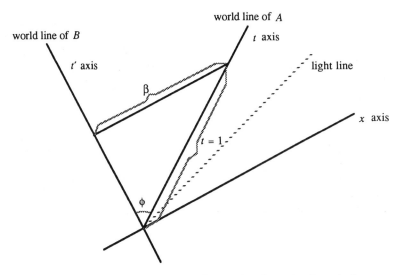

Figure 11.18. Gudermannian diagram that shows that $\beta = \sin \phi$.

Thus, if two inertial frames have relative velocity β (in natural units), and if one inertial observer holds a ruler of length λ, the other inertial observer thinks that the length of the ruler has been contracted to length $\lambda\sqrt{1 - \beta^2}$.

In this chapter we have described the Gudermannian diagram, which is a special case of the Minkowski diagram. The Gudermannian diagram can only have two sets of axes, but it has the advantage that the same scale (i.e., unit of measurement) applies to both sets of axes. This feature is in contrast to the general case of the Minkowski diagram, which can have any number of sets of axes, but generally a different scale for each set. The unit of measurement on each set of axes in the Minkowski diagram is determined by the unit calibration hyperbolas.

Each of the two sets of axes in the Gudermannian diagram represents an inertial frame, and let β be the relative velocity between these two Gudermannian frames. Define the Gudermannian angle ϕ as $\beta = \sin \phi$. With respect to a reference frame in a Minkowski diagram, let one Gudermannian frame have the relative velocity $\tan(-\phi/2)$ (so that the t axis of this Gudermannian frame makes an angle $-\phi/2$ with respect to the reference time axis). Let the other Gudermannian frame have the velocity $\tan(\phi/2)$ (so that the t' axis of this other Gudermannian frame makes an angle $\phi/2$ with respect to the reference time axis). By the symmetry exhibited in Figure 11.19, it follows that both Gudermannian frames have the same scale (i.e., the same unit of measurement on each Gudermannian axis).

Finally we must verify that the relative velocity between the two Gudermannian frames as constructed in Figure 11.19 is indeed β. With respect to the reference frame their relative velocities are $\tan(\phi/2)$ and $\tan(-\phi/2)$. Using the

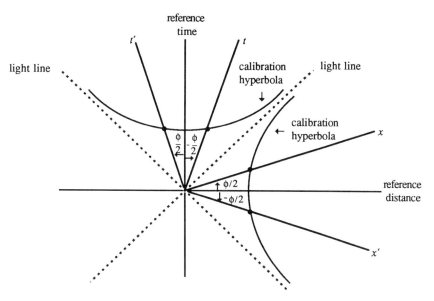

Figure 11.19. Because of their symmetry with respect to the reference frame, all four Gudermannian axes x, t, x', t' strike the calibration hyperbolas at the same time or same distance. Thus both Gudermannian frames have the same scale.

physical velocity addition formula, the relative velocity between the two Gudermannian frames is

$$\frac{\tan(\phi/2) - \tan(-\phi/2)}{1 - \tan(\phi/2)\tan(-\phi/2)}$$

By the use of trigonometric identities, this expression does reduce to $\sin\phi$. But the Gudermannian angle ϕ was defined by $\beta = \sin\phi$. Thus we have verified that the Gudermannian construction does produce the relative velocity $\beta = \sin\phi$ as required.

CHAPTER 12

THE GUDERMANN WORLD

Stand still, you ever moving spheres of heaven,
That time may cease, and midnight never come;
Fair Nature's eye, rise, rise again, and make
Perpetual day; or let this hour be but
A year, a month, a week, a natural day.

Christopher Marlowe (1564–1593)
Doctor Faustus, Scene 14

HYPERBOLIC SINE AND COSINE

When the physics start to get difficult, then it is time to turn back and do some mathematics in order to get our feet on the ground again. In Chapter 8, we introduced hyperbolic functions, and in this chapter we wish to develop some more

Hyperbolic Sine and Cosine

Plate 12.1. Sir Joseph Larmor (1857–1942)

properties about them. Two of these functions, the hyperbolic sine and the hyperbolic cosine of a variable u, are defined by the equations

$$\sinh u = \frac{e^u - e^{-u}}{2}, \qquad \cosh u = \frac{e^u + e^{-u}}{2}$$

Let us show that the point (x, t) with $x = \sinh u$ and $t = \cosh u$ lies on the unit hyperbola $t^2 - x^2 = 1$. We simply substitute the defining relations into the equation of the hyperbola, and we see that these coordinates do satisfy the equation:

$$t^2 - x^2 = (1/4)(e^{2u} + 2 + e^{-2u}) - (1/4)(e^{2u} - 2 + e^{-2u}) = 1$$

We have thus established the basic identity

$$\cosh^2 u - \sinh^2 u = 1$$

The hyperbolic tangent is defined as

$$\tanh u = \frac{\sinh u}{\cosh u} = \frac{e^u - e^{-u}}{e^u + e^{-u}}$$

and the hyperbolic secant is defined as

$$\text{sech } u = \frac{1}{\cosh u}$$

From the definitions, we find that

$$\cosh u + \sinh u = e^u$$

$$\cosh u - \sinh u = e^{-u}$$

from which it follows that any combination of the exponentials e^u and e^{-u} can be replaced by a combination of sinh u and cosh u, and conversely. Also, since e^{-u} is positive, we see that cosh u is always greater than sinh u. For large values of u, e^{-u} is small, and cosh $u \approx$ sinh u, tanh $u \approx 1$. The graphs of the hyperbolic functions are shown in Figure 12.1.

At $u = 0$, cosh $u = 1$, sinh $u = 0$, tanh $u = 0$, so the hyperbolic functions all have the same values at 0 that the corresponding trigonometric functions have. The hyperbolic cosine is an even function; that is,

$$\cosh(-u) = \cosh u$$

which means that the curve is symmetric about the vertical axis. The hyperbolic sine and hyperbolic tangent are odd functions; that is,

$$\sinh(-u) = -\sinh u,$$

$$\tanh(-u) = -\tanh u$$

which means that each is symmetric with respect to the origin. The symmetric properties of the hyperbolic functions are similar to the corresponding circular trigonometric functions.

Hand-held scientific calculators have keys to compute hyperbolic sine, cosine,

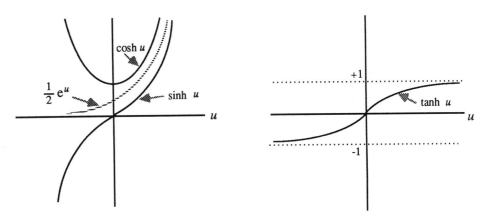

Figure 12.1. Graphs of hyperbolic sine, cosine, and tangent.

and tangent, as well as their inverses. This ready availability of the hyperbolic functions makes them useful in solving problems in relativity theory.

Certain major differences between the hyperbolic and circular functions should also be noted. The circular functions are periodic: $\sin(u + 2\pi) = \sin u$, $\tan(u + \pi) = \tan u$, and so on, but the hyperbolic functions are not periodic. Again, they differ greatly in the range of values they assume:

$\sin u$ varies between -1 and 1, oscillates;

$\sinh u$ varies from $-\infty$ to ∞, steadily increases.

$\cos u$ varies between -1 and $+1$, oscillates;

$\cosh u$ varies from $+\infty$ to $+1$ to $+\infty$.

$\tan u$ varies from $-\infty$ to ∞;

$\tanh u$ varies from -1 to $+1$.

The following identities require only the definitions and some algebraic calculations to be established:

$$\sinh(u + u') = \sinh u \cosh u' + \cosh u \sinh u'$$

$$\cosh(u + u') = \cosh u \cosh u' + \sinh u \sinh u'$$

These identities can be used to establish the *physical velocity addition formula* that was treated in Chapters 4 and 9. In fact, this addition formula is nothing more than the identity

$$\tanh(u + u') = \frac{\tanh u + \tanh u'}{1 + \tanh u \tanh u'}$$

which can be established from the preceding two identities. We also have

$$\tanh(u - u') = \frac{\tanh u - \tanh u'}{1 - \tanh u \tanh u'}$$

THE GUDERMANNIAN

The function $\sin^{-1} \tanh u$, which occurs frequently in mathematics, is called the *Gudermannian* of u, named after the German mathematician Christoph Gudermann (1798–1852). His papers were published in 1830. The symbol used is gd u (read: Gudermannian of u). Thus

$$\text{gd } u = \sin^{-1} \tanh u$$

From this definition, we have

$$\text{gd } 0 = 0, \quad \text{gd}(-u) = -\text{gd } u, \quad \text{gd}(+\infty) = \pi/2, \quad \text{gd}(-\infty) = -\pi/2$$

When u increases, gd u increases. Its value lies between $-\pi/2$ and $\pi/2$. Some values are given in the following table:

tanh u	u	gd u (in Radians)	gd u (in Degrees)
0.000	0.0	0.000	0.0°
0.462	0.5	0.480	27.5°
0.7616	1.0	0.864	49.5°
0.9052	1.5	1.132	64.9°
9.964	2.0	1.302	74.6°
0.9866	2.5	1.407	80.6°
0.9951	3.0	1.471	84.3°
0.9982	3.5	1.510	86.5°
0.9993	4.0	1.534	87.9°
0.9998	4.5	1.549	88.8°
0.9999	5.0	1.557	89.2°

To find the inverse of the Gudermannian, define the *Gudermannian angle* ϕ as

$$\phi = \text{gd } u = \sin^{-1}(\tanh u), \quad -\pi/2 < \phi < \pi/2$$

and solve for u. The result is

$$u = \text{gd}^{-1} \phi = \tanh^{-1}(\sin \phi)$$

We see that the relationship of u and ϕ is

$$\tanh u = \sin \phi$$

We can depict this relationship by means of the right triangle with hypotenuse 1 and opposite leg tanh u, as shown in Figure 12.2. Because

$$1 - \tanh^2 u = 1 - (\sinh^2 u / \cosh^2 u) = (\cosh^2 u - \sinh^2 u)/\cosh^2 u$$
$$= 1/\cosh^2 u$$

we see that the adjacent leg of the triangle is $1/\cosh u = \text{sech } u$, as shown in Figure 12.2.

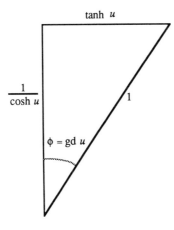

Figure 12.2. Right triangle showing relationship between Gudermannian angle φ and hyperbolic parameter u.

In relativity theory, it is usual to define the Greek trio α, β, and γ as

$$\alpha = \cos \phi = \operatorname{sech} u$$
$$\beta = \sin \phi = \tanh u$$
$$\gamma = \sec \phi = \cosh u$$

The right triangle of Figure 12.2 can be relabeled as shown in Figure 12.3.

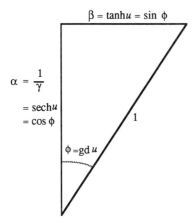

Figure 12.3. Right triangle showing relationship between the Gudermannian angle φ and the Greek trio α, β, γ.

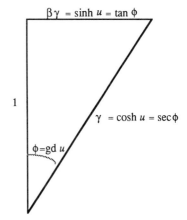

Figure 12.4. A similar right triangle showing a further relationship between the Gudermannian angle φ and the Greek trio α, β, γ.

If we multiply each side of the right triangle in Figure 12.3 by cosh u, we get the similar right triangle shown in Figure 12.4. Each of the right triangles in Figures 12.2, 12.3, and 12.4 is a Euclidean triangle and, hence, obeys the Pythagorean theorem:

$$\alpha^2 + \beta^2 = 1, \quad \text{or } (1/\cosh u)^2 + \tanh^2 u = 1$$

$$1 + \beta^2 \gamma^2 = \gamma^2, \quad \text{or } 1 + \sinh^2 u = \cosh^2 u$$

In conclusion, the Gudermannian $\phi = \text{gd } u$ is a function that converts the hyperbolic parameter u to the Gudermannian angle ϕ. This gives the Greek trio a simple geometric interpretation: $\alpha = \cos \phi$, $\beta = \sin \phi$, $\gamma = \sec \phi$.

REASON FOR THE GUDERMANNIAN

At this point, one might ask "Why do we introduce the Gudermannian?" Suppose that frame x, t and frame x', t' have relative velocity β. We thus construct the Minkowski diagram shown in Figure 12.5. Here arc DP is a part of the calibration hyperbola

$$t^2 - x^2 = t'^2 - x'^2 = 1$$

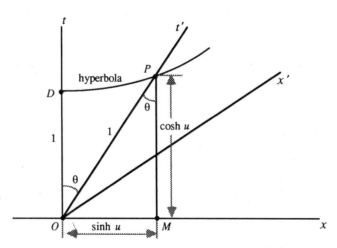

Figure 12.5. Minkowski diagram with affine right triangle OPM.

Thus $OP = OD = 1$, as we have already labeled them. Also because

$$\cosh^2 u - \sinh^2 u = 1$$

we see that P has coordinates

$$x = \sinh u \quad t = \cosh u$$

as shown in the figure. By construction, line OP has slope (with respect to the t axis) given by β; that is,

$$\beta = \text{slope of } OP = \tan \theta = \frac{x}{t} = \frac{\sinh u}{\cosh u} = \tanh u$$

Thus we have the famous relation $\beta = \tan \theta = \tanh u$. The parameter γ is defined as

$$\gamma = \frac{1}{\sqrt{1 - \beta^2}} = \frac{1}{\sqrt{1 - \tanh^2 u}} = \cosh u$$

So far, so good. But now comes the trouble. In the diagram, we have labeled the angle OPM as θ. From the diagram we might write down (but, as we will soon see, incorrectly) that

$$\sin \theta = \sinh u, \quad \cos \theta = \cosh u \quad \textbf{(WRONG!)}$$

So, why do we bother with the hyperbolic functions at all? Why not just use the circular trigonometric functions of θ? Why introduce the angle $\phi = \text{gd } u$? Why not just use the angle θ instead? The answer is that the right triangle OMP in the diagram is not Euclidean, but affine. Triangle OMP does not obey the Pythagorean theorem, but instead it obeys the pseudo-Pythagorean theorem

$$OP^2 = PM^2 - OM^2$$

which is

$$1 = \cosh^2 u - \sinh^2 u$$

If we let $\sin \theta = \sinh u$ **(WRONG!)** and $\cos \theta = \cosh u$ **(WRONG!)**, then we would have the nonsense

$$1 = \cos^2 \theta - \sin^2 \theta \quad \textbf{(WRONG!)}$$

In conclusion, the hyperbolic parameter u cannot be interpreted as the angle of any triangle, and for this reason we must introduce bona fide angles such as θ defined by $\beta = \tan \theta = \tanh u$ and ϕ defined by $\beta = \sin \phi = \tanh u$. Thus θ and ϕ are related by the equation

$$\tan \theta = \sin \phi$$

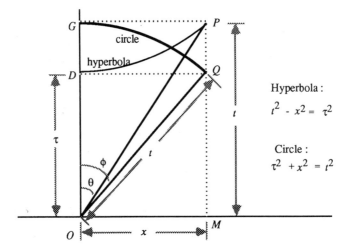

Figure 12.6. Relationship between θ and φ.

Let us now derive the geometric relationship between the affine angle θ with the Euclidean angle φ. (Figure 12.6.) Let P be a point on a line that makes an angle θ with the t axis. Because the point P is timelike, the angle θ must be less than 45°. Draw a horizontal line from P to intersect the t axis at G. Next draw a circular arc with center O and radius OG. Let the circular arc intersect the vertical line through P at Q. Thus OQ is equal to OG. Draw the horizontal line from Q to intersect the t axis at D. Thus $DQ = GP$. Finally, define φ as the angle DOQ. Then the desired result is given by

$$\sin \phi = \frac{DQ}{OQ} = \frac{GP}{OG} = \tan \theta$$

GEOMETRIC INTERPRETATION FOR THE TIMELIKE HYPERBOLA

Let us now give a geometric interpretation of the Gudermannian. First, we consider the *time like case*. Figure 12.7 shows the circle $x^2 + t^2 = \tau^2$ and the timelike equilateral hyperbola $t^2 - x^2 = \tau^2$ in the first quadrant.

The parametric equations for the circle are

$$x = \tau \sin \phi, \quad t = \tau \cos \phi$$

and the parametric equations for the hyperbola are

$$x = \tau \sinh u, \quad t = \tau \cosh u$$

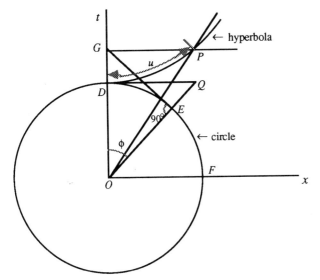

Figure 12.7. Timelike hyperbola and associated circle. Angle OEG is a right angle.

Let P be any point on the hyperbola. Draw a horizontal line through P and let it intersect the t axis at G. From G draw a tangent to the circle, and denote the point of tangency by E. Finally, define ϕ to be the angle DOE. Let us mark the scale of u along the hyperbolic arc DP in order to show visually the connection of u with the sector DOP, even as we show ϕ as marking the angle DOE. We now want to prove that this is a visual picture of the relationship of circular calibration ϕ with hyperbolic calibration u, as expressed by the Gudermannian relationship $\phi = \operatorname{gd} u$.

Let us now prove that $\phi = \operatorname{gd} u$. First, draw a horizontal line from D and let it intersect the extension of radius OE at Q. Now triangles QOD and GOE are each right triangles, with respective sides OD and OE equal (as they are both radii of the circle). Moreover, both right triangles contain the angle ϕ; that is,

$$\phi = \text{angle } QOD = \text{angle } GOE$$

Therefore, the two right triangles are congruent. Thus, their hypotenuses are equal: $OQ = OG$.

Since P is a point on the equilateral hyperbola, its coordinates are

$$x = GP = \tau \sinh u, \quad t = OG = \tau \cosh u$$

Hence $OQ = OG = \tau \cosh u$. Since the circle has radius τ, we have

$$OD = OE = \tau$$

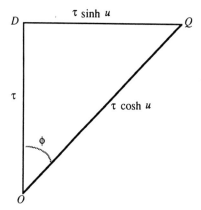

Figure 12.8. Right triangle QOD extracted from Figure 12.7.

Thus we can label right triangle QOD as shown in Figure 12.8. We have labeled side DQ is equal to $\tau \sinh u$, because

$$DQ = \sqrt{OQ^2 - OD^2} = \sqrt{\tau^2 \cosh^2 u - \tau^2} = \tau \sinh u$$

(Because GP is also equal to $\tau \sinh u$, we see that the points P and Q lie on a vertical line, that is, a line parallel to the t axis.) From Figure 12.8, we see that

$$\sin \phi = \frac{DQ}{OQ} = \frac{\tau \sinh u}{\tau \cosh u} = \tanh u$$

which gives

$$\phi = \sin^{-1}(\tanh u)$$

But the right-hand side is the definition of gd u. Therefore this equation is $\phi = \text{gd } u$, which is what we wanted to prove.

The Gudermannian of u can also be defined by the equation

$$\phi = \text{gd } u = \tan^{-1}(\sinh u)$$

The result can be seen by noting that from Figure 12.8 we have

$$\tan \phi = \frac{\tau \sinh u}{\tau} = \sinh u$$

Figure 12.8 can be used to explain *time dilation*. Side OD represents the proper time (i.e., the period of a stationary clock). Hypotenuse OQ then represents the dilated time (i.e., the period as it appears to an observer who sees the

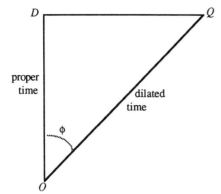

Figure 12.9. Time dilation.

clock moving with speed β). Hence the formula for time dilation becomes

dilated period of clock = (period of clock at rest) sec φ

= (period of clock at rest) sec (gd u)

where the hyperbolic parameter u is defined by the equation $\beta = \tanh u$. This dilation formula is illustrated in Figure 12.9.

GEOMETRIC INTERPRETATION FOR THE SPACELIKE HYPERBOLA

Now we wish to consider the spacelike case. Figure 12.10 shows the circle $x^2 + t^2 = \sigma^2$ and the spacelike hyperbola $x^2 - t^2 = \sigma^2$ in the first quadrant. Let R be a point on the hyperbola with hyperbolic parameter u_c. Draw a vertical line through R, and let it intersect the x axis at H. From H draw a tangent to the circle, and denote the point of tangency by E. Finally, define ϕ_c to be the angle HOE. The parametric equations for the circle are

$$x = \sigma \cos \phi_c, \quad t = \sigma \sin \phi_c$$

and the parametric equations for the hyperbola are

$$x = \sigma \cosh u_c, \quad t = \sigma \sinh u_c$$

We want to show that $\phi_c = \text{gd } u_c$. First, draw a vertical line from F and let it intersect the extension of radius OE at S. Now triangles SOF and HOE are each right triangles, with respective sides OF and OE equal (as they are both radii of the circle). Moreover, both right triangles contain the angle ϕ_c; that is,

$$\phi_c = \text{angle } SOF = \text{angle } HOE$$

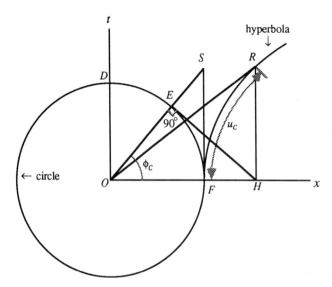

Figure 12.10. Spacelike hyperbola and associated circle. Angle OEH is a right angle.

Therefore the two triangles are congruent, which means that their hypotenuses are equal: $OS = OH$. But $x = OH = \sigma \cosh u_c$ so $OS = \sigma \cosh u_c$. Thus we can label the right triangle SOF as shown in Figure 12.11.

We see that side FS is equal to $\sigma \sinh u_c$, because

$$FS = \sqrt{\sigma^2 \cosh^2 u_c - \sigma^2} = \sigma \sinh u_c$$

(Because HR is also equal to $\sigma \sinh u_c$, it follows that R and S lie on a horizontal line in Figure 12.10.) From Figure 12.11 we see that

$$\sin \phi_c = \frac{FS}{OS} = \frac{\sigma \sinh u_c}{\sigma \cosh u_c} = \tanh u_c$$

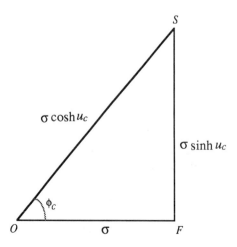

Figure 12.11. Right triangle SOF extracted from Figure 12.10.

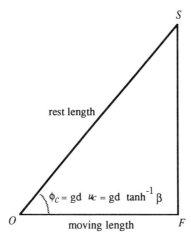

Figure 12.12. Length contraction.

Thus

$$\phi_c = \sin^{-1}(\tanh u_c) = \text{gd } u_c$$

Figure 12.11 can be used to explain *length contraction*. Hypotenuse OS represents the rest length (i.e., the length of a stationary object). Side OF then represents the moving length (i.e., the length it appears to an observer who sees the object moving with speed β). Hence the formula for length contraction becomes

moving length of ruler = (rest length of ruler) $\cos \phi_c$

= (rest length of ruler) $\cos (\text{gd } u_c)$

where the hyperbolic parameter u_c is defined by the equation $\beta = \tanh u_c$. This contraction formula is illustrated in Figure 12.12.

In terms of ϕ_c and u_c, the Greek trio is

$$\alpha = \cos \phi_c = \text{sech } u_c = \frac{1}{\cosh u_c}$$

$$\beta = \sin \phi_c = \tanh u_c$$

$$\gamma = \frac{1}{\sin \phi_c} = \sec \phi_c = \cosh u_c$$

In other words, the Greek trio has the same expression in terms of ϕ_c and u_c as it has in terms of ϕ and u. Whenever the context is clear, there is no need to carry the subscript c on ϕ_c and u_c, and so we can use ϕ and u both for the case of the timelike hyperbola and the case of the spacelike hyperbola.

Plate 12.2. Albert Einstein on a sailboat in 1936.

A MATTER OF PERSPECTIVE

As we know, the geometry of the Minkowski space-time world is affine geometry, and affine geometry is like projective geometry except that parallel lines do indeed look parallel. In my hand, I am holding a 1-foot ruler. If I hold it broadside, it looks like 1 foot (i.e., 12 inches). If I hold it end-on, it looks like a point (neglecting, of course, the thickness of the ruler). By holding it at an angle, I can make it look like any distance between 0 and 1 foot. At a gentle angle, it looks like 10 inches; at a steep angle it looks like 2 inches. We easily understand that the apparent length of an object to our eye depends upon our perspective; the same type of thing holds in the theory of relativity. It is fortunate because it has led to an unending supply of puzzles and paradoxes which are the delight of relativity theory. However, by leaving the affine world of Minkowski and going to the Euclidean world of Gudermann, the resolution of many of these paradoxes becomes much easier.

Foremost among these pleasantries is the *garage paradox*, which appears in almost every book on relativity. Imagine a garage with an automatic door at each end. Door *A* (the in door) is open, but it automatically closes as soon as a moving car is inside the garage. Door *B* (the out door) is closed, but it automatically opens just before the front of a moving car would hit it, thereby letting the car exit from the garage. We are given the following three facts.

1. The garage is 10 feet long in its inertial frame.
2. A car is traveling at the speed $\beta = \sqrt{3}/2$.
3. In its inertial frame, the car is 20 feet long.

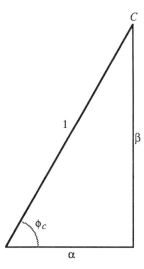

Figure 12.13. Right triangle with angle ϕ_c and opposite leg β.

We now want to find the contraction factor $\alpha = \gamma^{-1}$ between the two frames. The relative velocity β is

$$\beta = \tanh u_c = \sin \phi_c$$

where u_c is the hyperbolic parameter, where ϕ_c is the circular angle and where $\phi_c = \text{gd } u_c$. The right triangle with acute angle ϕ_c is shown in Figure 12.13. The hypotenuse is 1 and the leg opposite ϕ_c is β. Hence the adjacent leg is

$$\alpha = \sqrt{1 - \beta^2} = \cos \phi_c$$

which is the required contraction factor. Hence, in this case, with $\beta = \sqrt{3}/2$ the contraction factor is

$$\alpha = \sqrt{1 - (3/4)} = 1/2$$

The angle $\phi_c = \text{gd } u_c$ is equal to $\phi_c = \sin^{-1} \beta = \sin^{-1} \sqrt{3}/2 = 60°$.

As a result, the observer in the garage frame (called the garage man) will see the speeding car shortened by the contraction factor $\alpha = 1/2$. That is, the garage man will see the length of the car as $20(1/2) = 10$ feet. Because for him the garage is also of length 10 feet, for an instant he sees the car entirely enclosed in the garage; door A behind the car closes just as door B in front of the car opens. However, from the point of view of the car's reckless driver, it is the garage that gets shortened. Her car retains its rest length of 20 feet, but the garage moving at her shrinks to $10(1/2) = 5$ feet. In other words, to her the garage appears to be contracted to half its length. How can her 20-foot car fit into a 5-foot garage? As she sees the process, the car for a certain period of time must stick out at both ends, so both doors must be open at once during those times. In other words, door B (the out door) must open before door A (the in door) closes; that is, the event of door B opening must come before the event of door A closing. Common sense tells us that the garage man or the driver must be wrong. Either the car was at one point in a closed garage, or it was not. How can this be? This is called the *paradox of the reckless driver and the garage*.

The following explanation can be given as a resolution of this paradox. Relativity insists that this is not a question with an unambiguous answer. The secret lies in the difference in the sequence of door openings and closings, as seen by the garage man and the driver:

Garage man (Unprimed garage frame)	Driver (Primed car frame)
Door A closes and door B opens at the same instant, so the car is inside the garage at this instant.	Door B opens first, and then later door A closes, so the car sticks out at both ends during this time period.

The garage man is sure he knows that both doors operated at the same time, so at this instant the car actually fits into the garage. But the driver is equally sure that both doors were open for a certain time period, which means that door B opened before door A closed.

Let A be the event that door A closes, and we will let this event be the origin for both inertial frames. We let the garage inertial frame have coordinates (x, t) and the car inertial frame have coordinates (x', t'). Thus event A has the coordinates

$$A: \quad (x_A = 0, t_A = 0), \quad (x'_A = 0, t'_A = 0)$$

Let event B be the event that door B opens. The garage has length $x = 10$ in its inertial frame, and the moving (contracted) car has length $x = 10$. Thus, the x coordinate of event B is $x_B = 10$, and event B happens in the garage frame at the same time as event A, so the time of event B is also $t_B = 0$:

$$B: \quad (x_B = 10, t_B = 0)$$

We can now use the Lorentz transformation to get the coordinates of event B in the car's inertial frame. We have (since $\gamma = 1/\alpha = 2$, $\beta = \sqrt{3}/2$)

$$x'_B = \gamma(x_B - \beta t_B) = 2[x_B - (\sqrt{3}/2)t_B]$$

$$t'_B = \gamma(-\beta x_B + t_B) = 2[-(\sqrt{3}/2) x_B + t_B]$$

Since B has coordinates $(x_B = 10, t_B = 0)$ in the garage frame, it will have the following coordinates in the car frame:

$$x'_B = 2[10 - (\sqrt{3}/2)0] = 20$$
$$t'_B = 2[-(\sqrt{3}/2)10 + 0] = -10\sqrt{3} = -17.3$$

Thus we have

$$B: \quad (x'_B = 20, t'_B = -17.3)$$

As a check, we see that the square of the space-time interval between events A (the origin) and B is the same in both frames:

$$x_B^2 - t_B^2 = 10^2 - 0^2 = 100$$

$$x'^2_B - t'^2_B = 20^2 - (-10\sqrt{3})^2 = 400 - 300 = 100$$

We therefore have the table

Garage Man	Driver
Door A at $x_A = 0$ closes at $t_A = 0$ and door B at $x_B = 10$ opens at $t_B = 0$ (car inside garage at $t = 0$)	Door B at $x'_B = 20$ opens at $t'_B = -17.3$ and then door A at $x'_A = 0$ closes at $t'_A = 0$ (car sticks out at both ends of garage from $t'_B = -17.3$ to $t'_A = 0$)

The closing of door A and the closing of door B are simultaneous ($t_A = 0$ and $t_B = 0$) in the garage frame, but are not simultaneous ($t'_A = 0$ comes after $t'_B = -17.3$) in the car frame. This is an application of the general principle that two events simultaneous in one frame cannot be simultaneous in any other frame.

One of Einstein's profound observations on the significance of relativity is that it removes the separateness of the age-old concepts of distance and time. This view is reflected by the term *space-time* or the more ominous sounding *fourth dimension* used in H. G. Well's book, *The Time Machine*. This is illustrated by our driver. To her, the car has a perfectly reasonable length of 20 feet, and time t' is the same throughout its length. To the garage man, however, the car has shortened, but, as if in compensation, the car has been spread out in time. We can see this by the right triangle with acute angle $\phi_c = 60°$, hypotenuse 20, and sides $20 \cos 60° = 10$ and $20 \sin 60° = 17.3$, as shown in Figure 12.14.

To the driver, time t' is the same at C as at A, so AC is equal to the rest length of 20 feet for the car. To the garage man, time t' is different (by 17.3)

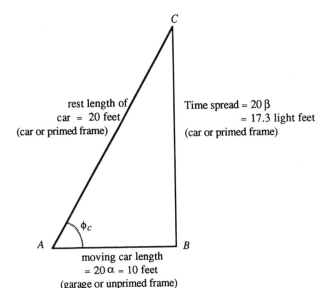

Figure 12.14. The rest length is the hypotenuse of a right triangle with legs the moving length and the time spread.

from C to B, and AB is equal to the car's moving length of 10 feet. To the driver (at rest in the car) the car is an object that extends 20 feet purely in space, but to the garage man, the speeding car becomes an object that extends 10 feet in space but also 17.3 light-feet in time (i.e., the amount of time it takes light to travel 17.3 feet). In other words the moving car has a space extent of 20α and a time extent of 20β light-feet. As we know, these quantities are the two legs of a right triangle with acute angle $\phi_c = \operatorname{gd} u_c$, where u_c is defined by $\beta = \tanh u_c$, so by the Pythagorean theorem

$$10^2 + (17.3)^2 = (20\alpha)^2 + (20\beta)^2$$
$$= 20^2(\alpha^2 + \beta^2) = 20^2(1 - \beta^2 + \beta^2) = 20^2$$

That is, the two combined in this way gives us back the original length 20 feet of the car.

The significance of this space-time approach may now be appreciated. If we regard time as a dimension on a par with any of the three spatial ones, then, by using a right triangle defined by the angle ϕ_c, the combined length of an object in space and time will be the same for both inertial observers. For an observer at rest with respect to the object, the object will be purely spatial in extent (car is 20 feet). But to the other inertial observer, the car's full extent will be part in space and part in time. If we measure the disagreement (17.3) of clocks at the two ends of the object in length units (light-feet) and treat this as a fourth dimension to apparent spatial length (10 feet), then the Pythagorean space-time length of the object remains the same (20 feet):

$$20 = \sqrt{10^2 + 17.3^2}$$

What one observer calls a pure length, the other sees as a combination of a length and a time spread. Space and time thus lose their separateness, becoming a unified concept in a Euclidean world in which the Pythagorean theorem holds with acute angle $\phi_c = \text{gd } u_c$. Minkowski also found that space and time lose their separateness and become a unified concept in an affine world (called the Minkowski world) in which a pseudo-Pythagorean theorem holds (i.e., the invariance of time-space interval).

The point is that the existence of an object, those properties of it that are independent of an observer's frame of reference, can only be described in the four-dimensional realm called space-time. Suppose we hold a book up to a bright light and look at its shadow on the wall. Edge on, it produces a thin shadow. Held with its cover to the light, it produces a shadow in the form of a broad rectangle. The two-dimensional world of shadows on a wall cannot represent the full solid reality of the book all at once; it exists in its own natural three-dimensional world. As a metaphor, the shrunken shadows of moving four-dimensional objects are all we see in our three-dimensional world. The different positions of the book corresponds in this metaphor to the different vantage points of moving observers.

EXAMPLE OF THE GUDERMANN WORLD

Let us suppose that we have two inertial frames x, t and x', t' with relative velocity β. As usual, we are using natural units of measurement in which the ultimate velocity (that of light) is $c = 1$, and hence β is necessarily less than one in magnitude. We now want to construct a space-time diagram based on the Gudermannian, and this construction is known as the Gudermann world. We have already introduced this Gudermannian space-time diagram in Chapter 2 and Chapter 11. As we know, the usual space-time diagram is called the Minkowski world, and it is one in which the calibration is made by unit hyperbolas and in which the pseudo-Pythagorean theorem holds. The genius of Christoph Gudermann lies in the fact that in his space-time diagram, the calibration is made by the unit circle and the Pythagorean theorem holds. The price we pay for this Euclidean world of Gudermann is that we can only depict two frames in one diagram, whereas in the affine world of Minkowski we can depict any number of frames in one diagram. However, for most problems this is not a serious shortcoming.

In this section, we continue the example of the garage and speeding car given in the previous section. Let us actually construct the Gudermann space-time diagram for the example of the garage frame x, t and the car frame x', t', where $\beta = \sqrt{3}/2$. Most pocket scientific calculators (such as the Sharp Scientific Calculator EL-506) have keys that compute sine, cosine, and tangent, and hyperbolic sine, hyperbolic cosine, hyperbolic tangent, as well as the inverses of all these functions. It also has keys for the other usual functions, such as square root and the like. Since the Gudermannian angle ϕ_c is defined as

$$\phi_c = \text{gd } u_c = \sin^{-1} \tanh u_c = \sin^{-1} \beta$$

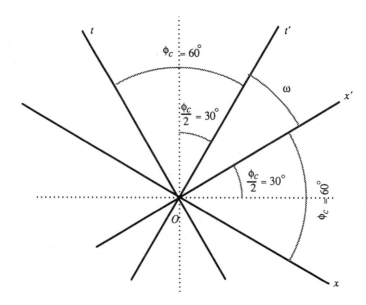

Figure 12.15. Gudermannian plane.

we can immediately calculate the Gudermannian angle as

$$\phi_c = \sin^{-1} \beta = \sin^{-1}(\sqrt{3}/2) = 60°$$

Refer to Figure 12.15. To construct this figure, first draw the vertical and horizontal axes as dotted lines. One half of the Gudermannian is $\phi_c/2 = 60°/2 = 30°$. Draw the t axis as a line 30° to the left of the vertical, the t' axis 30° to the right of the vertical, the x' axis 30° up from the horizontal, and the x axis 30° down from the horizontal. Then forget the horizontal and vertical axes, as they were only used as aids to construction.

Thus, we have constructed two sets of axes, such that the angle between t and t', as well as the angle between x and x', are each equal to the Gudermannian angle ϕ_c. But such a construction has a very interesting property, namely, the t axis is at right angles to the x' axis and the t' axis is right angles to the x axis. [To prove this, let ω be the angle between the t' and x' axis. The angle between the vertical and horizontal is then $(\phi_c/2) + \omega + (\phi_c/2) = \phi_c + \omega = 90°$. But $\phi_c + \omega$ is also the angle between the t and x' axes, as well as the angle between the t' and x axes.] Because of these right angles between frames, the geometry between these frames is Euclidean, and the Pythagorean theorem holds. Axes constructed in this way make up the *Gudermann world*.

Let x, t represent the garage frame, and x', t' the car frame. (See Figure 12.16.) In both frames, we let event A be the origin: $(x_A = 0, t_A = 0)$ and $(x'_A = 0, t'_A = 0)$. The rest length of the garage is 10 feet, so we will depict the garage on the x axis from event A $(x_A = 0, t_A = 0)$ to event B $(x_B = 10, t_B =$

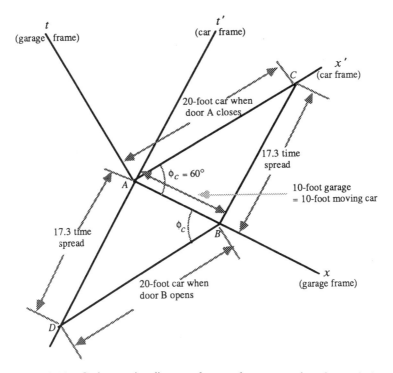

Figure 12.16. Gudermannian diagram of garage frame x, t and car frame x', t'.

0). We made both occur at $t = 0$, for we want A and B to be simultaneous in the garage frame. The rest length of the car is 20 feet, and we will depict the car in the x', t' frame in two positions DB and AC. Position DB represents the car just when door B opens, whereas position AC represents the car just when door A closes. Event D is chosen to have the same t' coordinate as B (i.e., line DB is drawn through B parallel to the x' axis). In this way, events D and B are simultaneous in the car frame, so the distance between them represents the rest length of the car. Likewise events A and C are simultaneous in the car frame, so AC is also the rest length of the car.

Because the Gudermann world is Euclidean, we can use ordinary (i.e., circular) trigonometry to find both space and time coordinates. In the car frame, the space coordinate x' of B is given by line AC and the time coordinate t' of B by line AD. By construction $AC = DB = 20$, and thus

$$AD = DB \sin \phi_c = 20 \sin 60° = 17.3$$

Because D lies on the t' axis below the origin A, we must put a negative sign in front of 17.3. Thus, in the car frame, the coordinates of B are $x'_B = 20$, $t'_B =$

−17.3. In summary we have

Event	Garage Frame	Car Frame
A	($x_A = 0, t_A = 0$)	($x'_A = 0, t'_A = 0$)
B	($x_B = 10, t_B = 0$)	($x'_B = 20, t'_B = -17.3$)

Because $t = 0$ for both A and B in the garage frame, the garage man sees event A (door A closes) at the same time as event B (door B opens). However, in the car frame, $t' = -17.3$ for event B, whereas $t' = 0$ for event A, so the car driver sees event B (door B opens) before event A (door A closes). Thus the garage man sees the car totally within the garage because events A and B are simultaneous for him. The car driver, on the other hand, sees the car sticking out from both ends of the garage between the nonsimultaneous events A and B for her.

As we know, the space-time interval between events is invariant in all frames. The squared space-time interval between A and B in the garage frame is

$$(x_B - x_A)^2 - (t_B - t_A)^2 = x_B^2 - t_B^2 = x_B^2$$

since $x_A = 0, t_A = 0$ is the origin, and since $t_B = t_A = 0$ (A and B simultaneous). The squared space-time interval between A and B in the car frame is

$$(x'_B - x'_A)^2 - (t'_B - t'_A)^2 = x'^2_B - t'^2_B$$

since $x'_A = 0, t'_A = 0$ is the origin. Because of the invariance, we have

$$x_B^2 = x'^2_B - t'^2_B$$

which is

$$x'^2_B = x_B^2 + t'^2_B$$

or in terms of the Gudermann diagram (Figure 12.16)

$$(DB)^2 = (AB)^2 + (DA)^2$$

or

$$(\text{rest length of car})^2 = (\text{moving length of car})^2 + (\text{time spread})^2$$

or

$$20^2 = 10^2 + 17.3^2$$

Therefore, we have shown that the invariance of the space-time interval in relativity theory is the same as the validity of the Pythagorean theorem in the Gudermann world.

CHAPTER 13

CONCLUSION

If the dull substance of my flesh were thought,
Injurious distance should not stop my way;
For then, despite of space, I would be brought,
From limits far remote, where thou dost stay.

William Shakespeare (1564–1616)
Sonnet 44

THE LONG ROAD WITH NO TURNING

In his satirical *History of Lunar States and Empires* (1656), the French writer Cyrano de Bergerac describes this amazing journey which supposedly he had taken. One day he was lifted up into the air. On landing several hours later, he was astonished

Plate 13.1. George Francis FitzGerald (1851–1901)

to find that he was not in his own land of France nor even in Europe. Instead he was in Canada. He claimed that his trans-Atlantic flight was quite possible, for he believed that while he was up in the air, the Earth had continued to rotate eastward. This was the reason why he would land in Canada and not France.

We know that even if we were able to ascend above the atmosphere, we would not be able to benefit by the fanciful method of travel that the French satirist imagined. Indeed, when we rise above the surface of the spinning Earth, we continue to move by inertia with the same speed, that is, the speed with which the Earth is moving beneath us. Landing again on Earth, we would find ourselves where we were before we went up. The situation is the same as if we made a jump inside a moving train. We jump up and land at the same place inside the train. The Earth is the first example, and the train is the second; each acts as an inertial frame of reference, a vital concept in the theory of relativity.

When all measurements are made with respect to only one inertial frame, then our usual perception of space and time is valid. Consider now a second inertial frame moving at uniform speed with respect to the first. The theory of relativity asserts that the perception of space and time in the second frame is quite different from the perception in the first frame. However, these two contrary perceptions are not random or chaotic, but are in fact connected by a mathematical law known as the Lorentz transformation. This transformation is so important that it has been said that all the precepts of special relativity are but applications of this transformation, or more precisely the Poincaré group of transformations of which the Lorentz transformation is a part.

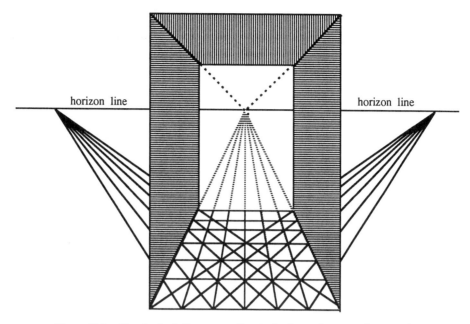

Figure 13.1. Sketch of a hallway according to the focused system of perspective.

What is the essence of the Poincaré group? The essence is one of *projection* and *horizon*. (See Figure 13.1.) To explain these two terms in a qualitative way, we present the following example, an analogy: *it is a long road that has no turning*. This analogy is based upon a straight set of railway tracks on a flat Earth and extending infinitely far in both directions. Let person A stand at a given point on these tracks and survey their length in both directions. (See Figure 13.2.)

In looking to the right, A does not see a set of parallel lines extending to infinity; instead A sees two lines coming together at a point on the horizon. The actual railroad tracks run on to infinity, but A's perception of them is that of two slanting lines which come together at a finite point. Person A does not see a plan view of the parallel tracks. Instead, A sees the results of a projection process that takes place in the retinas of her eyes. Her perception is that of a perspective drawing similar to what is done by artists. The true point at infinity is replaced by a finite point corresponding to her horizon. In this way, the set of tracks to the right is compressed by means of perspective into a finite segment to the right. Similarly, the infinitely long extent of tracks to the left is also compressed by the same projection process into a finite segment to the left. Her two horizons represent finite limits that encompass the actual infinite length of the actual set of parallel tracks. In conclusion, the projection process used by artists allows a person to compress in the mind's eye an infinitely long set of parallel railroad tracks into a finite segment between her horizons, with the person at the center point, or zero

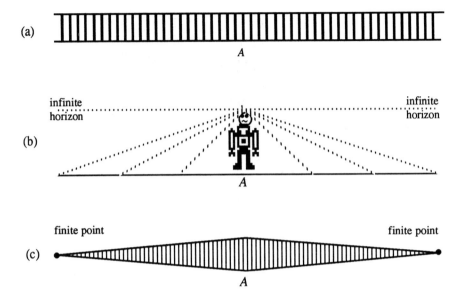

Figure 13.2. The analogy of the long road with no turning.

(a) A plan view of a set of parallel railway tracks extending to infinity in both directions, with the position of A somewhere on them.

(b) Looking in both directions, A sees the tracks as a projection, with the tracks ending on her horizon.

(c) The perception of A of the tracks in each direction is a set of finite slanting lines meeting at a point on her horizon. The infinite length of parallel tracks has been compressed in the projection to a finite length of nonparallel tracks meeting at her horizons to the right and left.

point. This finite interval corresponds to the inertial frame of reference of the person in question.

We have thus outlined the essential features in the analogy of the road with no ending. The road is infinitely long, but a person's perception of it is a finite projection with herself at the center, and the end points at her horizons.

A person's perception of the long road with no turning is the same perception no matter where she is located on the road. This is the key point in the analogy. Suppose she is lost in Siberia and finds her way to the tracks of the trans-Siberia railroad. She still does not know where she is, for she would see the same thing if she were 50 miles to the east or 100 miles to the west, or anywhere on this road with no ending.

The set of infinitely long railroad tracks on the flat Earth corresponds to the space-time continuum. These tracks are the same tracks no matter where one is located on the tracks. There is only one set of tracks despite the fact that there can be many people located at different points on the tracks. Each person by means of the perspective projection described earlier creates his or her own inertial frame of reference. Each frame corresponds to a finite line segment running from horizon to horizon, with the particular person at the center point. Although each person is located at a different point on the railroad tracks, his or her perception of the tracks has the same form as that of any other person. The sameness of all inertial frames expresses a fundamental symmetry of the Poincaré group, namely, one based on hyperbolic space-time invariance.

In considering this analogy, we realize that the space-time symmetry inherent in relativity theory is no stranger to the human mind. The idea of projection with reference to our location and our horizons have governed human action since the time of earliest man. Each of us must judge other things in relationship to our own values, which is equivalent to saying that we consider our location as the center point of the road with no ending. A cave man knew that his world extended from where he stood to his horizon; so do we. All his world, as well as ours, is projected into that finite extent allowed by the human mind. We cannot see infinitely far. Everything seems to come together at the same point on the horizon, no matter where we stand on the long road with no turning and no ending.

SO CLOSE, YET SO FAR

Sir Isaac Newton spent many years pondering the nature of space and time. He asked, Do we need absolute space and absolute time? Translated into the terms of this analogy, Do we need the long road with no turning, or is it enough that we consider as real only each person's perception of it? No human ever sees the plan view of the infinite road. She only sees her finite projection bounded by her finite horizons. The absoluteness of the long road might be but an illusion. Is absolute space and absolute time required for the purposes of physics?

In 1687, Newton's work *Principia* was published. Experts agree that this represented the birth of modern science. Historians of science are unanimous in this assessment. The *Principia* is the most important single work in scientific history. Newton, accordingly, is widely acclaimed as the greatest scientist of all time. The three parts of the *Principia* set forth Newton's laws of motion, the law of universal gravitation, and the rules for reasoning in physics. Those rules include the ideas of keeping explanatory principles to a minimum, relying on experiment to determine scientific truth, and regarding natural laws observed on Earth to be typical of the universe unless experiment showed otherwise.

Einstein wrote that Newton "determined the course of western thought, research, and practice like no one else before or since." Building on the realization that the same force pulling an apple to the ground held the moon in its orbit around the Earth, Newton conceived the idea of universal law, namely, that the principles operating on Earth applied throughout the cosmos. He unified the physics of the Earth with the physics of the heavens. Practicing physicists today, trying to unify the description of nature further, still marvel at Newton's accomplishment. The key is the idea that we can measure in the laboratory the forces responsible for astrophysics and cosmology. This took a great leap of the imagination.

In the *Principia*, Newton described an elegant experiment in which he rotates a pail of water. Initially the water is flat, but as he starts rotating the pail, the surface of the water becomes concave, and the concavity increases until the water is rotating at the same rate as the pail. Newton asks what causes the concavity of the surface of the water? His answer was that the concavity must be revealing the water's absolute rotation, and that could be only with respect to absolute space.

Bishop George Berkeley (1685–1753) and Baron Gottfried Wilhelm von Leibniz (1646–1716), both contemporaries of Newton, were highly critical of Newton's absolute space and absolute time. Leibniz believed that space and time have no independent existence but are merely relations among pieces of matter. The philosophical mistrust of the concepts of absolute space and absolute time were further discussed by Ernst Mach (1838–1916) in the nineteenth century. He rejected the concept of absolute space, and with it absolute motion, and said that the inertia of a body arises from its interaction with all other matter in the universe. According to Mach, the concavity of the water in Newton's pail does not arise from absolute rotation of the water relative to absolute space, as Newton contended. Instead, Mach said it arose from the rotation of the water relative to the other masses in the universe. Mach argued that the same concavity would result if the water were not rotating, but instead the rest of the universe were rotating about the water. Mach's ideas had considerable influence on Einstein.

Despite his brilliant experiment of the rotating pail of water, Newton was still disturbed by a result that he deduced quite easily from his laws of motion. This result is now referred to as the principle of relativity, which in Newton's time would only apply to mechanics, as electricity and magnetism then were imperfectly understood. Newton's statement of this principle is: "The motion of bodies included in a given space are the same among themselves, whether the space is at rest, or moving uniformly in a straight line without circular motion." Newton's use of the word "space" in this context does not refer to absolute space but to the spatial coordinates of the given inertial frame. In other words, Newton's statement says that within a given inertial frame, no mechanical experiment will reveal the velocity of the frame. All mechanical processes inside the inertial frame will take place as if the frame were at rest.

The reason why the principle of relativity was upsetting to Newton is that it showed that there was no actual distinction between rest and uniform motion. According to his laws, rest and uniform motion are relative; that is, there is no absolute difference between an inertial frame at rest and one in uniform motion. Thus the principle of relativity makes the concepts of absolute space and absolute time redundant.

It is a long road that has no turning. Why did Newton fail to discover the fundamental space-time symmetry of nature? In the analogy, we saw that two vital elements were involved, namely, projection and horizon. *Projection in the analogy corresponds to the principle of relativity in nature.* Each person can locate herself anywhere on the railroad, and she will see in her mind the same projection of the tracks. Each person can locate herself in any inertial frame, such as her living room chair, or a seat on a jet airplane, and her coffee pours the same way. The second element in the analogy is horizon. The infinite extent of the railroad tracks is limited to a finite perception in her mind's eye by her horizon. *What does the horizon in the analogy correspond to in nature?* This is the question for whose answer Newton was striving in his search for an understanding of the nature of space and time.

Newton's final answer was given when he was over 70 in a special section, the general scholium, inserted at the end of later editions of the *Principia*:

> The Supreme God is a Being eternal, infinite, absolutely perfect. . . . He is eternal and infinite, omnipotent and omniscient; that is, his duration reaches from eternity to eternity; his presence from infinity to infinity. . . . He is not eternity and infinity, but eternal and infinite; he is not duration or space, but he endures and is present. He endures forever, and is everywhere present; and by existing always and everywhere, he constitutes duration and space.

In plain words, the horizon in the analogy corresponds to our view of perfection in nature, or in Newton's words, the Being eternal, infinite, absolutely perfect. And how is absolute perfection manifested in nature? The twentieth-century answer is that nature's perfection is manifested in light. Thus today we believe we have the answer to what Newton sought, namely, *horizon in the analogy corresponds to the principle of light in nature*. The principle of relativity and the principle of light make up Einstein's theory of special relativity and give us the space-time symmetry of the Poincaré group. As we have related in this book, Newton by 1680 knew of Roemer's discovery of the finite velocity of light and the relativistic Doppler effect. A simple mathematical analysis of Roemer's results was all that was required for Newton to have arrived at the principle of light, and thus the special theory of relativity. Newton was so close, yet so far. His general scholium is actually an extremely clear and accurate statement of the principle of light. It is the missing link to the understanding and appreciation of Newton's failure to grasp relativity theory, when it was figuratively within his hands.

If we translate Newton's general scholium into modern terminology, we obtain the principle of light (in vacuum). A rough translation is as follows:

> There is neither absolute space, nor absolute time. There is a supreme entity, absolute space-time. Absolute space-time is governed by the symmetry brought on by the invariance of the velocity of light (in vacuum). Light is eternal, light is infinite, light is absolutely perfect. Light is eternal because its duration reaches from eternity to eternity. Light is infinite because its presence goes from infinity to infinity. Light is absolutely perfect because it has no horizon. Light can find no rest point on the long road with no turning; that is, light belongs to no inertial frame. As a result light has no rest mass, but it is pure energy of motion. Belonging to no inertial frame, light in vacuum must have the same velocity in every inertial frame. Being perfect,

light must have the ultimate velocity. Thus light in vacuum must always be moving at the same and ultimate velocity, no more, no less. Light is not duration and space, but light endures and is present. For a beam of light, all times are the same time and all places are the same place. Time does not pass for light, for light is ageless. Space does not extend for light, for light is without space. A beam of light is at all times at the same place, and at all places at the same time. Light in vacuum endures forever and is everywhere present. Light is not limited or affected by time or by distance.

Newton said it all in this metaphor, but he was not understood. At the critical point the greatest mathematician of all time forsook the mathematics which would have made clear the space-time symmetry expressed by his metaphor. This book has been about metaphor and mathematics, and this great failing of Newton shows that both are required in science. In the words of Einstein: "Enough of this. Newton, forgive me. You found the only way that, in your day was at all possible for a man of the highest powers of intellect and creativity."

THREE FLASHES OF INSIGHT

Exactly 200 years after the publication of Newton's *Principia*, the Michelson-Morley experiment was performed in Cleveland. The year was 1887. Michelson, a brilliant experimental physicist came to the Case School of Applied Science in 1882 and soon began a collaboration with Morley, a chemist at nearby Western Reserve College. Michelson had already achieved some fame for his accurate measurements of the speed of light. His interest in light extended to the problem of the aether, a hypothetical substance that supposedly filled all of space. It was then assumed that such a substance was necessary to transmit light waves, just as sound waves required air or some other material medium. But calculations showed that the aether had to be lighter than anything known, as firm as steel to transmit the transverse waves of light, and nearly impossible to detect.

Michelson and Morley constructed an extraordinarily sensitive apparatus designed for the detection of the aether, but could find no sign of it. Michelson considered his effort a failure, but later others realized that the significance of the Michelson-Morley experiment was that it decisively established that the aether is undetectable, so in effect the aether does not exist. Michelson and Morley did not appreciate the true value of their so-called failure. As it turned out, the Michelson-Morley experiment was the beginning of our modern understanding of the nature of space and time. Newton's work in 1687 formed the program of research in physics for two centuries. The Michelson-Morley experiment in 1887 was the first major crack that appeared in the Newtonian edifice. Soon other cracks began to appear.

Some of these other cracks came from Heinrich Hertz's discovery in the same year, 1887, of the photoelectric effect, and his subsequent proof of the physical existence of electromagnetic radiation in forms other than light. The photoelectric

Plate 13.2. Albert Einstein (1879–1955)

effect is due to the emission of electrons by certain substances when exposed to light. Such substances have important technological uses in light meters and electronic eyes.

In 1905, Einstein's theory of special relativity was published. It drastically changed the ordinary concepts of space and time, and it led to the concept that mass and energy are two manifestations of the same thing. Einstein's work explicitly says that relativity theory made the idea of the aether no longer necessary. However, the theory of relativity made up only half the revolution of twentieth-century physics. Quantum physics represented the other half. It describes the workings of nature in the realm of the subatomic world. Einstein also played an important role in the early development of quantum physics, for in 1905 he used the quantum hypothesis to explain the photoelectric effect. The quantum hypothesis had been introduced in 1900 by Max Planck. Einstein's 1905 explanation of the photoelectric effect was verified by experiments a decade later. This was an important step in making scientists take the new quantum theory seriously. Einstein's explanation required treating light as made up of small particles, or photons. This was a major innovation, because for the previous 100 years light had been considered as a wave.

Both in quantum theory and relativity theory, light plays a key role. In relativity, there are two principles. The principle of light says that light travels at the ultimate velocity. Here it is easy to visualize light as a stream of *photons*, each traveling at the same constant velocity. The principle of relativity can be interpreted in terms of the relativistic Doppler effect, first detected in 1676 by

Roemer. Here it is convenient to regard light as a *wave* whose wavelength in each inertial frame is governed by the Doppler shift.

In quantum theory scientists faced the dilemma of reconciling Einstein's requirement that light behave like a stream of particles with the established requirement that light also behaves like wave motion. Actually, this was the same dilemma faced by Newton in the seventeenth century, who also treated light as made up of small particles in some situations and as waves in other circumstances.

The resolution of the dilemma came in 1925 when a Danish physicist proposed the principle of complementarity. Niels Bohr (1885–1962) said that light was sometimes waves, sometimes particles. It literally depends upon how you look at it. It was the nature of the experiment that determined the form of light. It could act either as a wave or a particle, but never both at the same time. The same principle applied to subatomic particles. An electron, for example, could either appear as a particle or as a wave.

Science recognizes the work of the grand unifiers, those who bring together various pieces in the scientific puzzle and put them together in a unified and connected whole. Newton and Einstein are recognized universally for this role. But before this can be done, all the missing pieces in the puzzle must be found. Without these missing pieces, the grand unifiers could not complete the finished product in their methodical way. However, the missing pieces are often found by people with flashes of insight and a touch of imagination which do not fit easily into the general scientific scheme.

It is an interesting facet of history that the three great leaps forward required to find the missing pieces both in quantum theory and in relativity theory were made by three Irish physicists. It seems that they were not bound by the strict disciplines and conventions of the English and continental physicists. Instead, their minds were free and open, so they could make the great discoveries that went beyond reality, and into the realm of fantasy. The same quality is found in the poetical works of William Butler Yeats (1865–1939) (Nobel Prize, 1923) and the prose of James Joyce (1882–1941), both Irish writers.

The Irish physicist and mathematician William Rowan Hamilton (1805–1865) originated the idea that the same entity could be treated both as a particle and a wave. Moreover, he worked out the mathematics for this concept. His idea of wave-particle duality was considered so strange that people remembered it only as one of the great curiosities of science. However, his mathematics was so elegant that it was preserved in some of the classic books of physics. When it came time in 1925 to put Bohr's quantum ideas into mathematical form, Erwin Schrödinger (1887–1961) and Werner Heisenberg (1901–1976) only had to turn to the pages containing Hamilton's equations, and they saw that all the essential mathematics had already been worked out for them by this eccentric Irishman. Thus the modern theory of quantum mechanics, originated in 1925–1927, abounds with the name of Hamilton, who 100 years previously discovered the essential missing piece of the quantum mechanics puzzle.

In relativity theory there were two essential missing pieces which had to be found. These two missing pieces drastically changed the ordinary concepts of space and time. With these two missing pieces, Einstein was able to complete the puzzle and so form a logically consistent theory. The two missing pieces concern the concept of space and the concept of time. From the time of Newton, both these concepts were regarded as absolute and independent of each other. But then came the Michelson-Morley experiment in 1887.

If you think of Earth as moving around the sun, then it is moving in different directions at different times, and at different velocities, with respect to the absolute space of Newton. It was thought that the aether resided in this absolute space. The Michelson-Morley experiment measured the difference in the velocity of light in each of two perpendicular directions. As the Earth traveled through the aether, the concept of absolute space and absolute time required that the velocity of light in one direction would be different from that in the perpendicular direction. But they were not. Michelson found that the velocity in one direction was always the same as that in the perpendicular direction. To explain this mystery, Irish physicist George Francis FitzGerald (1851–1901) invented the outrageous idea that moving bodies actually contract in the direction of their motion. Next, Irish physicist Joseph Larmor (1857–1942) invented the equally outrageous idea that moving clocks go slower. He said that when you think you have measured the velocity of light, you are fooled by your clocks having changed their rates. These things happen in such a way as to make you think that light is still moving with the same velocity in both directions. The concepts of FitzGerald space contraction and Larmor time dilation were so preposterous they were initially met with utter disbelief. However, they were the two missing pieces that destroyed Newton's concept of absolute space and his concept of absolute time, respectively. However, physicists would not accept them until Einstein took a hand, and with one stroke of wizardry in 1905 put the puzzle together and resolved all the difficulties in his special theory of relativity.

Science had labored for two centuries after the work of Newton, but two flashes of insight, namely, FitzGerald's space contraction and Larmor's time dilation demolished the idea of absolute space and absolute time, and provided the missing pieces from which Einstein's theory of relativity sprang. Likewise, the flash of insight in Hamilton's wave-particle duality provided the missing piece for quantum physics. These three moments of inspiration go beyond metaphor and mathematics, and it is believed that no satisfactory explanation of this process of creativity can ever be given.

CHRONOLOGICAL TABLE

EGYPTIANS, BABYLONIANS, AND CHINESE

B.C.

4000–500　　Calendar: Reconciliation of year and month.
　　　　　　Recognition of zodiac as path of Sun, Moon, and five planets.
　　　　　　Pythagorean theorem.

GREEK

600　　Anaximander: The Earth is a sphere.
　　　　Pythagoras: Spherical shell universe. The Pythagorean theorem. Musical harmony.
　　　　Euclid: Geometry.

300　　Heraclides: Retrograde motions of planets as loops described.

250　　Eratosthenes: Circumference of the Earth calculated.
　　　　Apollonius: Eccentric planetary orbits to account for irregular motions conceived.

150　　Hipparchus: Measuring apparatus and observations. Stereographic projection.

A.D.

150　　Ptolemy: *Almagest*. Improvement on Apollonian theory. *Geographia*. World map.

MODERN

1419　　Prince Henry of Portugal: Navigation school established.
1474　　Toscanelli: Western route described to the Portuguese.
1492　　Columbus: Discovery of America.
1543　　Copernicus: Heliocentric theory published.
1547　　Dee: Speaks to Mercator.
1569　　Mercator: World map on Mercator projection.
1589　　Tycho Brahe: Astronomical observatory.
1599　　Wright: Mathematical theory of Mercator projection.
1605　　Kepler: Orbits of planets are ellipses.
1610　　Galileo: Telescopic discoveries of mountains on moon, sunspots, moons of Jupiter.
1634　　Galileo: Law of falling bodies. Law of inertia.
1672　　Newton: Spectrum and theory of colors.
1676　　Roemer: Movement of light.

1687	Newton: Law of gravitation.
1727	Bradley: Aberration of the stars explained.
1864	Maxwell: Light as an electromagnetic wave.
1887	Michelson: Michelson–Morley experiment.
1904	Lorentz: Lorentz equations.
1904	Poincaré: Light as signal with ultimate velocity.
1905	Einstein: Special theory of relativity. Equivalence of mass and energy.
1915	Einstein: General theory of relativity (or law of gravitation).

PLATE CREDITS

ASHMOLEAN MUSEUM, OXFORD UNIVERSITY
Plate: 8.4

CALIFORNIA INSTITUTE OF TECHNOLOGY
Plates: 7.2, 13.2

HAAGS GEMEENTEMUSEUM, THE HAGUE
Plate: 4.4

HEBREW UNIVERSITY OF JERUSALEM
Plate: 7.1

THE HISTORY OF SCIENCE COLLECTIONS OF THE UNIVERSITY OF OKLAHOMA LIBRARY
Plates: 0.2, 0.3, 1.1, 1.2, 1.3, 1.6, 1.7, 2.2, 2.4, 4.6, 6.8, 6.9, 8.2, 9.2, 9.3

HOUGHTON LIBRARY, HARVARD UNIVERSITY
Plate: 10.4

THE LIBRARY OF CONGRESS
Plates: 2.3, 2.5, 4.3, 4.5, 5.1, 5.2, 6.2, 6.3, 6.4, 6.5, 6.6, 6.7, 6.10, 6.11, 6.12, 6.13, 9.4, 11.1, 11.2

MUSEUM OF FINE ARTS, BOSTON
Plate: 10.2

THE NATIONAL PORTRAIT GALLERY, LONDON
Plate: 2.1

THE NATIONAL PORTRAIT GALLERY, SMITHSONIAN INSTITUTION, WASHINGTON
Plate: 3.2

THE NIELS BOHR LIBRARY OF THE AMERICAN INSTITUTE OF PHYSICS
Plates: 0.1, 1.4, 1.5, 3.1, 4.1, 4.2, 4.7, 6.1, 8.1, 9.1, 10.1, 10.3, 11.4, 12.1, 12.2, 13.1

THE ORIENTAL INSTITUTE OF THE UNIVERSITY OF CHICAGO
Plate: 11.3

TATE GALLERY, LONDON
Plate: 8.3

POETRY CREDITS

INTRODUCTION: William Shakespeare's *Macbeth*, act 1, scene 3. From *The Works of William Shakespeare*, vol. 8. The Mershon Co., New York, 1885.

CHAPTER 1: Henry Vaughan's *The World*. From *The Works of Henry Vaughan*, vol. 2. Clarendon Press, Oxford, 1914.

CHAPTER 2: William Wordsworth's *The Prelude*, III, 59–63. From *The Complete Poetical Works of William Wordsworth*. Houghton Mifflin, Boston and New York, 1904.

CHAPTER 3: John Milton's *Paradise Lost*, III, 1–6. From *The Poetical Works of John Milton*, 2nd Edition. J. Johnson, London, 1809.

CHAPTER 4: William Blake's *Auguries of Innocence*. From *The Poetical Works of William Blake*. Chatto & Windus, London, 1906.

CHAPTER 5: Walt Whitman's *Miracles*. From *The Complete Poems and Prose of Walt Whitman*. Ferguson Bros. & Co. Printers, Philadelphia, 1888.

CHAPTER 6: Walt Whitman's *Night on the Prairies*. From *Ibid*.

CHAPTER 7: Walt Whitman's *When I Heard the Learn'd Astronomer*. From *Ibid*.

CHAPTER 8: William Blake's *Mock on Mock on Voltaire Rousseau*. From *The Poetical Works of William Blake*. Chatto & Windus, London, 1906.

CHAPTER 9: William Shakespeare's *Sonnet 30*. From *The Works of William Shakespeare*, vol. 2. Chatterton-Peck Co., New York, 1885.

CHAPTER 10: Alexander Pope's *An Essay on Man*, II, 19–22. Effingham Maynard & Co., New York, 1890.

CHAPTER 11: Dante's *Paradiso*, XVII, 16–18. From *La Devina Commedia di Dante Alighieri*, Tom. III. Nella Stamperia de Romanis, Roma, 1822.

CHAPTER 12: Christopher Marlowe's *The Tragical History of Doctor Faustus*, scene 14. From *Christopher Marlowe*. American Book Co., New York, 1912.

CHAPTER 13: William Shakespeare's *Sonnet 44*. From *The Works of William Shakespeare*, vol. 2. Chatterton-Peck Co., New York, 1885.

BIBLIOGRAPHY

1. Historical books on special relativity and related subjects

ALIOTO, A. M. *A History of Western Science*. Prentice-Hall, Englewood Cliffs, N.J., 1987.

BANVILLE, J. *Doctor Copernicus*. David R. Godine, Boston, 1976.

BANVILLE, J. *Kepler*. David R. Godine, Boston, 1981.

BERSTEIN, J. *Einstein*. Viking, New York, 1973.

BORN, M. *The Restless Universe*. Dover, New York, 1951.

CLARK, R. W. *Einstein, The Life and Times*. World Publishing, New York, 1971.

D'ABRO, A. *The Rise of the New Physics*, 2 volumes. Dover, New York, 1951.

EINSTEIN, A. *Relativity, The Special and General Theory, A Popular Exposition*. Crown, New York, 1920.

EVERITT, C. W. F. *James Clerk Maxwell, Physicist and Natural Philosopher*. Scribners, New York, 1975.

FRENCH, A. P. (ed.). *Einstein, A Centenary Volume*. Heinemann, London, 1979.

FRIEDMAN, A. J., and DONLEY, C. C. *Einstein as Myth and Muse*. Cambridge University Press, Cambridge, 1985.

GOLDBERG, S. *Understanding Relativity, Origin and Impact of a Scientific Revolution*. Birkhauser, Boston, 1984.

HAAS-LORENTZ, G. L. DE (ed.). *H. A. Lorentz, Impressions of His Life and Work*. North Holland, Amsterdam, 1957.

HOFFMAN, B. *Albert Einstein, Creator and Rebel*. New American Library, New York, 1972.

HOFFMAN, B. *Relativity and Its Roots*. W. H. Freeman, New York, 1983.

HOLTON, G., AND ELKANA, Y. (eds.). *Albert Einstein, Historical and Cultural Perspectives*. Princeton University Press, Princeton, N. J., 1982.

PAIS, A. *Subtle Is the Lord, The Science and Life of Albert Einstein*. Oxford University Press, New York, 1982.

PYENSON, L. *Young Einstein, The Advent of Relativity*. Heyden, Philadelphia, 1985.

POWERS, J. *Philosophy and the New Physics*. Methuen, London, 1982.

RUSSELL, B. *The ABC of Relativity*. New American Library, New York, 1969.

RYAN, D. P. (ed.). *Einstein and the Humanities*. Greenwood Press, New York, 1987.

SWENSON, L. S. *Genesis of Relativity*. Burt Franklin, New York, 1979.

TELLER, E. *The Pursuit of Simplicity*. Pepperdine University Press, Malibu, Calif., 1980.

WHITTAKER, E. T. *A History of the Theories of Aether and Electricity*. Harper and Bros., New York, 1960.

2. Mathematical books on special relativity

AHARONI, J. *The Special Theory of Relativity*. Dover, New York, 1985.

ANDERSON, J. L. *Principles of Relativity Physics*. Academic Press, New York, 1967.

BERGMANN, P. G. *Introduction to the Theory of Relativity*. Dover, New York, 1976.
BONDI, H. *Relativity and Common Sense*. Dover, New York, 1980.
BOHM, D. *The Special Theory of Relativity*. Benjamin-Cummings, Reading, Mass., 1965.
BORN, M. *Einstein's Theory of Relativity*. Dover, New York, 1965.
BRIDGMAN, P. W. *A Sophisticate's Primer of Relativity*, Wesleyan University Press, Middletown, Conn., 1983.
BURKE, W. L. *Spacetime, Geometry, Cosmology*. University Science Books, Mill Valley, Calif., 1980.
EDDINGTON, A. S. *The Mathematical Theory of Relativity*, 2nd ed, Cambridge University Press, 1952.
FRENCH, A. P. *Special Relativity*. W. W. Norton, New York, 1968.
HELLIWELL, T. M. *Introduction to Special Relativity*. Allyn & Bacon, Boston, 1966.
KACSER, C. *Introduction to the Special Theory of Relativity*. Prentice Hall, Englewood Cliffs, N.J., 1967.
LAWDEN, D. F. *Elements of Relativity Theory*. John Wiley, New York, 1985.
MERMIN, N. D. *Space and Time in Special Relativity*, McGraw-Hill, New York, 1968.
PAULI, W. *Theory of Relativity*. Dover, New York, 1958.
RINDLER, W. *Essential Relativity, Special, General, and Cosmological*. Springer-Verlag, New York, 1977.
RINDLER, W. *Introduction to Special Relativity*. Clarendon Press, Oxford, 1982.
SHADOWITZ, A. *Special Relativity*. W. B. Saunders, Philadelphia, 1968.
SKINNER, R. *Relativity for Scientists and Engineers*. Dover, New York, 1982.
SYNGE, J. L. *Relativity, The Special Theory*. Interscience Publishers, New York, 1956.
TAYLOR, E. F., and WHEELER, J. A. *Spacetime Physics*. W. H. Freeman, San Francisco, 1966.
TOLMAN, R. C. *Relativity, Thermodynamics, and Cosmology*. Dover, New York, 1987.
UGAROV, V. A. *Special Theory of Relativity*. MIR Publishers, Moscow, 1979.
WEYL, H. *Space, Time, and Matter*. Methuen, London, 1922.

3. Books on special relativity with less mathematics
BERGMANN, P. G. *The Riddle of Gravitation*. Scribners, New York, 1968.
CALDER, NIGEL. *Einstein's Universe*. Penguin Books, New York, 1979.
COLEMAN, J. A. *Relativity for the Layman*. Viking-Penguin, New York, 1969.
DAVIS, P. *The Edge of Infinity*. Simon & Schuster, New York, 1981.
EPSTEIN, L. C. *Relativity Visualized*. Insight Press, San Francisco, 1985.
FRIEDRICHS, K. O. *From Pythagoras to Einstein*. Random House, New York, 1965.
GARDNER, M. *The Relativity Explosion*. Vintage Books, New York, 1976.
GIBILISCO, S. *Understanding Einstein's Theories of Relativity*. Tab Books, Blue Ridge Summit, Pa., 1983.
HAWKING, S. W. *A Brief History of Time*. Bantam Books, New York, 1988.
JEANS, SIR JAMES. *Physics and Philosophy*. Dover, New York, 1981.

KAHAN, G. $E = mc^2$: *Picture Book of Relativity*. Tab Books, Blue Ridge Summit, Pa., 1983.

KAUFMANN, W. J. *Relativity and Cosmology*. Harper & Row, New York, 1977.

LANDSBERG, P. T. *The Enigma of Time*. Adam Hilger Ltd., Bristol, 1984.

LILLEY, S. *Discovering Relativity for Yourself*. Cambridge University Press, Cambridge, 1981.

MARDER, L. *Time and the Space Traveller*. University of Pennsylvania Press, Philadelphia, 1974.

MURCHIE, G. *Music of the Spheres*. Dover, New York, 1967.

NARLIKAR, J. V. *The Lighter Side of Gravity*. W. H. Freeman, New York, 1982.

PARK, D. *The Image of Eternity, Roots of Time in the Physical World*. New American Library, New York, 1980.

REICHENBACH, H. *The Philosophy of Space and Time*. Dover, New York, 1958.

ROTHMAN, T. *Frontiers of Modern Physics*. Dover, New York, 1985.

RUCKER, R. V. B. *Geometry, Relativity, and the Fourth Dimension*. Dover, New York, 1977.

SCHWARTZ, J. T. *Relativity in Illustrations*. New York University Press, New York, 1962.

WALD, R. M. *Space, Time, and Gravity*. University of Chicago Press, Chicago, 1977.

4. Specialized references for Chapter 1

CHRISTIANSON, G. E. *This Wild Abyss, The Story of the Men Who Made Modern Astronomy*. The Free Press, Macmillan, New York, 1978.

GALILEI, GALILEO. *Dialogues Concerning Two New Sciences*, trans. by H. Crew and A. de Salvio. Northwestern University Press, Evanston, Ill., 1950.

EINSTEIN, A. Zur Elektrodynamic bewegter Körper. *Ann. Phys*, Vol. 17, pp. 891–921, 1905.

NEWTON, I. *Mathematical Principles of Natural Philosophy*. London, 1687 (reprinted in *Great Books of the Western World* series, Vol. 34, of the *Encyclopaedia Britannica*, Chicago, 1952).

POINCARÉ, H. L'état actuel et l'avenir de la Physique mathématique, lecture delivered on September 24, 1904 to the International Congress of Arts and Science, St. Louis, Missouri. (*Bull. Sci. Mat.*, Vol. 28, pp. 302–324, 1904.)

POINCARÉ, H. Sur la dynamique de l'électron. *C. R. Acad Sci Paris*, pp. 1504–1508, 1905.

STEWART, A. B. The discovery of stellar aberration. *Scientific American*, Vol. 210, p. 100, Mar. 1964.

5. Specialized references for Chapter 2

BRILLOUIN, L. *Relativity Reexamined*. Academic Press, New York, 1970.

ESSEN, L. *The Special Theory of Relativity, A Critical Analysis*. Clarendon Press, Oxford, 1971.

FRIEDMAN, M. *Foundations of Space-Time Theories, Relativistic Physics and Philosophy of Science*. Princeton University Press, Princeton, N.J., 1983.

FRISCH, D. H., and SMITH, J. H. Measurement of the relativistic time dilation using μ-mesons. *American Journal of Physics*, Vol. 31, p. 342, 1963.

KENNEDY, R. J., AND THORNDIKE, E. M. Experimental establishment of the relativity of time. *Phys. Rev.*, Vol. 42, pp. 400–418, 1932.

LARMOR, J. On the ascertained absence of effects of motion through the aether in relation to the constitution of matter and the FitzGerald-Lorentz hypothesis. *Philos. Mag*, Vol. 7, pp. 621–625, 1904.

LORENTZ, H. A. The relative motion of the Earth and the ether. *Versl. Kon. Akad. Wetensch., Amsterdam 1*, Vol. 74, 1892.

NORDENSON, H. *Relativity, Time and Reality, A Critical Investigation of the Einstein Theory of Relativity from a Logical Point of View.* Allen and Unwin, London, 1969.

PENMAN, S. The muon. *Scientific American*, Vol. 205, p. 46, July 1961.

WILL, C. M. *Was Einstein Right?* Basic Books, New York.

6. Specialized references for Chapter 3

APOLLONIUS OF PERGA. *On Conic Sections*, Ancient Greek book circa 200 B.C. (reprinted in *Great Books of the Western World* series, Vol. 11, of the *Encyclopaedia Britannica*, Chicago, 1952).

GOLDMAN, M. *The Demon in the Aether, The Story of James Clerk Maxwell, Father of Modern Science.* Heyden, Philadelphia, 1983.

MAXWELL, J. C. *Treatise on Electricity and Magnetism*, (2 volumes), Oxford University Press, Oxford, 1873 (reprinted Dover, New York, 1952).

MEHRA, J. *Einstein, Hilbert, and the Theory of Gravitation.* Reidel, Boston, 1974.

NEUGEBAUER, O. *A History of Ancient Mathematical Astronomy.* Springer-Verlag, New York, 1975.

ROSSER, W. G. V. *Classical Electromagnetism via Relativity, An Alternative Approach to Maxwell's Equations.* Plenum, New York, 1968.

SMITH, C. *Conic Sections by the Methods of Coordinate Geometry.* Macmillan, London, 1930.

7. Specialized references for Chapter 4

ALVAGER, T., FARLEY, F., KJELLMAN, J., AND WALLIN, I. Test of the second postulate of special relativity in the GeV region. *Physics Letters*, Vol. 12, p. 260, 1964.

FOX, J. G. Experimental evidence for the second postulate of special relativity. *American Journal of Physics*, Vol. 30, p. 297, 1962.

GILL, T. P. *The Doppler Effect.* Academic Press, New York, 1965.

HUYGENS, C. *Treatise on Light.* Paris, 1678 (reprinted in *Great Books of the Western World* series, Vol. 34, of the *Encyclopaedia Britannica*, Chicago, 1952). [In this book, Huygens gives an account of Roemer's work on the movement of light.]

NEWTON, I. *Optics*, London, 1704 (reprinted in *Great Books of the Western World* series, Vol. 34, of the *Encyclopaedia Britannica*, Chicago, 1952).

ROBERTSON, H. P. Postulate versus observation in the special theory of relativity. *Reviews of Modern Physics*, Vol. 21, p. 378, 1949.

ROTHMAN, M. A. Things that go faster than light. *Scientific American*, Vol. 203, pp. 142–152, July 1960.

VOIGT, W. Uber das Doppler'sche Prinzip. *Gotting Nachr*, Vol. 14, 1887 (reprinted in *Phys. Z.*, Vol. 16, pp. 381–386, 1915).

8. Specialized references for Chapter 5

JAFFE, B. *Michelson and the Speed of Light.* Doubleday, Garden City, N.Y., 1960.

LIVINGSTON, D. M. *The Master of Light, A Biography of Albert A. Michelson.* Scribners, New York, 1973.

MICHELSON, A. A., and MORLEY, W. E. On the relative motion of the Earth and the Luminiferous Ether. *Amer. J. Sci.*, Vol. 34, pp. 333–345, 1887.

SHANKLAND, R. S. Conversations with Albert Einstein. *American Journal of Physics*, Vol. 31, p. 47, 1963.

SHANKLAND, R. S. Michelson-Morley experiment. *American Journal of Physics*, Vol. 32, p. 16, 1964.

SHANKLAND, R. S. The Michelson-Morley experiment. *Scientific American*, pp. 107–114, Nov. 1964.

SWENSON, L. S. *The Etheral Aether, A History of the Michelson-Morley-Miller Aether-Drift Experiments, 1880–1930.* University of Texas Press, Austin, 1972.

WEISSKOPF, V. F. The visual appearance of rapidly moving objects. *Physics Today*, Vol. 13, 1960.

9. Specialized references for Chapter 6

BURKE, J. *The Day the Universe Changed.* Little Brown, Boston, 1985.

GHIM, W. *Life of Mercator.* Duisburg, 1595.

KELLAWAY, G. P. *Map Projections.* Methuen, London.

KLINE, M. *Mathematics in Western Culture.* Oxford University Press, New York, 1953.

OSLEY, A. S. *Mercator, A Monograph on the Lettering of Maps in the 16th Century Netherlands.* Watson-Guptill, New York.

ROBINSON, A. H. *Elements of Cartography.* John Wiley, New York.

WRIGHT, E. Certaine Errors in Navigation Arising either of the ordinarie erroneous making or using of the Sea Chart, Compasse, Crosse Staffe, and Tables of the Declination of the Sunne, and Fixed Starres detected and corrected, Printed at London by Valentine Sims, 1599 (facsimile reprint by Walter J. Johnson, Norwood, N.J., 1974).

10. Specialized references for Chapter 7

BERTOZZI, W. Speed and kinetic energy of relativistic electrons. *American Journal of Physics*, Vol. 32, p. 551, 1964.

EINSTEIN, A. Ist die Tragheit eines Korpers von seinem Energie inhalt abhangig? *Ann. Phys.*, Vol. 18, pp. 639–641, 1905.

JARNAKER, J. *The Double Solution of the Theory of Relativity, A Study of Ontological Symmetry.* Almqvist & Wiksell, Uppsala, 1971.

PROKHOVNIK, S. J. *Light in Einstein's Universe, The Role of Energy in Cosmology and Relativity.* Reidel, Boston, 1985.

VAN DER WALLS, J. D. Energy and mass. *Proc. R. Acad. Amsterdam*, pp. 821–831, 1912.

WIGNER, E. P. Violations of symmetry in physics. *Scientific American*, Vol. 213, p. 28, Dec. 1965.

ZEE, A. *Fearful Symmetry, The Search for Beauty in Modern Physics.* Macmillan, New York, 1986.

11. Specialized references for Chapters 8 through 12

BORN, M. *Physics in My Generation.* Springer-Verlag, New York, 1969.

BREHME, R. W. A geometric representation of Galilean and Lorentz transformations. *American Journal of Physics*, Vol. 30, p. 489, 1962.

FEYNMAN, R. P., LEIGHTON, R. B., AND SANDS, M. *The Feynman Lectures on Physics.* Addison-Wesley, Reading, Mass., 1963.

GELFAND, I. M., MINLOS, R. A., AND SHAPIRO, Z. YA. *Representation of the Rotation and Lorentz Groups and Their Applications.* Pergamon Press, Oxford, 1963.

GOLDBLATT, R. *Orthogonality and Spacetime Geometry.* Springer-Verlag, New York, 1987.

KITTEL, C., KNIGHT, W. D., AND RUDERMAN, M. A. *Mechanics, Berkley Physics Course*, Vol. I, McGraw-Hill, New York, 1965.

KLEIN, F. *Elementary Mathematics from an Advanced Standpoint. Geometry.* Dover, New York, 1939.

LYUBARSKII, G. YA. *The Application of Group Theory in Physics.* Pergamon Press, Oxford, 1960.

MINKOWSKI, H. *Das Relativitatsprinzip*, lecture delivered to the Math. Ges. Gottingen, Nov. 5, 1907. *Ann. Phys.*, Vol. 47, pp. 927–938, 1915.

MINKOWSKI, H. *Raum und Zeit*, lecture delivered to the 80th Naturforscherversammlung at Cologne, Sept. 21, 1908. *Phys. Z.*, Vol. 20, pp. 104–111, 1909.

PEDOE, D., *Geometry and the Visual Arts.* Dover, New York, 1976.

SEARS, F. W. Some applications of the Brehme diagram. *American Journal of Physics*, Vol. 31, p. 269, 1963.

SINGH, J. *Great Ideas of Modern Mathematics.* Dover, New York, 1959.

SNAPPER, E., AND TROYER, R. J. *Metric Affine Geometry.* Academic Press, New York, 1971.

INDEX

A

Abelian group, 282
Absolute geometry, 326
Addition equation for physical velocities, *See* Physical velocity addition formula
Adiabatically isolated system, 216
Aether concept, 156–157
Aether reference frame, 147–51
Affine geometry, 326, 328
    FitzGerald length contraction, 346–49
    Gudermannian space-time diagram, 340–43
    metric, 327, 328
    Minkowski coordinate systems, 337–39
    Minkowski plane, 330–34
    parallelism of, 328
    proper/improper times, 343–46
    time and distance by radar measurements, 334–37
al-Haytham, Ibn (Alhazen), 20
Alighieri, Dante, 8–10
Ampère, André Marie, 86
Analysis Situs (Poincaré), 17
Analytic geometry, 243
Ant metaphor, 287–89
Apparent rating, 73
Apparent velocity, 106–12
Apparent velocity addition formula, 111
Aristarchus of Samos, 166
Aristotle, 20, 148
Astronomia Nova (Kepler), 70–71

B

Bacon, Roger, 20
Bergerac, Cyrano de, *See* de Bergerac, Cyrano
Berkeley, Bishop George, 378
Blake, William, 143
Bohr, Niels, 382
Bolyai, Janos, 326
Born, Max, 299
Bradley, James, 22
Brahe, Tycho, 12–13, 19, 21, 70, 194
Bridgman, P. W., 113–14
Bunsen, Robert Wilhelm, 118

C

Calibration, 319
Cartesian coordinate system, calibration of, 32
Cartography, relativity theory and, 171–76
Catenary:
    definition of, 287
    derivation of, 294–97
    intrinsic equation of, 296
    relativity and, 286–97
Circle, in plane geometry, 251–52
Color of light, 132–36
Columbus, Christopher, 191–94, 197–201
    Toscanelli's maps and, 175–76
Commutative group, 282
Conformal mapping, 171
Conservation of energy, 214–16, 224–25, 234
Conservation laws, 233–37
    conservation of mass, 234
    conservation of momentum, 211–12
    generality of, 233
Contraction factor, 62, 219
Contraction of a moving ruler, 322–23
Contraction of distance, 62
Contraction of length, equation for, 220
Copernicus, Nicolaus, 12–13, 19, 43–44, 97
Copernicus angle, 93, 138
Coulomb, Charles Augustin de, 86

D

Dante, *See* Alighieri, Dante
de Bergerac, Cyrano, 373–74
Dee, John, 180, 194
de Nemore, Jordanus, 231–32
De Revolutionibus Orbium Coelestium (Copernicus), 12
Descartes, René, 216, 243–44
Des Relativitatsprinzip (Minkowski lecture), 299–300
Dialogue Concerning Two Chief World Systems (Galileo), 17
Dilation factor, 60, 81, 219
Dilation of mass, 220
Dilation of time, 60, 233, 323
Direct formula, 76
Directrix, 91
Discourse on the Method for Properly Guiding the Reason and Finding Truth in the Sciences (Descartes), 243
Dispersion, 126
Displacement current, 86
Distance, and time, 104–6
Doppler, Christian Johann, 119, 127
Doppler effect, 119–26, 127, 132–35, 379
    See also Relativistic Doppler factor
Dynamical Theory of the Electromagnetic Field, A (Maxwell), 86

E

Earth:
   flat map representation of, 170
   as an inertial frame, 158
Eccentric angle, 93
Eccentric anomaly, 71
Einstein, Albert, 12, 19, 20, 35, 36–39, 84–85, 293, 377–78
   equivalence of energy and mass equation, 19, 227–31
   field equation, 84
   relativistic energy, 223–27
Electromagnetic spectrum, 116–19
Electromagnetic waves, 46, 157
Elements, The (Euclid), 35, 240–41, 326
Elliptic geometry, 242
Elliptic orbit, 14
Energy:
   conservation of, 214–16, 224–25, 234
   kinetic, 215–16, 224, 227
   relativistic, 223–27
   rest, 229
Epistolarium Astronomicarum (Brahe), 13
Equivalence of energy and mass, 19, 227–31
Eratosthenes, 166
Erlangen program, 328
Euclid, 35, 240, 242–43, 325–26
Euclidean geometry, 240–43, 326–27
   orthogonal axes in, 272
Euclidean space, 329
Euler, Leonhard, 324–25
Exaggerated latitude, 207

F

Fable of the King's Messengers, 99–101
Fable of the Muon and the Mountain, 63–68, 217–20
Fable of the Two Eskimos, 58–62
Faraday, Michael, 86
Faucault, Leon, 22
Fermi, Enrico, 26
FitzGerald, George Francis, 16, 155, 299, 383
FitzGerald length contraction, 67, 155–56, 159–64, 321–23, 346–49
   light clocks and, 159–64
Fixed stars, definition of, 150
Fizeau, A.H.L., 22
Flat maps, 170–71
Fresnel, Augustin, 148
Freud, Sigmund, 78
Fundamental law of dynamics, 30

G

Galilean principal of relativity, 17, 18, 46
Galilean transformation, 40–41
Galilei, Gallileo, 12, 16–17, 18, 20, 22, 29–30, 35, 127, 211, 231, 287
Garage paradox, 364–66
Geography (Ptolemy), 171, 173
Geometry:
   absolute, 326
   affine, 326, 328
   analytic, 243
   elliptic, 242
   Euclidean, 240–43, 326–27
   hyperbolic, 242, 256–59
   metric, 327
   metric affine, 327, 328
   Minkowski, 315–21
   non-Euclidean, 240–43, 283, 326
   projective, 326–27
   Riemannian, 19
Globes, spatial attributes, 203
Gravitation theories, 12–20
Graviton, 231
Group theory, 18, 281–85
   definition of, 17, 282
Group velocity, 111
Gudermann, Christoph, 329–30, 353, 369
Gudermannian angle, 52–55, 60, 137, 342, 347–48, 354
Gudermannian of u (the Gudermannian), 353
   geometric interpretation for spacelike hyperbola, 361–63
   geometric interpretation for timelike hyperbola, 358–61
   reason for, 356–58
Gudermannian space-time diagram, 52, 53, 340–43
Gudermann world, 350–72
   example, 369–72
   hyperbolic sine and cosine, 350–53

H

Half-offset, 336
Hamilton, William Rowan, 382
Harvey, William, 97
Heisenberg, Werner, 118, 382
Heraclitus, 209
Hertz, Heinrich, 87, 88, 299, 380
Hilbert, David, 84–85, 328
Hipparchus, 166–68, 207
Hipparchus projection, 183, 203–6
History of Lunar States and Empires (de Bergerac), 373–74
Hubble, Edwin, 126
Huygens, Christian, 22, 128, 131, 148, 243
Hyperbola, 252, 319, 351
Hyperbolic functions, compared to circular functions, 353
Hyperbolic geometry, 242, 256–59
Hyperbolic parameter, 258, 296

Hyperbolic radian, 259
Hyperbolic trigonometry, 256–60

I

In Cabin'd Ships at Sea (Whitman), 81–82
Inertia, 30
Inertial frames of reference, 30–33, 43, 98–99, 106, 120–21, 123, 142, 148, 232, 309
Inertia lines, 277
Inner product, 272–78, 329
   space-time inner product, 277
   time-space inner product, 275
International System of Units (Système Internationale d'Unités), *See* SI units
Intrinsic equation of the catenary, 296
Invariance of distance, 271
Invariance of time-space interval, 275
Invariants, symmetry and, 232–33
Inverse equation, 77
Isotropy of space, 234

J

Jet lag illustration, 2–6

K

Kepler angle, 138
Kepler, Johannes, 12–15, 16, 19–20, 194
Kepler ellipse, 91
Kepler's equations, 70–72, 77, 138–40
   derivation of, 91–95
Kepler's formula, 94–95, 109, 145
Kinetic energy, 215–16, 224, 227

L

Langevin, Paul, 18
Larmor, Sir Joseph, 60, 299, 383
Larmor time dilation, 156, 294, 321–23
Latitude angle, 137
Law for the composition of Doppler factors, 139
Law of action and reaction, 30
Law of conservation of angular momentum, 30
Law of conservation of momentum, 30
Law of inertia, 29–30, 31, 157–158, 212–14
Law of the composition of velocities, 139
Law of the conservation of energy, 225, 231–32
Leibniz, Baron Gottfried Wilhelm von, 378
Length contraction, 363
Length of time, 27
Light:
   time and, 10
   traveling through a vacuum, 149
Light clocks, and the FitzGerald length contraction, 159–64

Light lines, 37, 277, 331
Light-meter, 28
Light principle, 34–36, 148, 157, 379–80
Light-second, 24, 28
Light-year, 24
Linear algebra, 328
Linear transformations, 260–65
Longfellow, Henry Wadsworth, 285
Lorentz, Hendrik, 12, 15–16, 19–20, 194, 291
Lorentz equations, 79–81, 140
   reflexive, 80–81
Lorentz transformation, 16, 36–39, 40–41, 109, 140–41, 201–3, 232, 269–70, 278–80, 290–92
   two-dimensional, 284–85

M

Mach, Ernst, 378
Macrobius, 150
Mapping:
   conformal, 171
   flat maps, 170–71
   Toscannelli's map, 171–76
Mass, 30, 209–11
   relativistic mass, 220–22
Mathematical model of relativity, 78–81
Maxwell, James Clerk, 12–13, 19, 35, 85, 116–19, 131–32, 149–51, 170, 293
   Dynamical Theory of the Electromagnetic Field, A, 86
   electromagnetic theory, 87, 170
   Maxwell's equations, 85–88, 141, 230–31
   theory of electricity, magnetism, and light, 42–43
Measurement units, 26
Menelaus, 166
Mercator, Gerard, 177–80, 194
Mercator projection, 136–37, 181–88, 208
   as one-to-many transformation, 190–91
   popularity of, 188–89
   transverse Mercator projection, 188–91, 196
   underlying principle, 181, 184–85
   Universal Transverse Mercator (UTM) grid, 190
Metaphor of ants, 287–89
Metric affine geometry, 327, 328, 332
Metric geometry, 327
Michelson, Albert A., 12–13, 19, 26, 43, 87–88, 151, 161, 380
Michelson-Morley experiment, 13, 15, 116, 132, 147–64, 380, 383
   null result, 155–56
   recognition principle, 156–59
Midpoint, 336
Milton, John, 6–7
Minkowski, Herman, 298–99
Minkowski world, 298–323
   definition of, 300

Minkowski world, (*cont.*)
  geometrical properties of, 300–301
  Minkowski coordinate systems, 337–39
  Minkowski geometry, 315–21
  Minkowski plane, 330–34
  Minkowsi space, 328, 331
  proper time/proper place, 313–15
  relativity of simultaneity, 305–6
  relativity of the same place, 309
  timelike/spacelike events, 310–12
  world lines, 301–5
Momentum, 221
  conservation of, 211–12
  definition of, 211
  law of inertia and, 212–14
  Newton's second law of motion and, 214
Morley, Edward, 13, 43, 87–88, 151, 161, 380
  See also Michelson-Morley experiment

N

Natural units, 27–29
Negative reciprocals, 268
Nemore, Jordanus de, *See* de Nemore, Jordanus
Neutrino, 231
Newton, Isaac, 12, 18, 19, 20, 34, 35, 44–45, 74, 126–30, 194, 211, 231, 293, 377–78
  laws of motion, 29–30
    fundamental law of dynamics, 30
    law of action and reaction, 30
    law of inertia, 29–30, 31, 157–158, 212–14
    second law of motion, 19, 83, 214
  Newtonian dynamics, 212–14, 216
Newton rainbow, 118
Non-Euclidean geometry, 240–43, 283, 326
Nuñez, Pedro, 177

O

Oblique Mercator projection, 188–91
Ockham's razor, 78, 90–91, 141
  definition of, 74
Oersted, Hans Christian, 86
Olympic metaphor of relativity, 72–78
One-clock velocity, 112
Optical wave propagation, 15
Orthogonal axes in Euclidean geometry, 272
Orthogonality, definition of, 272
Orthogonal lines in space-time geometry, 275
Orthogonal vectors, 329

P

Paradiso, 8–9
Paradox of the reckless driver and the garage, 366–67
Peirce, Benjamin, 325
Perpendicular vectors, 329
Philosophiae Naturalis Principia Mathematica (Newton), 29, 216, 241, 377–80
Photons, 20, 50–51, 118–19, 278, 331
  gamma factor, 231
  passage of time and, 8
  relativistic mass, 231
  rest mass, 231
  spin angular momentum, 231
  time/space and, 278
Physical theory of relativity, 88–91
Physical velocity, 106–12
Physical velocity addition formula, 16, 48, 109, 139, 294
  derivation from Lorentz transform, 144–45
Physics in My Generation (Born), 299
Planck, Max, 118–19
Planck's constant, 118–19
Planck's equation, 119, 134
Poetic metaphors, 6–10
Poincaré group, 18, 136, 141–43, 232–33, 281–85, 379
  displacements in space and time in, 236–37
  symmetry, 96–146
  of space-time, 96–99
Poincaré, Henri, 12, 17–19, 20, 34, 194, 236, 242, 299, 300
Polar equation for an ellipse, 92
Pope, Alexander, 7–8, 298
Positive reciprocals, 268
Primed system, 75
Principle of perseverance, 210
Principles of forbiddenness, 234
Proclus, 240, 250
Projection, 326–27, 375
Projective geometry, 326–27
Propagation of light, 23–24
  compared to propagation of sound, 24
Proper mass, 221
Protagoras, 104
Pseudo-Pythagorean theorem, 317, 369
Ptolemy, 166, 171, 176
Pure color light rays, 132–33
Pythagoras, 82–83, 250–51
Pythagorean theorem, 248–50, 317

Q

Quantum theory, 118, 282, 381–82

R

Radar measurements, of time and distance, 334–37
Reaction time, 22
Recognition principle, 65, 156–59
Reflection through the origin, 263–64
Reflexive equation, 77, 109

INDEX

Reflexive formula, 76
Reflexive Lorentz equations, 80–81
Relativistic Doppler factor, 101–4, 124, 129–31, 137, 208
　derivation from Lorentz transform, 145–46
　See also Doppler effect
Relativistic energy, 223–27
Relativistic frame of reference, 33
Relativistic kinetic energy, 229
Relativistic mass, 220–22
　photons, 231
Relativity:
　catenary and, 286–97
　mathematical model, 78–81
　physical theory, 88–91
　as a shear in space-time, 36–39
Relativity, The Special and the General Theory (Einstein), 306
Relativity of simultaneity, 305–6
Relativity of the same place, 309
Relativity principle, 34–36, 148
Rest energy, 229
Rest mass, 221
　neutrinos, 231
　photons, 231
Rhumb lines, 183
Riemannian geometry, 19
Robinson, Edward Arthur, 34
Roemer, Olaf, 20–21, 35, 127–31, 137, 150, 379
　Doppler effect and, 127–28
　Roemer's multiplicative equation, 139–40, 143
Rotation, 254, 265–71
　of axes, 267
　versus shear, 278–80
　of space coordinates, 271–74
　in three dimensions, 283
Rotation group, 284
Rotation symmetry, 143
Round globe, spacing of parallels on, 168–70
Round-trip velocity, 112

S

Scale, 319
Schrödinger, Erwin, 118, 382
Science, as fable, 286–87
Second law of motion (Newton), 19, 83
　compared to conservation laws, 234
　momentum and, 214
Section of the projection, 327
Separation lines, 277
Shear, 264, 265–71
　versus rotation, 278–80
　of space-time coordinates, 274–78
Shear group, 284–85
Shear symmetry of space-time, 38, 142

Shear transformation, *See* Lorentz transformation
SI units, 27, 28, 105
　unnatural, 114
Sizes and Distances of the Sun and Moon (Aristarchus), 168
Sophisticate's Primer of Relativity, A (Bridgman), 113–14
Space and Time (Minkowski lecture), 300
Space coordinates, rotation of, 271–74
Spacelike hyperbola, geometric interpretation, 361–63
Space-time concept, 96–99, 197–98
Space-time coordinates, shear of, 274–78
Space-time inner product, 277
Space-time interval, 25, 318, 338
Space-time symmetry, 136–41
　connection between conservation laws and, 234–35
Special Galilean group, 17
Special theory of relativity, 33, 106
　invariance/symmetry and, 236
　length/time and, 106
　mathematics and, 43
　summary of requirements, 35
Spectrum, 126
Speed, 106–7
Speed of light, 8, 23, 132–36
Stereographic polar projection, 187, 203–6
Stereographic projection, 136, 171
Symmetric objects, definition of, 141–42
Symmetry:
　invariants and, 231–33
　laws of physics and, 141
　principle of, 232

T

Taylor, Brook, 83–84
Theory of relativity:
　Newtonian mechanics and, 221
　reference speed, 6
Thomson, J. J., 211
Thoreau, Henry David, 74, 77
Time and distance, 104–6
　by radar measurements, 334–37
Time dilation, 104, 360
Timelike hyperbola, geometric interpretation, 361–63
Timelike/spacelike events, 310–12
Time Machine, The (Wells), 367
Time of length, 27
Time-of-length units, 24–25
Time-space inner product, 275
Time-space interval, 25, 39–40, 268, 279, 289–90, 301, 317–18, 338
Todd, David Peck, 150–51
Toscanelli, 194

Toscannelli's map, 171–76
Total energy, 224
Total mass, 223
Transformation, definition of, 261
Translation in space symmetry, 142
Transverse Mercator projection, 188–91, 196
Treatise on Electricity and Magnetism (Maxwell), 12–13, 23
Trigonometric functions, 254
    geometric method of defining, 167–68
Trigonometry, 165–68, 253–55
    hyperbolic, 256–60
True anomaly, 71
True rating, 73
Two-clock velocity, 112
Two-dimensional Lorentz transformation, 284–85
"Two Rivers, The" (Longfellow), 285
Two-way velocity of light, 112–16
Tycho, See Brahe, Tycho
"Tyger, The" (Blake), 143

U

Uniformity of space, 234
Uniformity of time, 234
Universal Transverse Mercator (UTM) grid, 190
Unprimed system, 75

V

Vector space, 328
Velocity of light, 20–24, 106, 110, 230–31
    natural units based on, 27–29
    two-way, 112–16
    in a vacuum, 24

W

Wave equation, 82–85
Wave motion, 81–82
Wave theory, 148–49
Wells, H. G., 367
Whitman, Walt, 81–82
Wigner, Eugene, 231
World lines, 99–101, 301–5, 331
Wright, Edward, 131–32, 136, 180–81, 183, 194, 200, 208
    discovery of the law of the scale of the map, 180
    mathematical theory of navigation, 194
Wright's additive equation, 139–40
Wright's equation, 137, 187, 194, 207–8

Y

Young, Thomas, 148